MX

Bioinformatics

METHODS EXPRESS

The **METHODS EXPRESS** series

Series editor: B. David Hames

Faculty of Biological Sciences, University of Leeds, Leeds LS2 9JT, UK

Bioinformatics

Biosensors

Cell Imaging

DNA Microarrays

Expression Systems

Genomics

Immunohistochemistry

PCR

Protein Arrays

Proteomics

Whole Genome Amplification

MX

Bioinformatics

METHODS EXPRESS

edited by **Paul H. Dear**

MRC Laboratory of Molecular Biology,

Cambridge, UK

Scion

© Scion Publishing Ltd, 2007

First published 2007

A CIP catalogue record for this book is available from the British Library.

ISBN: 978 1 904842 16 3 (paperback)
ISBN: 978 1 904842 23 1 (hardback)

Scion Publishing Limited
Bloxham Mill, Barford Road, Bloxham, Oxfordshire OX15 4FF
www.scionpublishing.com

Important Note from the Publisher

Dedication

To Felicity

Typeset by Phoenix Photosetting, Chatham, Kent, UK
Printed by Ajanta Offset and Packagings Ltd, Delhi, India

Cover image by Paul H Dear, representing fruiting bodies of the amoeba *Dictyostelium discoideum*.

Contents

Contributors

Bateman, Alex Wellcome Trust Sanger Institute, Wellcome Trust Genome Campus, Hinxton, Cambridge, CB10 1SA, UK. E-mail: agb@sanger.ac.uk

Clark, Melody S. British Antarctic Survey, Natural Environment Research Council, High Cross, Madingley Road, Cambridge, CB3 0ET, UK. E-mail: mscl@bas.ac.uk

Foster, Peter G. Department of Zoology, Natural History Museum, London, UK. E-mail: p.foster@nhm.ac.uk

Griffiths-Jones, Sam Wellcome Trust Sanger Institute, Wellcome Trust Genome Campus, Hinxton, Cambridge, CB10 1SA, UK. Current address: Faculty of Life Sciences, University of Manchester, Michael Smith Building, Oxford Road, Manchester, M13 9PT, UK. E-mail: sam.griffiths-jones@manchester.ac.uk

Gruber, Arthur Department of Parasitology, Institute of Biomedical Sciences, University of São Paulo, Av. Prof. Lineu Prestes 1374, São Paolo SP, Brazil, 05508-000. E-mail: argruber@usp.br

GuhaThakurta, Debraj Rosetta Inpharmatics LLC, Merck & Co., Research Genetics Department, 401 Terry Avenue North, Seattle, WA 98109, USA. E-mail: debraj_guhathakurta@merck.com

Hall, Neil The Institute for Genomic Research, 9712 Medical Center Drive, Rockville, MD 20850, USA. Current address: University of Liverpool, School of Biological Sciences, Biosciences Building, Crown St, Liverpool, L69 7ZB, UK. E-mail: neil.hall@liv.ac.uk

Heringa, Jaap Centre for Integrative Bioinformatics, Vrije Universiteit De Boelelaan 1081a, 1081 HV Amsterdam, The Netherlands. E-mail: heringa@few.vu.nl

Lesk, Arthur M. Department of Biochemistry and Molecular Biology, The Pennsylvania State University, University Park, PA 16802, USA. E-mail: aml25@psu.edu

Lopez, Rodrigo EMBL Outstation – Hinxton, European Bioinformatics Institute, Wellcome Trust Genome Campus, Hinxton, Cambridge, CB10 1SD, UK. E-mail: rls@ebi.ac.uk

Morgenstern, Burkhard Universität Göttingen, Institut für Mikrobiologie und Genetik, Abteilung für Bioinformatik, Goldschmidtstr. 1, D-37077 Göttingen, Germany. E-mail: burkhard@gobics.de

Pirovano, Walter Centre for Integrative Bioinformatics, Vrije Universiteit De Boelelaan 1081a, 1081 HV Amsterdam, The Netherlands. E-mail: pirovano@few.vu.nl

Rajandream, Marie-Adele Wellcome Trust Sanger Institute, Wellcome Trust Genome Campus, Hinxton, Cambridge, CB10 1SA, UK. E-mail: mar@sanger.ac.uk

Sangar, Vineet Department of Biochemistry and Molecular Biology, The Pennsylvania State University, University Park, PA 16802, USA. E-mail: vus102@psu.edu

Schlitt, Thomas British Antarctic Survey, Natural Environment Research Council, High Cross, Madingley Road, Cambridge, CB3 0ET, UK. Current address: Department of Medical and Molecular Genetics, King's College London School of Medicine, 8th Floor Guy's Tower, London, SE1 9RT, UK. E-mail: thomas.schlitt@genetics.kcl.ac.uk

Schriml, Lynn M. The Institute for Genomic Research, 9712 Medical Center Drive, Rockville, MD 20850, USA. E-mail: lschriml@tigr.org

Stormo, Gary D. Washington University School of Medicine, Department of Genetics, Campus Box 8510, Room 5410, 4444 Forest Park Parkway, St. Louis, MO 63108, USA. E-mail: stormo@genetics.wustl.edu

Preface

In 1984, our total knowledge of DNA sequence amounted to 2 825 441 bases — enough to be printed in a modest book. In fact, such a book was printed — as a two-volume paperback (*Nucleotide Sequences 1984*. IRL Press) — and a copy of it still sits in my lab's library. At a pinch, you could do bioinformatics by finding the right page and then marking up the restriction sites or stop codons with a pencil.

Twenty-odd years later, there is about 170 billion bases (50 human genomes' worth) of sequence data in GenBank alone, most of it heavily annotated with experimental data or the results of computational analysis. Finding the right page has become correspondingly harder.

To the honest wet-bench scientist who just wants to know whether *rabC* has a homolog in *Drosophila*, or whether there are any transcription factors on human chromosome 14q23, there is now a bewildering glut of bioinformatic resources. Data may be duplicated and scattered; a protein might have a dozen different names in different databases; a sequence won't upload because it's in EMBL format instead of FASTA format. This apparent impenetrability is a great shame, because the data resources and computational tools available to biologists are by far the most extensive and sophisticated available to scientists in any discipline.

This book is not a comprehensive guide to all facets of bioinformatics. Instead, it aims to guide you — the honest wet-bench scientist — through a selection of the more accessible and user-friendly resources, showing how to answer the sort of questions that you are most likely to ask. In each chapter, an overview of the subject is followed by a series of worked examples with step-by-step instructions. After following these marked paths, you'll find that the resources out there are immensely powerful and, with a little experience, not so bewildering after all.

Paul H. Dear
May 2007

Acknowledgements

My thanks go to the authors for their excellent chapters. Thanks also to David Hames for inviting me to edit this volume; to all at Scion for the astonishing elasticity of their deadlines; and to Jane Hoyle for moderating the quirkier aspects of my grammar and spelling. I'm grateful to many colleagues at LMB (especially Alan Bankier, Sarah Teichmann, and Paul Hart) for help with a range of technical matters and to the staff at many bioinformatics help desks, particularly Giulietta Spudich at the EBI. Special thanks go to my wife, Denise, for her inexhaustible patience and to my daughter, Felicity, whose tree house would otherwise have been finished by now.

Before you begin

Computer hardware and software

Wherever possible, the protocols in this book use web-based resources that will work with most current web browsers on Macs or PCs. In some cases, your browser may need plug-ins (additional modules that add specific functions to the browser) such as Java; most web sites will tell you whether you need these and where they can be downloaded from (usually at no cost).

In general, web pages should look and behave similarly on all platforms. However, there are a few differences. For instance, some 'standard' button-names are set by the browser and not by the web site – a button that is called 'Upload' when viewed in one browser may be called 'Choose file' in another.

A few of the protocols require software (usually free) to be installed and run on your own machine; in these cases, instructions are given in the relevant chapter. It is also helpful to have some software on your own computer for viewing or manipulating files. In particular, a basic text editor (the simpler the better) is very useful. Microsoft Word or other word processors can be used at a pinch, but take care to save files as 'Text only' (see below, under 'File formats, editing and saving').

Typefaces used in this book

Throughout this book, underlining is used to indicate URLs (for instance, http:// www.ncbi.nlm.nih.gov/) and also for the names of example files, which can be downloaded from the book's web site (for example, ABCC9fasta.txt). **Bold typeface** is used for the names of buttons, menus, links, menu items, and other 'active' features on web pages or in software. A `monospaced font` is used for inputs or commands that you should enter verbatim and to show the output of programs. The ↻ symbol indicates that a single line of text has been 'wrapped around' to fit on one page. The names of programs (though not the names of websites which give access to them) are indicated in SMALL CAPITALS.

Online resources to accompany this book

A web site accompanies this book at www.scionpublishing.com/bioinformatics. Before you start reading a chapter you should download, and unpack, the zip file for the chapter you are working on. These zip files contain example files (datasets, copies of the expected results, etc.) which you will need to work through the corresponding protocols.

The web site also contains a compendium of links and data files, to save you having to type in each of the addresses referred to in the book. Throughout the text, you will find superscripts next to each URL, such as: http://www.ncbi.nlm. nih.gov/ [1.1] – simply click on the relevant link (in this case '1.1') to jump directly to the URL. The example files are similarly linked, such as: ABCC9fasta.txt[10.5] – to download these, right click on the link and select 'Save Target As...'.

File formats, editing, and saving

Text documents that need to be uploaded to the web sites used in these protocols (for example, files containing sequences for analysis) will need to be in specific formats appropriate to the site. These are almost always 'text only' formats so, if you view or edit a file using a word processor such as Microsoft Word, it is essential to save the file as a 'text only' document: the web server will not be able to cope with a Word '.doc' file. However, copying and pasting from a word processor document into a text-box on a web site is generally OK.

Another frustrating problem that you may encounter is that different text editors use different hidden symbols to mark the end of a line. This can cause problems when uploading a file (even 'simple' text-only file) to some web sites. If this happens, try using a different text editor (the simpler the better) to open and save the document before uploading it. Again, if the web site allows you to paste (instead of uploading) the file contents, this problem tends not to arise.

Some software, including e-mail programs, can introduce other changes into your text without your noticing. Watch out for the distinction between tabs and spaces, and between 'line wraps' (where a continuous line of text is displayed on two lines to fit into a window) and true line breaks (where an 'end of line' symbol breaks the line).

Finally, be aware that a few web servers that require an e-mail address will balk if your address contains hyphens ('e.g., joe@genome-lab.org') or some other characters.

The ever-changing web

Web sites and database contents are constantly changing – that is their point. I have tried to concentrate on sites that are likely to be stable, and to indicate cases where database updates may give results that differ from those in the protocols. If you find that things are not working exactly as described in this book, bear in mind the following points:

- A database search performed today might give slightly different results from those described here (for example, more hits if the database has expanded; or a different location for a gene if a genome sequence has been revised).
- Web sites may be updated, usually with additional features or cosmetic changes. If the button or link described in the text is not where you expect it to be, a little hunting will generally find it or its new equivalent.
- If a given web site is unavailable, try going back to the root address. For example, if http://www.hugelab.ac.im/bobslab/bioinformatics/tools/phylogeny/ annhialigner no longer works, try going back to http://www.hugelab.ac.im/ bobslab/ or even http://www.hugelab.ac.im and then follow links to the new location of the site you are after. If all else fails, a Google search for the name of the database or software will often lead you to its new home.

Explore!

Almost all of the web sites and resources described in this book offer many more functions, options, settings, and links than those described here. Spending a little time playing and trying 'nondefault' settings will yield unexpected (and sometimes even useful) results.

Abbreviations

AIC	Akaike information criterion
ASRV	among-site rate variation
CDD	Conserved Domains Database
CDR	complementarity-determining region
CDS	coding sequence
COGS	Clusters of Orthologous Groups of proteins
CPU	central processing unit
DAG	directed acyclic graph
DAS	distributed annotation system
E value	expectation value
EBI	European Bioinformatics Institute
EC	Enzyme Commission
EM	expectation–maximization
EMBnet	European Molecular Biology network
EMBOSS	European Molecular Biology Open Software Suite
EST	expressed sequence tag
EVD	extreme value distribution
GCG	Genetics Computer Group
GEO	Gene Expression Omnibus
GFF	generic file format
GHMM	generalized hidden Markov model
GI	GenInfo Identifier
GNN	Genome News Network
GO	Gene ontology
GOLD	Genomes On Line Database
GSS	genome survey sequence
GTR	general time reversible
HIV	human immunodeficiency virus
HMM	hidden Markov model
HSE	heat-shock element
HSP	high-scoring sequence pair
HTH	helix–turn–helix
IUPAC	International Union of Pure and Applied Chemistry
JGI	Joint Genome Institute
MAP	maximum a priori
MCMC	Markov chain Monte Carlo

MGI	Mouse Genome Informatics
ML	maximum likelihood
MP	maximum parsimony
MSA	multiple sequence alignment
MSD	Macromolecular Structure Database
MSP	maximum scoring pair
NCBI	National Center for Biotechnology Information
NIH	National Institutes of Health
NMR	nuclear magnetic resonance
NTF2	nuclear transport factor 2
OMIM	Online Mendelian Inheritance in Man
ORESTES	ORF ESTs
ORF	open reading frame
PDB	Protein Data Bank
PIR	Protein Information Resource
PRF	Protein Research Foundation
PSSM	position-specific scoring matrix
PTHLH	parathyroid hormone-like hormone
PWM	position weight matrix
RCSB	Research Collaboratory for Structural Bioinformatics
RT	reverse transcriptase
RT-PCR	reverse transcriptase polymerase chain reaction
SAGE	serial analysis of gene expression
SCPD	Promoter Database of *Saccharomyces cerevisiae*
SGD	*Saccharomyces* Genome Database
SNP	single-nucleotide polymorphism
SOAP	simple object access protocol
SRP	signal recognition particle
SSP	secondary structure profile
SSR	simple sequence repeat
STS	sequence-tagged site
TF	transcription factor
TFBS	transcription factor-binding site
TGI	TIGR Gene Indices
TIGR CMR	The Institute for Genomic Research Comprehensive Microbial Resource
UCSC	University of California Santa Cruz
UID	unique identifier
UTR	untranslated region
wwPDB	Worldwide Protein Data Bank
XBP-1	X-box-binding protein 1

Color section

Chapter 2. Navigating sequenced genomes

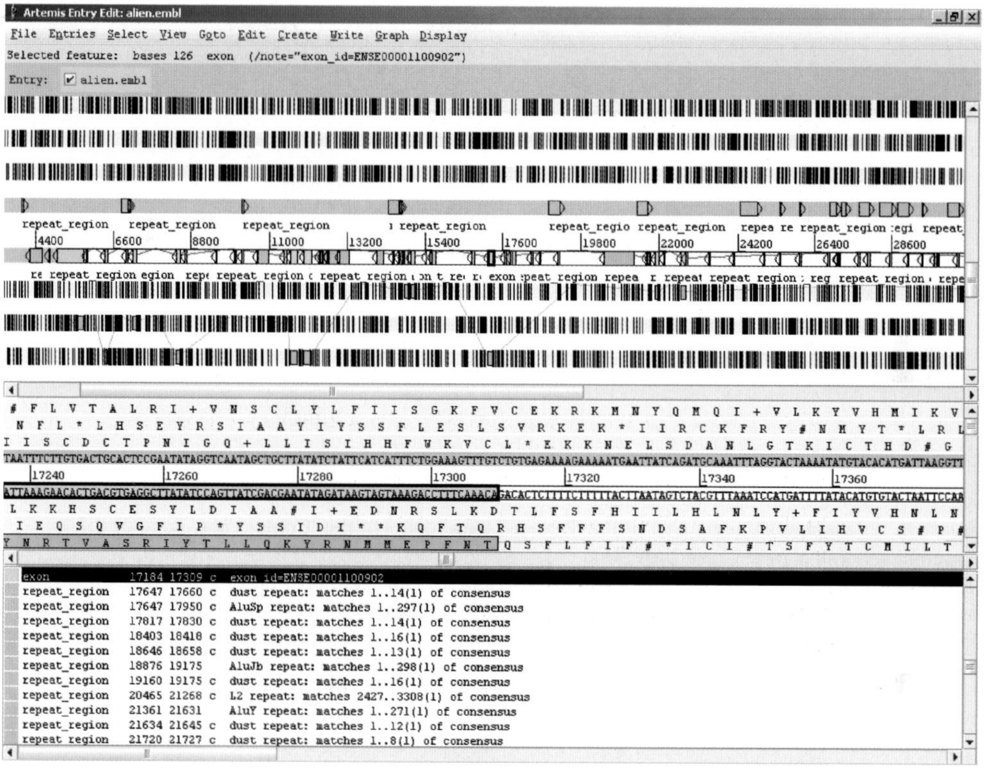

Figure 5. Screenshot of the ARTEMIS sequence viewer and annotation tool (see page 31).
The main window is divided into several sections. Below the main menu is information about the current selection and the sequences being viewed. Below this (and filling most of the top half of the screen), the 'overview' section shows stop codons in all six reading frames (short vertical black lines) and features on both strands (colored boxes, mainly in blue and yellow); the vertical scroll bar on the right controls the scale (zoom) of this window and the horizontal one scans along the sequence. The 'base view' (just below the overview) shows the sequence of both strands and, above and below these, the translation in all six frames; again, the scroll bars control the zoom and position of this window. The bottom third of the screen shows a list of annotated features. Many other aspects of the sequence can be displayed or hidden (see text).

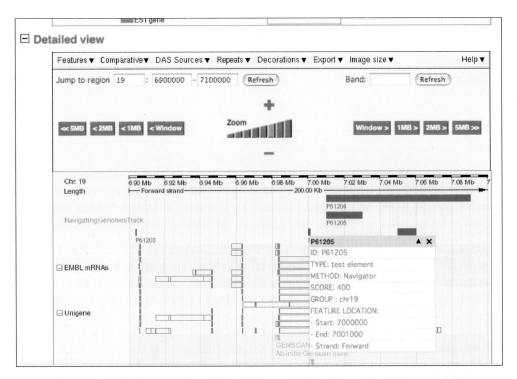

Figure 6. Screenshot showing part of the 'Detailed view' panel of the 'Contig View' page of the Ensembl genome browser (see page 36).
The data that was uploaded in *Protocol 7* is shown (dark bars just below the chromosome length scale) in the track called 'NavigatingGenomesTrack'. In this shot, the user has clicked on one of the uploaded features (P61205), and the small pop-up window displays information about this feature.

Chapter 3. Sequence similarity searches

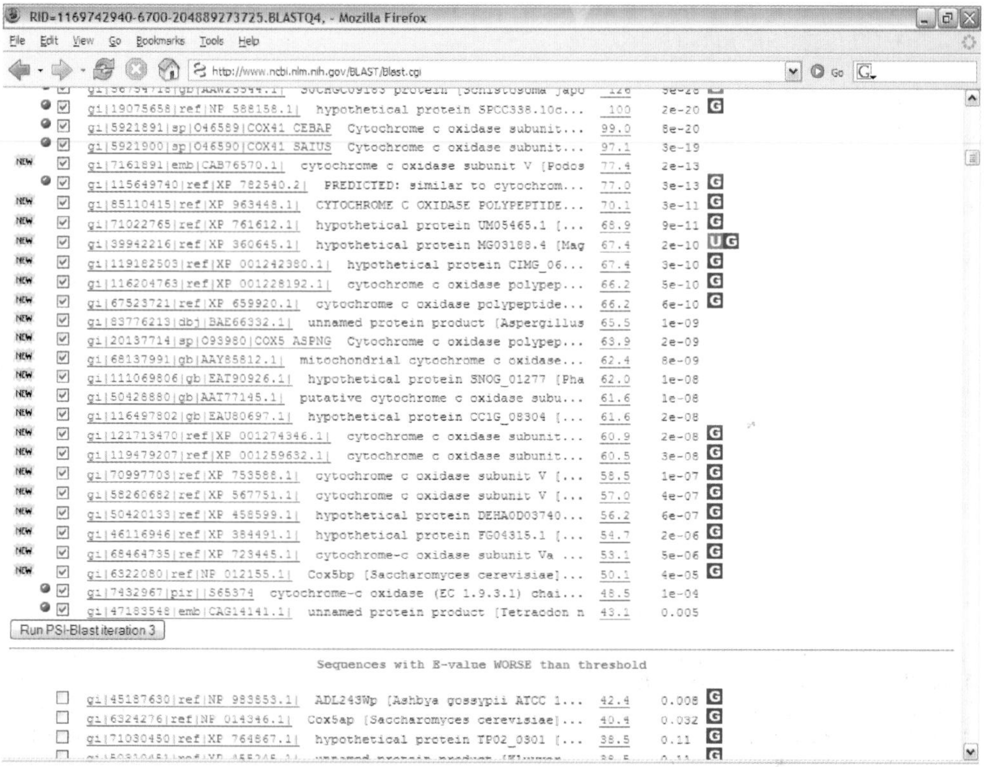

Figure 4. PSI-BLAST output (see page 64).
Part of the results after the second PSI-BLAST iteration are shown. The output format is essentially the same as for BLASTP (see *Fig. 2*), but new family members found in this iteration are indicated by 'NEW'. Family members found in the previous round are indicated by green dots. Sequences shown below the horizontal dividing line have too high an *E* value; only those above the line will be used in compiling the PSSM (or 'profile') of the protein family for the next iteration of the search.

Chapter 4. Gene prediction

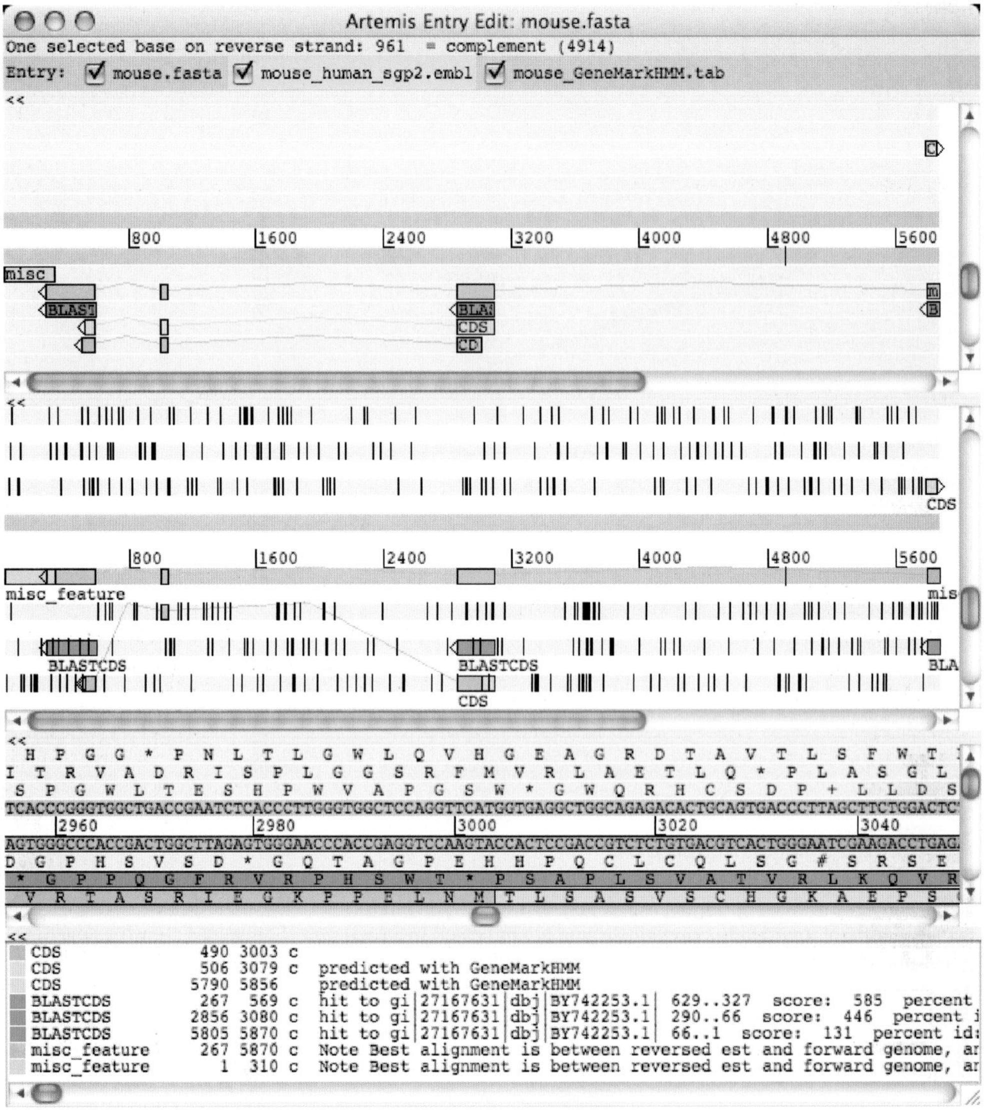

Figure 3. Mouse gene predictions viewed in ARTEMIS (see page 89).
The track immediately below the distance scale in the top section shows reversed alignment to EST
BB280527 (extreme left), produced by EST2GENOME (see *Protocol 11*). The track below this shows the
EST2GENOME alignment to BY742253, showing alignment of four parts of this EST to parts of the genomic
sequence. The track below that shows three BLASTN hits (labeled BLASTCDS) to *M. musculus* ESTs.
Beneath this, a three-exon gene predicted by GENEMARK.HMM is shown; a second GENEMARK prediction – a
small single-exon gene – appears at the extreme right on the forward strand (shown above the distance
scale). The lowest track in this section (just above the top-most horizontal scroll bar) shows the gene
predicted by SGP2. The second section of the screen (between the top two horizontal scroll bars) shows
the same region, but with the three reading frames represented as individual lines above (forward strand)
and below (reverse strand) the distance scale; features are shown in the relevant reading frame; vertical
lines are stop codons. The third section shows a close-up view of part of the sequence (nt 2960–3040)
including nucleotide and amino acid sequences. The bottom section lists the annotated features.

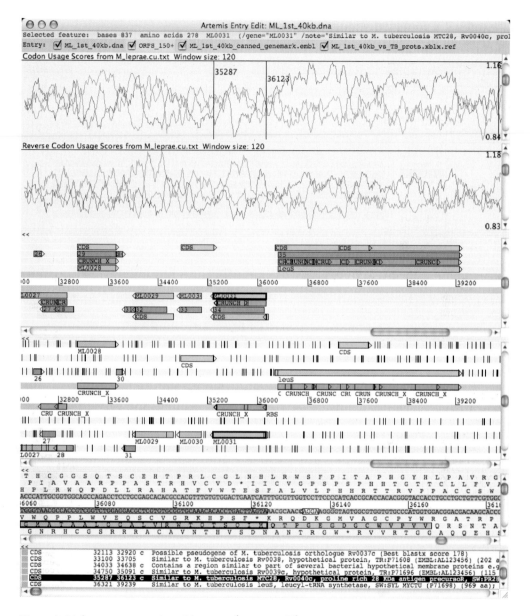

Figure 4. *M. leprae* genome viewed in ARTEMIS (see page 97).
The top section shows the codon usage plots for the forward (upper) and reverse (lower) strands in each frame. In the section below this, the tracks immediately above and below the distance scale show the CDSs of the published annotation (forward and reverse strands). The next tracks (working outwards from the distance scale) show the BLASTX matches (labeled 'CRUNCH X') to the *M. tuberculosis* protein set. The next tracks show the GENEMARK.HMM predictions (numbered) and then the trimmed ORFs of >150 amino acids (labeled CDS). The lower sections of the screen show the same information in greater detail, with the annotations listed in the bottom section.

Chapter 7. Expressed sequence tags

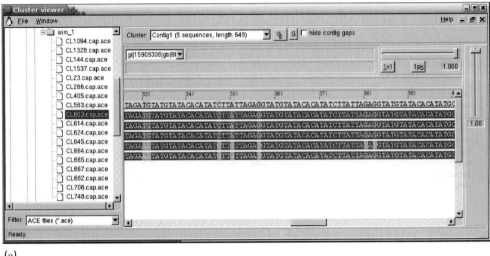

(a)

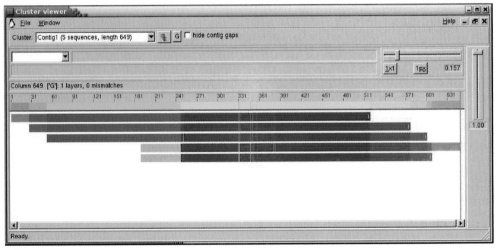

(b)

Figure 6. Visualizing cluster assemblies using CLVIEW (see page 160).
CLVIEW presents cluster assemblies in a zoomed view (*a*), displaying a directory tree on the left of the window and the aligned DNA sequences on the right; or in an overview of the assembled reads (*b*), displaying the consensus sequence with a yellow background, followed by a pile of aligned reads marked with a blue background. Base discrepancies are labeled in red and may represent potential SNPs.

Chapter 8. Protein structure, classification, and prediction

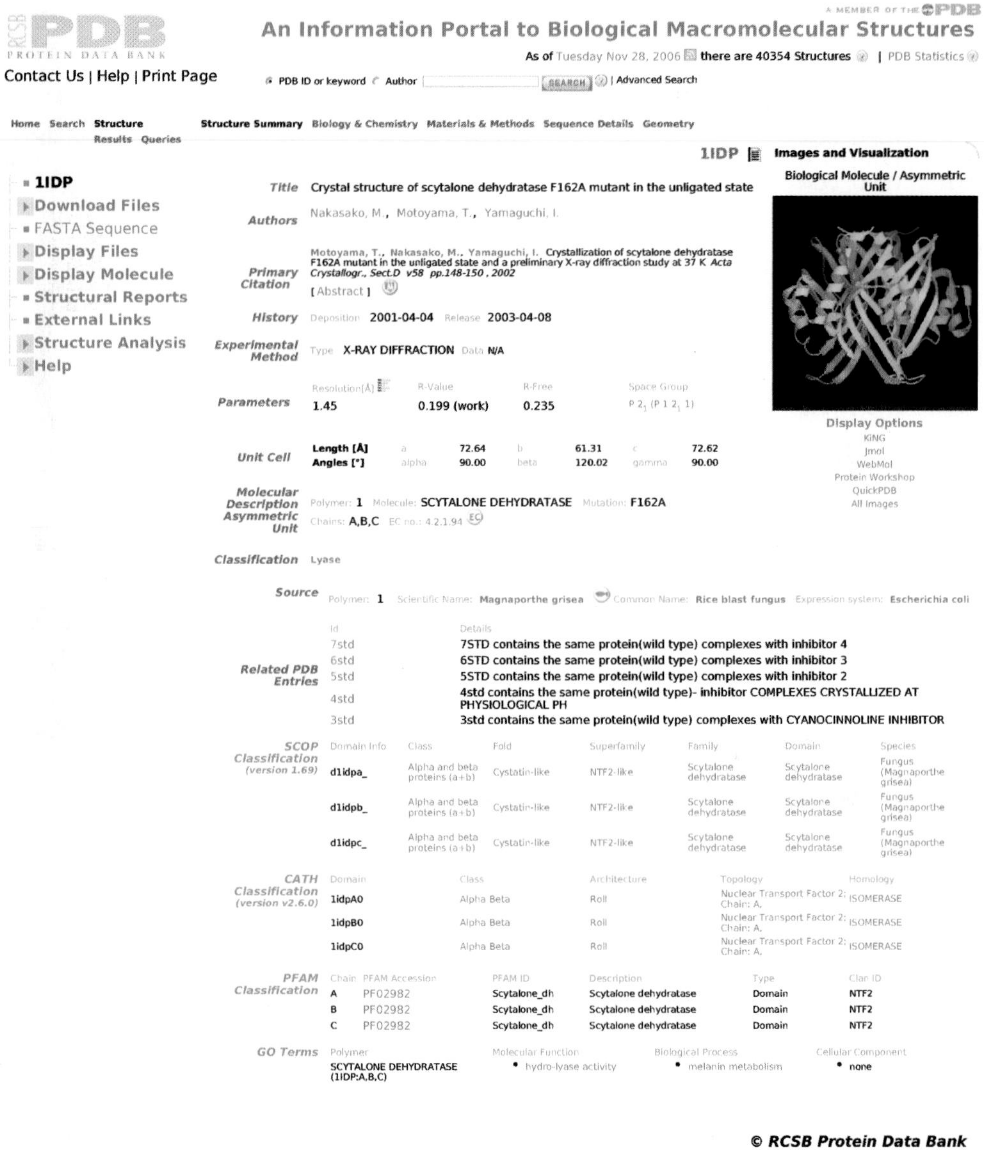

A MEMBER OF THE PDB

RCSB PDB PROTEIN DATA BANK

An Information Portal to Biological Macromolecular Structures

As of Tuesday Nov 28, 2006 there are 40354 Structures | PDB Statistics

Contact Us | Help | Print Page

PDB ID or keyword Author SEARCH | Advanced Search

Home Search Structure Results Queries **Structure Summary** Biology & Chemistry Materials & Methods Sequence Details Geometry

- **1IDP**
- Download Files
- FASTA Sequence
- Display Files
- Display Molecule
- Structural Reports
- External Links
- Structure Analysis
- Help

1IDP

Title Crystal structure of scytalone dehydratase F162A mutant in the unligated state

Authors Nakasako, M., Motoyama, T., Yamaguchi, I.

Primary Citation Motoyama, T., Nakasako, M., Yamaguchi, I. Crystallization of scytalone dehydratase F162A mutant in the unligated state and a preliminary X-ray diffraction study at 37 K *Acta Crystallogr., Sect.D v58 pp.148-150 , 2002*
[Abstract]

History Deposition 2001-04-04 Release 2003-04-08

Experimental Method Type X-RAY DIFFRACTION Data N/A

Parameters

Resolution[Å]	R-Value	R-Free	Space Group
1.45	0.199 (work)	0.235	$P\ 2_1$ ($P\ 1\ 2_1\ 1$)

Unit Cell

Length [Å]	a	72.64	b	61.31	c	72.62
Angles [°]	alpha	90.00	beta	120.02	gamma	90.00

Molecular Description Asymmetric Unit Polymer: 1 Molecule: SCYTALONE DEHYDRATASE Mutation: F162A
Chains: A,B,C EC no.: 4.2.1.94

Classification Lyase

Source Polymer: 1 Scientific Name: Magnaporthe grisea Common Name: Rice blast fungus Expression system: Escherichia coli

Related PDB Entries

Id	Details
7std	7STD contains the same protein(wild type) complexes with inhibitor 4
6std	6STD contains the same protein(wild type) complexes with inhibitor 3
5std	5STD contains the same protein(wild type) complexes with inhibitor 2
4std	4std contains the same protein(wild type)- inhibitor COMPLEXES CRYSTALLIZED AT PHYSIOLOGICAL PH
3std	3std contains the same protein(wild type) complexes with CYANOCINNOLINE INHIBITOR

SCOP Classification (version 1.69)

Domain Info	Class	Fold	Superfamily	Family	Domain	Species
d1idpa_	Alpha and beta proteins (a+b)	Cystatin-like	NTF2-like	Scytalone dehydratase	Scytalone dehydratase	Fungus (Magnaporthe grisea)
d1idpb_	Alpha and beta proteins (a+b)	Cystatin-like	NTF2-like	Scytalone dehydratase	Scytalone dehydratase	Fungus (Magnaporthe grisea)
d1idpc_	Alpha and beta proteins (a+b)	Cystatin-like	NTF2-like	Scytalone dehydratase	Scytalone dehydratase	Fungus (Magnaporthe grisea)

CATH Classification (version v2.6.0)

Domain	Class	Architecture	Topology	Homology
1idpA0	Alpha Beta	Roll	Nuclear Transport Factor 2: Chain: A.	ISOMERASE
1idpB0	Alpha Beta	Roll	Nuclear Transport Factor 2: Chain: A.	ISOMERASE
1idpC0	Alpha Beta	Roll	Nuclear Transport Factor 2: Chain: A.	ISOMERASE

PFAM Classification

Chain	PFAM Accession	PFAM ID	Description	Type	Clan ID
A	PF02982	Scytalone_dh	Scytalone dehydratase	Domain	NTF2
B	PF02982	Scytalone_dh	Scytalone dehydratase	Domain	NTF2
C	PF02982	Scytalone_dh	Scytalone dehydratase	Domain	NTF2

GO Terms

Polymer	Molecular Function	Biological Process	Cellular Component
SCYTALONE DEHYDRATASE (1IDP:A,B,C)	• hydro-lyase activity	• melanin metabolism	• none

Images and Visualization

Biological Molecule / Asymmetric Unit

Display Options
KiNG
Jmol
WebMol
Protein Workshop
QuickPDB
All Images

Figure 5. A page from the RCSB site, showing the Structure Explorer summary page for entry 1idp (*M. grisea* scytalone dehydratase) (see page 175).
Bibliographical information is shown, as well as some data about the structure and its determination, links to other databases, and a picture.

PolyPhobius prediction

Prediction of Q2PR35|Q2PR35_FUGRU

```
ID    Q2PR35|Q2PR35_FUGRU
FT    TOPO_DOM       1      25      NON CYTOPLASMIC.
FT    TRANSMEM      26      50
FT    TOPO_DOM      51      59      CYTOPLASMIC.
FT    TRANSMEM      60      80
FT    TOPO_DOM      81      97      NON CYTOPLASMIC.
FT    TRANSMEM      98     120
FT    TOPO_DOM     121     140      CYTOPLASMIC.
FT    TRANSMEM     141     162
FT    TOPO_DOM     163     196      NON CYTOPLASMIC.
FT    TRANSMEM     197     222
FT    TOPO_DOM     223     238      CYTOPLASMIC.
FT    TRANSMEM     239     260
FT    TOPO_DOM     261     272      NON CYTOPLASMIC.
FT    TRANSMEM     273     292
FT    TOPO_DOM     293     312      CYTOPLASMIC.
//
```

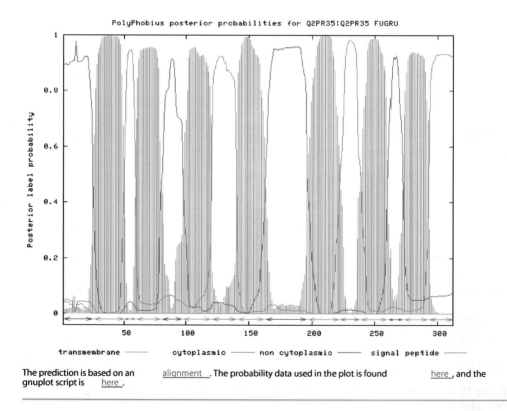

The prediction is based on an   alignment   . The probability data used in the plot is found   here , and the gnuplot script is   here .

Figure 13. Output of the PHOBIUS website, showing an example of the prediction of transmembrane regions and signal sequences (see page 191).

Coils output for FOS_CHICK
Proto-oncogene protein c-fos

[ISREC-Server] Date: Mon Nov 27 23:53:01 Europe/Zurich 2006

```
coils -def -in=../wwwtmp/.COILS.29000.7081.seq
-out=../wwwtmp/.COILS.29000.7081.out -mat=2
```

```
# COILS version 2.1
# using MTIDK matrix
# no weights
# Input file is ../wwwtmp/.COILS.29000.7081.seq
#>FOS_CHICK Proto-oncogene protein c-fos, 419 bases, 4DDEA701 checksum.
```

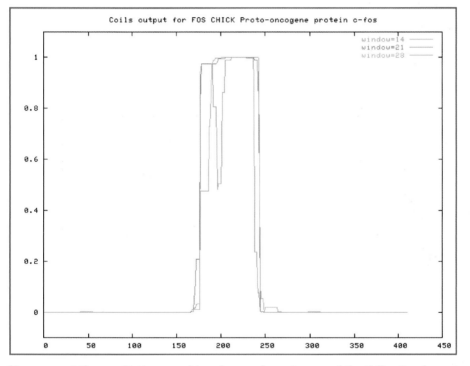

You can get the prediction graphics shown above in one of the following formats:

- GIF-format
- Postscript-format
- numerical format (window 14, 21, 28)

Back to ISREC home page

Figure 15. Output of the COILS website, showing an example of the rediction of coiled-coiled regions (see page 193).

Chapter 10. Prediction of protein function

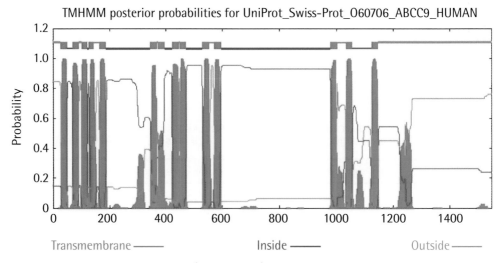

Figure 4. Graphical output from TMHMM (see page 219).
The lower part shows the probability that each part of the sequence lies inside or outside the cell or in a transmembrane helix, whilst the upper part predicts the organization of the protein.

Figure 7. Helical representation of the sequence of the human *fos* oncogene using PEPNET from the EMBOSS suite of programs (see page 223).
Note the leucine zipper between positions 165 and 193. (Only the first 231 amino acids of the protein are shown here.)

PPSearch Output

```
------------------------------------------------------------
|            ppsearch (c) 1994 EMBL Data Library           |
|        based on MacPattern (c) 1990-1994 R. Fuchs        |
------------------------------------------------------------

PROSITE pattern search started: Thu Jun 15 12:56:22 2006

Sequence file: /ebi/extserv/old-work/ppsearch-20060615-12562181829685.input

----------------------------------------
Sequence /ebi/extserv/old-work/ppsearch-20060615-12562181829685.input (260
residues):

Matching pattern PS00036 BZIP_BASIC:
   74: RRKLKNRVAAQTARDR
Total matches: 1

Total no of hits in this sequence: 1

==========================================

1311 pattern(s) searched in 1 sequence(s), 260 residues.
Total no of hits in all sequences: 1.
Search time: 00:00 min
```

Figure 10. Expected output of PPSEARCH **when used to search for patterns in 'USERSEQ1_fasta.txt' (see page 226).**

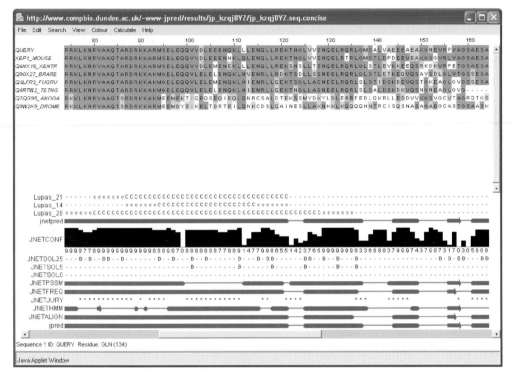

Figure 14. Jpred output shown in JALVIEW (see page 232).
The upper part of the screen shows the alignment of the query protein with others of significant similarity (shading indicates conservation) and contains a screen image of the results obtained when analyzing the sequence above using JALVIEW. Below this, the lines beginning 'Lupas' indicate predicted coiled-coil regions (at three different window sizes). The 'JNETSOL' lines indicate which residues are likely to be accessible to solvent ('B' indicates a buried residue – one that is less than 25%, less than 5% or 0% exposed to solvent in the three tracks). The remainder of the tracks all relate to secondary structure predictions: 'JNETPSSM', 'JNETFREQ', 'JNETHMM', and 'JNETALIGN' show predictions made by various methods (red tubes are α-helices; green arrows are β-sheets); asterisks in the 'JNETJURY' track show where these predictions disagreed and had to be resolved, whilst the 'jnetpred' track shows the consensus secondary-structure prediction. Finally, the histogram (and the corresponding numerical values beneath it) show the reliability of the prediction shown in 'jnetpred': for example, the large helix on the left includes one region towards its right end where the prediction is less certain, approximately at residues 113–116.

```
CLUSTAL (1.0) multiple sequence alignment


Sequence      MVVVAAAPNPADGTPKVLLLSGQPASAAGAPAAR-LPLMVPAQRGASPEAASGGLPQARK 59
XBP1_HUMAN    MVVVAAAPNPADGTPKVLLLSGQPASAAGAPAGQALPLMVPAQRGASPEAASGGLPQARK 60
XBP1_RAT      MVVVAAAPSAASAAPKVLLLSGQPASGG-----RALPLMVPGPRAAGSEAS--GTPQARK 53
XBP1_MOUSE    MVVVAAAPSAATAAPKVLLLSGQPASGG-----RALPLMVPGPRAAGSEAS--GTPQARK 53
              ********..* .:*************..    : ******. *.*..**:  * *****


Sequence      RQRLTHLSPEEKALRRKLKNRVAAQTARDRKKARMSELEQQVVDLEEENQKLLLENQLLR 119
XBP1_HUMAN    RQRLTHLSPEEKALRRKLKNRVAAQTARDRKKARMSELEQQVVDLEEENQKLLLENQLLR 120
XBP1_RAT      RQRLTHLSPEEKALRRKLKNRVAAQTARDRKKARMSELEQQVVDLEEENQKLQLENQLLR 113
XBP1_MOUSE    RQRLTHLSPEEKALRRKLKNRVAAQTARDRKKARMSELEQQVVDLEEENHKLQLENQLLR 113
              ************************************************* :** *******


Sequence      EKTHGLVVENQELRQRLGMDALVAEE--EAEAKGNEVRPVAGSAESAALRLRAPLQQVQA 177
XBP1_HUMAN    EKTHGLVVENQELRQRLGMDALVAEE--EAEAKGNEVRPVAGSAESAALRLRAPLQQVQA 178
XBP1_RAT      EKTHGLVIENQELRTRLGMNALVTEEVSEAESKGNGVRLVAGSAESAALRLRAPLQQVQA 173
XBP1_MOUSE    EKTHGLVVENQELRTRLGMDTLDPDEVPEVEAKGSGVRLVAGSAESAALRLCAPLQQVQA 173
              *******:****** ****::* .:*  *.*:**. ** ************ ********


Sequence      QLSPLQNISPWILAVLTLQIQSLISCWAFWTTWTQSCSSNALPQSLPAWRSSQRSTQKDP 237
XBP1_HUMAN    QLSPLQNISPWILAVLTLQIQSLISCWAFWTTWTQSCSSNALPQSLPAWRSSQRSTQKDP 238
XBP1_RAT      QLSPPQNIFPWILTLLPLQILSLISFWAFWTSWTLSCFSNVLPQSLLIWRNSQRSTQKDL 233
XBP1_MOUSE    QLSPPQNIFPWTLTLLPLQILSLISFWAFWTSWTLSCFSNVLPQSLLVWRNSQRSTQKDL 233
              **** *** ** *::*.*** **** *****:** ** **.*****  **.********


Sequence      VPYQPPFLCQWGRHQPSWKPLMN----------- 260
XBP1_HUMAN    VPYQPPFLCQWGRHQPSWKPLMN----------- 261
XBP1_RAT      VPYQPPFLCQWGPHQPSWKPLMNSFVLTMYTPSL 267
XBP1_MOUSE    VPYQPPFLCQWGPHQPSWKPLMNSFVLTMYTPSL 267
              ************ **********
```

Figure 15. A multiple alignment using DBCLUSTAL (see page 234).

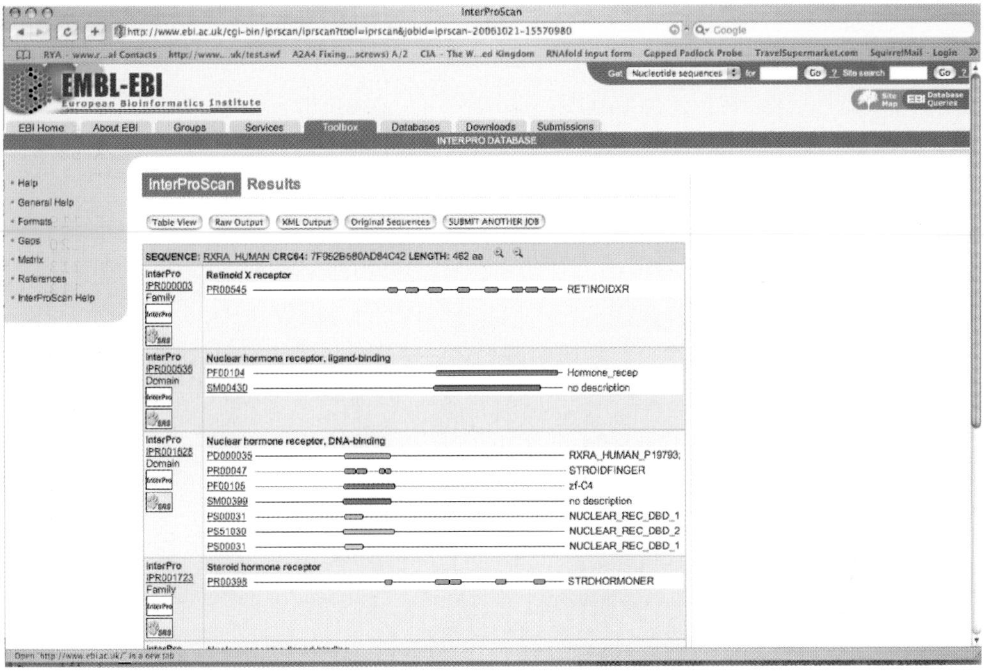

Figure 16. Part of the results of INTERPROSCAN for the protein RXRA_HUMAN (see page 237).

CHAPTER 1

Database resources for wet-bench scientists

Neil Hall and Lynn M. Schriml

1. INTRODUCTION

With the increasing amount of data being generated by genomic-scale studies, it has become much more important for biologists to store data in a structured way that makes it easily accessible and allows the integration of different data types from different sources. Hence, there are now hundreds of specialized databases available to biological researchers that cover a vast array of different data types from transcription factor binding sites to metabolic pathways and from protein domains to scientific journals. For these data resources to be properly exploited, one has to understand how the data is generated and structured, otherwise there is a danger that important data may be missed or, worse still, incorrect data blindly trusted.

Because of the number and diversity of data resources available, we cannot describe them all in a single chapter, so here we will describe the publicly available databases at the National Center for Biotechnology Information (NCBI) (1) (http://www.ncbi.nlm.nih.gov/ [1.1]) and provide examples of how you can query different information using their online tools. This chapter complements Chapter 2, which explores resources for navigating sequenced genomes with the emphasis on a second major group of tools – Ensembl. It should be noted also that there are many other tools available at different web sites and we will mention some of these later in this chapter.

1.1 Types of databases

The publicly available web resources described in this chapter can be divided into two types: primary and secondary databases. Whilst both types of database serve a useful purpose, one has to understand the distinction before making assertions based on a database query.

Bioinformatics: *Methods Express* (Paul H. Dear, ed.)
© Scion Publishing Limited, 2007

1.1.1 Primary databases

Primary databases include all of the repositories of the primary output from experimental work, such as GenBank (2) (http://www.ncbi.nlm.nih.gov/Genbank/index.html [1.2]), which contains nucleotide sequences, and ArrayExpress (3) (http://www.ebi.ac.uk/arrayexpress/ [1.3]) at the European Bioinformatics Institute (EBI), which contains microarray expression data. These repositories generally house information that is submitted by the scientist who generated it and little is done to process, curate, or provide quality control over what is entered. Therefore, they are usually very comprehensive, but one must always treat the data with caution.

1.1.2 Secondary databases

Secondary databases are less all-inclusive than the primary databases, but instead concentrate on data quality and include additional information and cross-referencing. Secondary databases usually draw on (and may be linked to) a number of primary databases, collecting together information centered on a particular topic. For example, Pfam (3) (http://www.sanger.ac.uk/Software/Pfam/ [1.4]) is a secondary database that curates protein domains and allows users to search proteins for known domains, whereas the Mouse Genome Informatics (MGI) (http://www.informatics.jax.org/ [1.5]) collects and curates genomic, genetic, and functional data associated with the laboratory mouse. There is a clear distinction between these two examples of secondary databases: Pfam covers one theme (protein domains) and does so for all organisms, whereas MGI (4) covers many types of data but for only one species.

1.2 Database resources at NCBI

At the time of writing, NCBI has over 20 databases, which can be searched either individually or en masse. Entrez (5) is a system that provides a common interface to all of the major NCBI databases, including PubMed (6), nucleotide and protein sequences, protein structures, complete genomes, taxonomy, and many others. It provides a consistent user interface and format, and allows queries to be made across multiple NCBI databases at once. The starting point for Entrez is http://www.ncbi.nlm.nih.gov/gquery/gquery.fcgi [1.6], and an overview, showing all of the Entrez-accessible databases and the connections between them, is available at http://www.ncbi.nlm.nih.gov/Database/datamodel/ [1.7].

Each of these databases (which are often called nodes) will contain entries with unique identifiers (UIDs). These identifiers are stable over time, whilst the data associated with them can change. For example, a gene will always have the same UID, but over time we may discover a new function for it or new splice sites so its annotation and sequence could change. These entries can be linked within and between nodes, which allows, for example, publication entries to be linked to protein entries. Each node has a specific entry format, although many features will recur among the different nodes. For example, there is a fully annotated GenBank record at http://www.ncbi.nlm.nih.gov/Sitemap/samplerecord.html [1.8]. Of particular interest are the three identifiers you will find in this single record: the *locus* name, the *accession number*, and the *GI*. The locus name (in this example

SCU49845) is unique to each entry and in many cases it may be identical or similar to the accession number (in this case, U49845). The accession number will always remain the same even if the entry is changed. There is also a *version number*, which indicates how many times the entry has updated: U49845.1 indicates that this is the first version of this entry. The GI is the 'GenInfo Identifier': if a sequence or protein translation changes in any way, a new GI number will be assigned, so GI sequence identifiers run parallel to the version numbers.

1.2.1 Nucleotide databases at NCBI

The main nucleotide database at NCBI (and, like the rest, accessible through Entrez) is GenBank (2), which contains all of the publicly available DNA sequences. However, there are subdivisions of GenBank that can be searched independently of the complete database. Depending on what you are looking for, it may be better to search a subdivision rather than the whole dataset. For example dbEST (7, 8) contains only that subset of sequence data and other information that relates to 'single-pass' cDNA sequences or expressed sequence tags (ESTs); similarly, dbGSS contains only single-pass genome survey sequences.

Further nucleotide databases (again, Entrez-accessible) exist outside GenBank. For example, dbSNP (9) (http://www.ncbi.nlm.nih.gov/entrez/query.fcgi?db=snp [1.9]) is a database of single-nucleotide polymorphisms, small insertions, and deletions. There is also a sequencing trace archive (http://www.ncbi.nlm.nih.gov/Traces/trace.cgi [1.10]), which contains sequences that have been submitted along with all of their underlying experimental data, so that you can view the original trace file that generated them. This data could be particularly useful if you wish to check the validity of a frameshift or insertion in a gene of interest.

1.2.2 Protein databases at NCBI

Protein databases can contain different levels of information from primary sequence to secondary and tertiary structures. As well as databases that contain entire peptide sequences, there are a number of resources dedicated to collecting and curating protein domains and motifs. The NCBI protein database is a concatenation of a number of subdatabases: it includes sequences from Swiss-Prot (10, 11), the Protein Information Resource (PIR) (12), the Protein Research Foundation (PRF), and the Protein Data Bank (PDB) (13), along with protein sequences translated from nucleotide sequences of RefSeq (10, 11) and GenBank. Therefore, when you search using the Entrez server, you will be doing a comprehensive search of all publicly available sequences. Like the nucleotide databases, these proteins will be based on variable data quality depending on their source. For example, Swiss-Prot has highly curated annotations and should be nonredundant, whereas the translations from GenBank will contain sequence errors and misannotations in a number of cases.

Additional protein-related information is available in NCBI's Structure database (http://www.ncbi.nlm.nih.gov/Structure/ [1.11]) and NCBI's Conserved Domains Database (CDD) (http://www.ncbi.nlm.nih.gov/Structure/cdd/cdd.shtml [1.12]). CDD contains domains from Pfam (14), Simple Modular Architecture Research Tool (SMART) (15), and Clusters of Orthologous Groups (COG) (16), as well as other domains curated at NCBI. The major utility of the domain database is for identifying domains in a protein sequence, which will allow the user to infer a function (also see Chapter 8).

1.2.3 Other databases at NCBI

As well as the major nucleotide and protein databases, NCBI houses a number of other related databases (nodes) that are linked to the sequence databases, as well as being themselves browsable and searchable through Entrez. One of the most commonly used databases is PubMed, which is a repository of biomedical journal articles. As well as being searchable using text queries, all of the articles in PubMed are linked to other NCBI entries related to them, such as nucleotide sequences. Similarly, there are databases of chemical structures (PubChem), microarray experiments (Gene Expression Omnibus, GEO) (17), taxonomy (Entrez Taxonomy), genes (Entrez Gene), maps (Map Viewer), and inherited diseases (Online Mendelian Inheritance in Man, OMIM) (18, 19) among others. Whilst some of these datasets may not seem obviously useful to your particular area of research, much of their functionality is derived from the fact that related records in each database are all linked so the user is able to traverse between datasets by following the links provided. For example, search the PubChem database for 'ethanol' and it will return not only a structure and description of the compound but also links to protein databases of enzymes that bind ethanol, as well as toxicology reports in the National Library of Medicine and relevant publications in PubMed.

2. METHODS AND APPROACHES

Here we discuss, in a little more detail, some of the tools available at NCBI through Entrez, tailored for searching their nucleotide, protein, and other databases. Additionally, we then provide a set of protocols, illustrating how to answer specific typical bioinformatics questions.

It is not possible to cover more than a tiny fraction of the resources, tools, and methods of query that are available through Entrez. However, we suggest that you start with the examples in the protocols and then take these as starting points to explore on your own.

2.1 Searching databases at NCBI

2.1.1 Text searches

The simplest and broadest search of NCBI databases is offered via the Entrez entry page: http://www.ncbi.nlm.nih.gov/gquery/gquery.fcgi [1.6]. Here, you can perform a simple text-based search across all databases or you can choose a specific NCBI database to search such as PubMed or Nucleotide. If you are searching across all databases, the simplest search you can perform is a text search of all fields. *Protocol 1* gives a simple example.

Protocol 1

A simple text search for *Plasmodium* across all NCBI databases using Entrez

1. Start at the NCBI home page: http://www.ncbi.nlm.nih.gov/ [1.1] and select **Entrez home** on the right to go to the Entrez cross-database search page: http://www.ncbi.nlm.nih.gov/gquery/gquery.fcgi [1.6].

2. In the text box towards the top of the screen ('Search across all databases'), type 'Plasmodium'. Click the adjacent **Go** button, or press 'Enter' on your keyboard.

3. The page will be updated. Next to each of the database names and icons in the lower part of the screen will be a number (or, in a few cases, 'none'). This indicates the number of entries in each of these databases that contained, anywhere within it, the word 'Plasmodium'.

4. Many of these records will not relate to *Plasmodium* itself. For example, some will describe proteins from other species that are noted as interacting with, or being similar to, *Plasmodium* proteins.

5. We therefore want to limit the search to entries in which the organism is *Plasmodium*. To do this, repeat the query, this time using the text 'Plasmodium[ORGANISM]'[a]

6. The page will be updated, this time giving the number of entries (in each of the databases) in which *Plasmodium* is in the 'Organism' field.

7. Clicking on any of the results will take you to the respective results page, listing all of the *Plasmodium* entries found. For example, clicking on **Nucleotide** or the adjacent icon will take you to the start of a list of over 200 000 *Plasmodium* nucleotide sequence entries.

Note

[a]A list of fields that can be searched is given here: http://www.ncbi.nlm.nih.gov/entrez/query/static/help/Summary_Matrices.html#Search_Fields_and_Qualifiers [1.13]. Many fields can be abbreviated; for example, 'ORGN' can be used in place of 'ORGANISM'.

Search terms can also be combined; for example, searching for 'malaria AND mosquito' will find all entries that contain (anywhere within the entry) both 'malaria' *and* 'mosquito'. Similarly, 'Plasmodium NOT Plasmodium[ORGANISM]' will find all entries that refer to *Plasmodium*, but that do not originate from the organism *Plasmodium* itself. More sophisticated searches can be made by querying each database individually, rather than globally. The advantage of this is that it will allow you to define your search using fields specific to that database. It is also possible to view a 'history' of previous searches and to combine these together to refine the search further. *Protocol 2* gives a simple example of searching a single database, using 'limits' and 'history' to build up a progressively more refined query.

Protocol 2

A search in PubMed using Limits and History

1. Navigate to the Entrez entry page (http://www.ncbi.nlm.nih.gov/gquery/gquery.fcgi [1.6]) and click on the **PubMed** link towards the top of the page.

2. In the text field at the top of the page, enter `malaria` and click the adjacent **Go** button.

3. The result is a list of over 40 000 (at the time of writing) entries that contain the word 'malaria' in any part of the entry.

4. We will repeat this search, restricting it to articles with 'malaria' in their title. Click on the **Limits** tab (just below the text entry box). Scroll to the bottom of the new page to find the pull-down menu **Default tag**. Select **Title** from the pull-down menu. Click the adjacent **Go** button (or scroll back up the page and click the one at the top).

5. The result is a list of about 20 000 articles, all with 'malaria' in their title.

6. We will now look at our previous searches and combine them to refine the search. Click on the **History** tab.

7. Into the text entry field, type `mosquito`. Also, click on the ticked box on the **Limits** tab to 'untick' it. This removes our previous limits settings.

8. Below, you will see a list of your most recent searches. The top-most one in the list will be:

    `#xx Search malaria Field: Title`

    where 'xx' is a number. Click on the number.

9. A pop-up menu of options will appear, asking how you want to combine your previous search (for articles with 'malaria' in the title) with the current search (for 'mosquito', not limited to the title). Click on **AND**.

10. The text box should now show:

    `(mosquito) AND (#xx)`

    meaning that we are about to search for records that contain mosquito in any part of the entry and that also contain 'malaria' in the title (from our previous search). Click **Go**.

11. The result is around 3500 entries (at the time of writing), each with 'malaria' in the title and with 'mosquito' somewhere in the entry.

NCBI provides a web page giving further details of how to search their databases at http://w ww.ncbi.nlm.nih.gov/entrez/query/static/help/helpdoc.html# Searching [1.14]. The following protocols give some examples of other ways to search the NCBI databases. The examples are in no way exhaustive, but they will introduce you to a range of search types that can form the basis of your own explorations.

Protocol 3

Determining the set of web resources available for a genome

1. Start at NCBI's home page (http://www.ncbi.nlm.nih.gov/ [1.1]) and click on the **Genomic Biology** link in the left blue bar. This link takes you to the genomic biology page: http://www.ncbi.nlm. nih.gov/Genomes/ [1.15].

2. Under **Genome resources** on the right, select **Eukaryotic** to go to an alphabetic list of genome projects, listed by species: http://www.ncbi.nlm.nih.gov/genomes/leuks.cgi [1.16].

3. Scroll down to find **Plasmodium falciparum 3D7** (at the time of writing, only one genome project is listed for this strain). On the same line, you will see a number of links, including a taxonomic identifier ('**36329**') and a link to the sequencing consortium's home page.

4. You will also see, at the right of the line, a series of colored abbreviations for different NCBI databases (**PM**, **R**, **G**, etc.). Clicking on any one of these will bring up data on *Plasmodium falciparum* 3D7 from the appropriate database. For example, clicking on **G** will show you all of the entries in the Genes database for this organism. If necessary, use your browser's 'back' button to return to http://www.ncbi.nlm.nih.gov/genomes/leuks.cgi [1.16].

5. Clicking on the organism name at the left will take you to the Genome Project database and display entries for *P. falciparum*, offering further links to data and resources for this organism.

NCBI Map Viewer provides one way to access positional genome information and to integrate it into searches. *Protocol 4* takes you through a typical use of Map Viewer.

Protocol 4

Finding sequence-tagged site (STS) markers on chromosome 3 of *P. falciparum* using the NCBI Map Viewer

1. Navigate to the Map Viewer home page (http://www.ncbi.nlm.nih.gov/mapview/ [1.17]) by the **Map Viewer** link from the NCBI home page (http://www.ncbi.nlm.nih.gov/ [1.1]).

2. On the Map Viewer home page, select **Plasmodium falciparum** (http://www.ncbi.nlm.nih.gov/ mapview/map_search.cgi?taxid=36329 [1.18]) from the pull-down **Search** menu (leave the text field empty) and click **Go**. You will be taken to the Map Viewer page for *P. falciparum*, including an ideogram of the karyotype.

3. Enter '**STS**' and '**3**' in the text fields **Search for** and **on chromosome(s)**, respectively, and click **Find**.

4. The results of your query are presented as hits on chromosome ideograms and in a tabular format. View the results in the Map Viewer graphical display by clicking on the **3** underneath the chromosome in the ideogram (to show all STSs) or by clicking on the blue links in the table below. Click on the first Map Element in the table (at the time of writing, this was **Pf2541**).

5. The resulting page will show STSs in a part of the chromosome, with the chosen STS (Pf2541) indicated. Clicking on its name will call up further information on that STS, including the polymerase chain reaction primer sequences.

Protocol 5

Searching for α-tubulin genes in *P. falciparum*

This search could be started from the Entrez Home page (searching all NCBI databases and then selecting those hits from the Gene database). Alternatively, as here, we can navigate to the Entrez Gene page to search only that database.

1. Navigate from the NCBI home page (http://www.ncbi.nlm.nih.gov/[1.1]) to the Entrez home page (http://www.ncbi.nlm.nih.gov/gquery/gquery.fcgi[1.6]) using the link on the right of the screen.

2. Click on the **Gene** link or adjacent icon (left side of screen) to go to the Entrez Gene page (http://www.ncbi.nlm.nih.gov/entrez/query.fcgi?db=gene[1.19]).

3. Into the text box at the top of the screen, type:

   ```
   Plasmodium falciparum[ORGANISM] alpha tubulin
   ```

   (see *Protocol 1* for an explanation of using '[ORGANISM]' to limit a search to entries originating from a species). Click **Go**.

4. The query should return 11 genes (correct at the time of writing) with α-tubulin in the annotation. Click on the link for **PFI0180w** to get a detailed summary of its annotation.

5. In this case, the PubMed reference for the gene (under **Links** on the right of the screen) links back to the genome project paper (Hall *et al.*), rather than to an original paper about α-tubulin. From this, we might assume that this gene's annotation has been predicted by homology to other tubulin genes, rather than having been verified experimentally.

6. Use your browser's 'back' button to return to the Entrez Gene page for PFI0180w. Scrolling down the page to the section headed 'General gene information' shows that all of the Gene Ontology terms relating to this protein (see Chapter 9 for an introduction to Gene Ontology) were assigned on the basis of evidence-code 'IEA' ('Inferred from Electronic Annotation'), confirming our assumption that the annotation was not verified experimentally.

7. Additional examples of gene searches are given on the Entrez Gene home page.

2.1.2 Sequence similarity searches and alignment of transcripts to genomic sequences

A common method of querying sequence databases is by similarity searching. The most well-known tool for similarity searching is BLAST (Basic Local Alignment Search Tool) (17), which allows you to search your query sequence against a database of your choosing. Chapter 3 gives detailed information about BLAST and related tools; here we introduce the use of such tools in the context of the NCBI databases.

Similarity searches of NCBI's nucleotide and protein sequence databases can be restricted to sequences from one or more species either by specifying the organisms in the Options section of the BLAST page or by submitting searches against databases on the organism-specific BLAST pages (http://www.ncbi.nlm.nih.gov/BLAST/[1.20]). Multiple query sequences can also be submitted in the same search using the organism-specific BLAST pages.

Protocol 6

Simple BLAST searches at NCBI

1. Navigate to the BLAST home page (http://www.ncbi.nlm.nih.gov/BLAST/ [1.20]) from the **BLAST** link included in the query bar found at the top of most NCBI pages.

2. The BLAST home page provides links to the suite of BLAST tools for comparisons between nucleotide or protein sequences. Searches may be conducted against highly divergent organisms (discontinuous megablast), the trace archive, the CDD, gene expression data in GEO, single-nucleotide polymorphisms, immunoglobulins, etc. In this case, we will search for proteins related to a *Plasmodium* α-tubulin, so select **protein blast** (under the 'Basic BLAST' heading).

3. On the 'Protein BLAST' page, leave all settings at their default values (note that we will be searching against the 'nr' or non-redundant database).

4. Open a new browser window and find the *P. falciparum* α-tubulin gene PFI0180w (see *Protocol 5*). When you have found the Entrez Gene page for this gene, scroll down to find the heading '**NCBI Reference Sequence (RefSeq)**' and click on the link to the gene product (**XP_001351911.1**). This should bring up the corresponding Entrez Protein page and, scrolling down, you will find the complete amino acid sequence for this protein. Copy this sequence (along with the numbers and spaces) and paste it into the **Search** box on the BLAST page.

5. Click on **BLAST**. You will be taken to a page saying that your request has been successfully submitted. The page will be automatically updated when your results are ready (BLAST searches can take some time to complete).

6. When the results are ready, you will see a diagram representing the best matches. The colored bars indicate the score of the match and the portion of your query sequence that it matches. In this case, there will be many full-length red bars indicating many close matches to the complete α-tubulin sequence.

7. Below this are listed the hits in order of BLAST score (best first). If you click on the accession number of the hit, you will go to the GenBank entry for that protein. If you click on the BLAST score, you will go to the alignment (remember that there may be more than one alignment per hit).

8. Now try repeating the search, but looking only for matches in *Arabidopsis thaliana*. To do this, navigate to the 'Protein BLAST' page (step 3 above), but this time, type `Arabidopsis thaliana` into the 'Organism' text field (under 'Choose Search Set'), before continuing as before. (Note: as you type the organism name, you will be prompted with a list of likely organisms—simply click on **Arabidopsis thaliana** to save typing the complete name.

9. The result this time is a smaller number of matches, some of them shorter or of lower score, to *Arabidopsis* sequences.

BLAST is not always the best tool for sequence alignment and NCBI provides other tools that may be more appropriate for your needs. Alignment of a mRNA or cDNA sequence to a genomic sequence can be computed using NCBI's SPIDEY (17) (http://www.ncbi. nlm.nih.gov/IEB/Research/Ostell/Spidey/ [1.21]) or SPLIGN (http://www.ncbi.nlm.nih.gov/ sutils/splign/splign.cgi [1.22]) alignment tools. Chapters 4 and 7 cover the alignment of transcripts with genomic sequence in more detail. *Protocol 7* gives a simple example of using SPLIGN to compare a cDNA sequence with genomic sequence.

Protocol 7

Aligning a cDNA sequence to genomic sequence

1. From the NCBI home page, select **Tools**. From the tools page (http://www.ncbi.nlm.nih.gov/Tools/[1.23]) click on the **Splign** link (scroll down to find it) to go to the SPLIGN page (http://www.ncbi.nlm.nih.gov/sutils/splign/splign.cgi[1.22]).

2. SPLIGN can be downloaded to run locally or you can submit a cDNA and genomic sequence to SPLIGN at NCBI, which we will do here. Click on the **click here** link (http://www.ncbi.nlm.nih.gov/sutils/splign/splign.cgi?textpage=online&level=form[1.24]) to submit an online job.

3. You will see text boxes to accommodate the cDNA sequence and the genomic sequence with which to align it. In the cDNA box, you can either paste a sequence or specify a sequence by its accession number. In this case, we will specify the cDNA for a chicken cDNA sequence: type the accession number 'AJ744697' into the cDNA box.

4. In the 'Genomic' box, you can again specify the sequence either by pasting it in or by giving an accession number. However, you can also select from a list of whole genome sequences using the pull-down menu underneath. Use this to select **Gallus gallus** (chicken).

5. Click on the **Align** button and wait until the results are ready.

6. The cDNA aligns to two places in the same genomic sequence contig (listed under 'Subject' at the top of the results page). Each alignment is a 'Model'.

7. For each model, the alignment of the cDNA to the genomic sequence is shown. This alignment will return six segments (six putative exons). The yellow boxes at the top represent the cDNA divided into the aligned segments; the genomic sequence is shown below. The vertical blue lines in the cDNA represent indels and the red lines mismatches in the alignments. To view the alignment for each segment, click on the graphical display or on the segment number.

Results of whole genome-to-genome pre-computed sequence comparisons are available in NCBI's Homologene resource, with highly similar sequences being represented as distinct Homologene groups. Text queries of the Homologene database yield results pages of matching Homologene groups containing highly related sequences from multiple organisms.

Pre-computed orthologs can be searched and browsed using Clusters of Orthologous Groups of proteins (COGS): http://www.ncbi.nlm.nih.gov/COG/new/[1.25]. These are genes that are predicted to be functional equivalents due to the fact that they are derived by vertical descent from a single ancestral gene in the last common ancestor of the compared species. You can also compare your gene to known COGS using the kognitor tool: http://www.ncbi.nlm.nih.gov/COG/grace/kognitor.html[1.26].

Results of pre-computed protein comparisons can be viewed at NCBI's BLink (BLAST Link) resource (http://www.ncbi.nlm.nih.gov/sutils/static/blinkhelp.html[1.27]). BLink results are provided by links on other Entrez database pages (e.g. **Entrez Protein**, **Entrez Gene**). BLink provides a tabular display of pre-computed highly related proteins for all organisms in the Entrez Protein database including hits to the CDD (http://www.ncbi.nlm.nih.gov/Structure/cdd/cdd.shtml[1.28]).

2.2 Downloading NCBI datasets

For most scientists, web sites such as NCBI will provide all of the functionality they will ever need for their research. However, for people who want to do a lot of data searches, web sites become impractical and it is sometimes necessary to download entire datasets and software tools for searching them, so that the analysis can be done on a local computer.

NCBI provides this at their FTP site: ftp://ftp.ncbi.nih.gov/ [1.29]. In the blast directory, you will find all of the protein and nucleotide databases available through the NCBI web site, as well as executable files that you can install on Windows, Mac OS X, or Linux operating systems. In the genomes directory, you can download individual genomes. For people who want to create their own scripts that incorporate NCBI datasets or NCBI software tools, there is a set of file standards and software tools called the NCBI toolbox that will allow you to process files, run searches, and format output on your own machine: http://www.ncbi.nlm.nih.gov/IEB/ToolBox/MainPage/index.html [1.30].

One important caveat with installing your own datasets is that you will need to update them constantly as they go out of date, whereas if you are running searches over the internet this is not a problem. You should also test the output of any executable that you install to make sure that it is running properly and that your databases are indexed correctly.

3. TROUBLESHOOTING

- **I can't find my genome at NCBI. Where else can I search?**
  This problem is becoming less widespread but it is not unusual. If a genome is published, then the rule is that it must be submitted to NCBI. However, if the genome project is still ongoing or it is unpublished, then it may not be submitted to GenBank but it may still be available. Many genome centers will submit ongoing projects to the trace archive at NCBI (http://www.ncbi.nlm.nih.gov/Traces/trace.cgi [1.31]), so check there first. If that does not work, many ongoing projects have links from the genome project page: http://www.ncbi.nlm.nih.gov/entrez/query.fcgi?db=genomeprj [1.32]. If you still cannot find it, then you will have to search yourself; the major academic genome centers are listed below as additional web resources.
- **How do I reconcile different versions of a genome at the various sites?**
  Unfortunately, you will find more than one version of a genome depending on where you look. This will happen if a genome project is ongoing but puts an intermediate version in GenBank in order to make the sequence widely available. The genome center may then continue to update their own version whilst leaving the old version in GenBank. GenBank will give a submission date in the top line of the entry. One must keep in mind that the GenBank version will be the 'official' version of the genome; this means that you will be able to give an accession number for this record in publications so that others can see the data. The genome center version is a transient file on a web site that

may disappear at any moment. So, whilst you may want to use the more recent version of the data, remember that it may not be there tomorrow.

4. ADDITIONAL WEB RESOURCES

Listed here is a selection of other web sites that are likely to be useful.

Major primary sequence generators

- The Wellcome Trust Sanger Institute: http://www.sanger.ac.uk/ [1.33]
- JGI Genomes: Eurkaryota, Archae, Bacteria: http://genome.jgi-psf.org/tre_home.html [1.34]
- Human Genome Sequencing Center, Baylor College of Medicine: http://www.hgsc.bcm.tmc.edu/ [1.35]
- The Broad Institute: http://www.broad.mit.edu [1.36]
- The Institute for Genomic Research: http://www.tigr.org [1.37]. Now renamed the J. Craig Venter Institute (JCVI; http://www.jcvi.org)
- Washington University Genome Sequencing Center: http://genome.wustl.edu/ [1.38]
- Genoscope: http://www.genoscope.cns.fr/ [1.39]

Bioinformatics institutes

- The European Bioinformatics Institute: http://www.ebi.ac.uk [1.40]
- National Center for Biotechnology Information: http://www.ncbi.nlm.nih.gov/ [1.1]
- Center for Information Biology and DNA Data Bank of Japan: http://www.cib.nig.ac.jp [1.41]

Genome annotation databases

- KEGG (Kyoto Encyclopedia of Genes and Genomes): http://www.genome.jp/kegg/ [1.42]
- GeneDB (Sanger Institute Pathogen Sequencing Unit annotation database): http://www.genedb.org/ [1.43]
- Ensembl Genomes: http://www.ensembl.org/index.html [1.44]
- CMR (Comprehensive Microbial Resource) Annotated Microbial Genomes: http://pathema.tigr.org/tigr-scripts/CMR/CmrHomePage.cgi [1.45]
- BRC Central (central web site of NIAID Bioinformatics Resource Centers, which houses databases of biodefense-related organisms): http://www.brc-central.org [1.46]
- Genome properties (a database of curated and calculated properties of microbial genomes): http://cmr.tigr.org/tigr-scripts/CMR/shared/GenomePropertiesHomePage.cgi [1.47]

Protein families, domains, and structures

- Pfam (curated protein domains): http://www.sanger.ac.uk/Software/Pfam/ [1.4]

- TIGRFAM (curated protein domains of microbes): http://www.tigr.org/TIGRFAMs/index.shtml [1.48]
- SMART: http://smart.embl-heidelberg.de/ [1.49]
- InterPro (protein families, domains, and functional sites in which identifiable features found in known proteins can be applied to unknown protein sequences): http://www.ebi.ac.uk/interpro/ [1.50]
- Protein data bank (provides a variety of tools and resources for studying the structures of biological macromolecules): http://www.rcsb.org/pdb [1.51]

Miscellaneous

- OBO (Open Biomedical Ontologies: an umbrella web address for well-structured controlled vocabularies for shared use across different biological and medical domains): http://obo.sourceforge.net/ [1.52]
- Amigo (a web interface for browsing gene ontologies, which will allow you to search for genes with specific functions or cellular locations, or that are involved in specific processes): http://www.godatabase.org/ [1.53]

5. REFERENCES

★ 1. **Wheeler DL, Barrett T, Benson DA, et al.** (2006) *Nucleic Acids Res.* **34**, D173–D180. – *This publication gives an overview of the NCBI databases and their associated tools. This reference is the most up to date at the time of writing, but NCBI publishes an overview of changes and updates in the Nucleic Acids Research database issue, which is published every year.*

2. **Benson DA, Karsch-Mizrachi I, Lipman DJ, Ostell J & Wheeler DL** (2006) *Nucleic Acids Res.* **34**, D16–D20.

3. **Parkinson H, Sarkans U, Shojatalab M, et al.** (2005) *Nucleic Acids Res.* **33**, D553–D555.

4. **Eppig JT, Bult CJ, Kadin JA, et al.** (2005) *Nucleic Acids Res.* **33**, 5.

★ 5. **Geer RC & Sayers EW** (2003) *Brief. Bioinform.* **4**, 5. – *A tutorial paper that covers some of the ground of this chapter but in more detail. It gives a useful overview of the concepts behind the Entrez tool using example tasks.*

6. **McEntyre J & Lipman D** (2001) *CMAJ,* **164**, 1317–1319.

7. **Boguski MS, Lowe TM & Tolstoshev CM** (1993) *Nat. Genet.* **4**, 332–333.

8. **Banfi S, Guffanti A & Borsani G** (1998) *Trends Genet.* **14**, 80–81.

9. **Sherry ST, Ward MH, Kholodov M, et al.** (2001) *Nucleic Acids Res.* **29**, 308–311.

10. **Boeckmann B, Bairoch A, Apweiler R, et al.** (2003) *Nucleic Acids Res.* **31**, 365–370.

11. **Boeckmann B, Blatter MC, Famiglietti L, et al.** (2005) *C. R. Biol.* **328**, 882–899.

12. **Barker WC, Garavelli JS, McGarvey PB, et al.** (1999) *Nucleic Acids Res.* **27**, 39–43.

13. **Sussman JL, Lin D, Jiang J, et al.** (1998) *Acta Crystallogr. D Biol. Crystallogr.* **54**, 1078–1084.

★ 14. **Bateman A, Coin L, Durbin R, et al.** (2004) *Nucleic Acids Res.* **32**, D138–D141. – *Pfam is possibly the most useful web resource available for gene function analysis. It is useful to understand how it is built and annotated before diving in and using it. This publication should be updated each year in the database issue of Nucleic Acids Research.*

15. **Letunic I, Goodstadt L, Dickens NJ, et al.** (2002) *Nucleic Acids Res.* **30**, 242–244.

★ 16. **Tatusov RL, Fedorova ND, Jackson JD, et al.** (2003) *BMC Bioinform.* **4**, 41. – *An in-depth description of how orthologous groups have been calculated for the COG database; it also describes the eukaryotic clusters (or KOGS). This database is an excellent tool for studying the phylogenetic coverage of genes.*

17. **Barrett T, Suzek TO, Troup DB,** *et al.* (2005) *Nucleic Acids Res.* **33**, D562–D566.
18. **Hamosh A, Scott AF, Amberger JS, Bocchini CA & McKusick VA** (2005) *Nucleic Acids Res.* **33**, D514–D517.
19. **Cantor MN & Lussier YA** (2004) *Medinfo*, **11**, 753–757.

CHAPTER 2

Navigating sequenced genomes

Melody S. Clark and Thomas Schlitt

1. INTRODUCTION

World sequencing capacity and technologies were fuelled by the race to complete the Human Genome Project and resulted in a massive investment in infrastructure, machinery, techniques, and personnel. The effective completion of this project has not seen a reduction in the amount of sequencing data generated. The techniques learnt, especially the use of shotgun sequencing, can efficiently sequence large vertebrate genomes, and the realization that low-density coverage of a genome could prove almost as useful to scientists as a completed genome, allied with a dramatic reduction in sequencing costs, brought about a liberation in the genomic science field. Therefore, the 'excess' capacity was not closed down but maintained so that, in theory, the DNA of any organism could be sequenced. A press release from the National Institutes of Health (NIH) in August 2005 announced that the public collections of sequence data (GenBank, EMBL, and DDBJ) had reached 100 gigabases from over 165 000 different organisms, and the sequence databases continue to grow at a tremendous rate. This continued production of sequencing data, much of which is for comparative analyses, has gradually led to a standardization of data presentation and genome viewers, some of which will be described here.

So what constitutes a 'sequenced' genome? This is either in the form of:

- A completely sequenced genome (to a genome-center standard of 99.9% accuracy, with no gaps).
- A draft genome with perhaps only three- to tenfold coverage, such that the genome is present in numerous contigs. The quality of the sequence data is not shown and may be variable, particularly at the ends of the contigs. Repeat sequences may be masked unless a single clone encompasses the whole repeat region with accurate sequence readthrough. Misassemblies can occur, so care should be taken when interpreting draft sequences (1).
- Ongoing genome sequencing data. This is very much 'work in progress', with sequence information made publicly available in ongoing draft form and with frequent updates. Again, repeat sequence data may be either masked or removed from the draft sequence.

Bioinformatics: *Methods Express* (Paul H. Dear, ed.)

This chapter will encompass all of these types of genomes and will explain how to access the data, appreciate the limitations of the associated annotation, identify sources of additional useful information, and download and manipulate the data. The chapter provides a number of worked examples and also lists several different genome viewers to use for both vertebrates and microbes. In general, the different genome browsers present identical data sets in slightly different ways, and some are easier to use than others. This chapter has focused on the use of Ensembl (for vertebrates) (2) and the related site Integr8 (mainly dealing with microbes) for several reasons:

- They combine the considerable resources of two major bioinformatics centers: The European Bioinformatics Institute and the Sanger Institute.
- Both are publicly funded with what appears to be a sustained commitment to future work and are not subject to the vagaries of commercial interest.
- They provide data as quickly as possible from the sequencing pipelines, allowing access to very early crude data releases.
- Numerous updates are provided.
- If you grasp Ensembl, then you do not need to relearn another browser when accessing Integr8.

Having said that, they may not suit everyone, and the best way to get to grips with genome viewers is to pick a gene, enter it into the different browsers, and decide which one you like the best with regard to data presentation and ease of use. Nevertheless, you would be well advised to explore Ensembl and Integr8 first, as understanding how they are used will help in understanding other browsers. You are also encouraged to explore references (3)–(6), which provide useful information on a number of browsers and related resources.

2. METHODS AND APPROACHES

2.1 Finding genome resources for an organism

If an organism has been (or is being) sequenced, then, generally speaking, every scientist who works in that community is aware of it. However, this does not necessarily mean that they will have privileged access to the data prior to the obligatory *Science* or *Nature* genome paper or know where the data is being stored. Also, the data may not be available immediately in one of the 'standard' public genome browsers. There may also be a requirement to try to identify similar genomes for comparative analyses. For this, you need to access a genome project-monitoring web site, which keeps an up-to-date log of what is being sequenced, with contact details (7). The most comprehensive site for this is the Genomes On Line Database (GOLD), which is maintained by Dino Liolios, Nektarios Tavernarkis, Phil Hugenholtz, and Nikos Kyrpides (8–10). *Protocol 1* is an example of a simple query to GOLD.

Protocol 1

The Genomes On Line Database (GOLD)

1. As an example, we will try to find out if there are any genome projects for the whiptail wallaby, *Macropus parryi*.

2. Go to http://www.genomesonline.org [2.1] and click on the **GOLD Tables** button.

3. You have a choice of the following buttons:
   - Published complete genomes
   - **Archaeal ongoing genomes**
   - **Bacterial ongoing genomes**
   - **Eukaryotic ongoing genomes**
   - **Metagenomes**

4. Click on **Published complete genomes** to call up a table of all published genome sequences. Several of the column headings at the top of the table are clickable links; clicking on one of these will sort the data according to that criterion.

5. Click **Organism** to sort the table by genus and species, and scroll down to look for *M. parryi*. There is no such entry (at the time of writing).

6. Perhaps there is an ongoing genome project for this species. We could go back and look through the list of 'Eukaryotic ongoing genomes', but instead we will use the 'search' function. Go back to the front page of GOLD and click on the **Search GOLD** link.

7. On the resulting page, you can specify what type of information you want to retrieve (using check boxes at the top of the screen; leave this at its default with the **All fields** box checked) and your search criteria (menus and fields in the lower part of the screen).

8. Under the **Type** menu on the left, select **species** and, in the adjacent text box, type 'parryi'. Press 'Enter' to start the search (or click the **Submit search** button at the bottom of the screen).

9. The query returns no results (at least, at the time of writing). Alas, there does not appear to be any whiptail wallaby genome project. However, are any other members of the genus being sequenced?

10. Use the 'back' button to return to the search form and this time search for the genus 'Macropus' (setting the **type** menu to **genus**).

11. This gives two results – both for *Macropus eugenii* genome projects. (For other species, different types of project such as expressed sequence tag (EST) projects may be listed.)

12. Clicking on the **Taxonomy** link for either of these two entries will take you to the appropriate entry in the National Center for Biotechnology Information (NCBI) Taxonomy Browser, where you will see that *M. eugenii* is the tammar wallaby. Additional information about – and links to – available genome and other data for this species are given.

13. Return to the GOLD page displaying the two *M. eugenii* genome projects. Links to relevant funding bodies, institutions, databases, and other resources are given for each of the genome projects. The identifiers listed under **GOLDSTAMP** (column on the extreme left) are unique to each genome project and link to a summary page or 'Gold Card' for the project.

14. The GOLD search page offers many options for searching, not only by organism type or by genome properties, but also by funding body, country, researcher, and many other factors.

GOLD will direct you to specific genome project information and also to an NCBI flat file (if it exists). A 'flat file' is a data file that contains records with no structured relationship. In the case of sequence files, these usually list minimal data on the source of the data, the full sequence, and (not always) annotation on coding regions. There are no links to additional sources of information and it is necessary to extract the data manually to be able to analyze it further. If, for example, you wanted to examine a particular gene from a sequenced microbe, then using the flat file you would have to scroll manually through the entire sequence to find it, if the annotation was in place, or else perform a BLAST search on the whole sequence, work out where the bit you wanted was, and then extract that particular piece of sequence. In a nutshell, this is why genome browsers are so useful.

There are distinct advantages in accessing the information via generic genome viewers. They present the data in a standard format and often link sequence data between different genomes, allowing easier comparisons and data handling. They are also not restricted in the organisms they list (some institutes only list those organisms being sequenced 'in house'), as long as the data is in the public domain.

It should be mentioned that GOLD does not necessarily direct you to a generic genome browser for the organism in question, even if the information is available there. For example, both the dog (*Canis familiaris*) and cow (*Bos taurus*) genomes are available in Ensembl, but a direct link to the relevant Ensembl pages is not available under the GOLD 'Organism' listing. In general, you will need to discover which of the generic genome browsers gives access to the data for the species you are interested in.

In the following sections, we will focus on the use of the Ensembl and Integr8 browsers; other popular resources will be covered later.

2.2 Browsing vertebrate genomes with Ensembl

Ensembl was developed as a browser for the Human Genome Project, but is now available for an increasing number of species (over two dozen at the time of writing). It is continually being updated with an increasing array of features. Most importantly, it is not restricted to the genome sequence produced by any one institute and it also allows easy comparison of data from different genomes. *Protocol 2* describes a typical use of Ensembl – browsing for information related to a given gene and its genomic environment. Gene symbols and descriptions can also be used instead of gene names.

Protocol 2

A short tour of Ensembl starting from the human parathyroid hormone-related protein gene

1. Go to Ensembl via http://www.ensembl.org [2.2] and click on **Homo sapiens** under the **Popular genomes** heading. In the text box next to the **Search e! Human** box at very top right of the screen, enter 'parathyroid hormone-related protein'. Set the adjacent pull-down menu to **Anything**. Press 'Enter' or click on the **Go** button.

2. There are only three results[a], starting with two Ensembl protein coding genes and, below it, an Ensembl gene family. Click on the link (**Ensembl protein coding gene: ENSG00000087494...**) for the gene.

3. This will call up an Ensembl Human Gene View page, starting with the Ensembl Gene Report for this gene (see *Fig. 1*). A range of information about the gene is displayed, along with links to further annotation and resources. Note that if you hover the cursor over the links on the left of the screen, brief explanations of the links will appear.

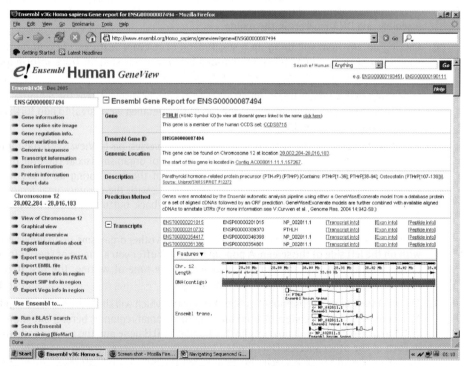

Figure 1. Ensembl gene report for parathyroid hormone-related protein.

4. To download directly the sequence data and some associated information for your own analyses, you may use **Export information about region**, **Export sequence as FASTA** or **Export EMBL file**. In each case, you will be offered various options such as which file format to use, which features of the annotation to include in the output, and how much flanking sequence to include. Each of these options will export a flat file, but with only minimal annotation. Much more information is displayed graphically using the other browser features. If you try

these export options, use the 'back' button on your browser to return to the Gene View page before continuing.

5. On the right-hand part of the screen, the **Genomic Location** box gives both the location of the gene in the genome (the chromosome and the start and end positions of the gene sequence) and the sequence contig in which the start of the gene lies. Access the genomic location of the gene by clicking on the link **28,002,284–28,016,183** (the exact coordinates may change in later versions of the genome assembly).

6. This takes you to an Ensembl Contig View page, showing an ideogram of chromosome 12 and, in the 'Overview' area below, an expanded view of the region surrounding the gene at 28 Mb in band 11p22. The parathyroid hormone-related protein gene (also known as parathyroid hormone-like hormone or PTHLH) is shown in the middle (the short brown bar above the name shows its extent on the chromosome) and neighboring genes are displayed to the left and right. A red box delineates the PTHLH gene alone; this region is shown in greater detail in the 'Detailed view' area below. Scroll down to bring the entire 'Detailed view' area on screen[b].

7. The 'Detailed view' area displays the data that were used to infer the Ensembl gene transcript for PTHLH. The data is displayed in 'tracks': each track is named towards the left of the box (for example 'Genscan'). The default tracks include matches with ESTs, Unigene, and mRNA sequences, and the transcripts predicted by the Genscan automated gene prediction software. Also shown are details of the sequence contigs spanning this region in the current sequence assembly and DNA markers in the region.

8. Clicking on any item will bring up a small menu of relevant features. Clicking on the track names (for example, **Unigene**) will bring up a help menu; clicking on **Track information...** in this menu will then bring up an explanation of that track.

9. In addition to the default tracks, many other tracks can be displayed for the region of interest. Use the **Features** pull-down menu (at the top of the 'Detailed view' area) to add further tracks to (or remove tracks from) the view.

10. To get an overview of the syntenic regions in other species, go to **View syntenic regions** on the left side of the screen. You will be offered a choice of organisms (six at the time of writing, starting with *Bos taurus*), and clicking on any one of these will bring up a 'classical' map showing the overall synteny of human chromosome 12 with the chromosomal regions in the other organism (see *Fig. 2*). After viewing this, use the 'back' button on your browser to return to the Contig View page.

11. To view a more detailed version of the syntenic region in other organisms, go to **View alignment with** on the left side of the screen. A pop-up menu will offer you a choice of species (or groups of species) to align with the human sequence. At the time of writing, the first item in the menu is **5 eutherian mammals** – select this item.

12. This will take you to an Ensembl AlignSlice View page, which will show the human chromosome 12 ideogram and an overview of the human chromosomal region as before, but the 'Detailed view' area below these will now display the human PTHLH gene and the corresponding regions in each of the other species. Only limited information (sequence contig and EMBL transcript tracks) are shown for each species by default, but the pull-down **Features** menu at the top of the 'Detailed view' area can be used to add tracks displaying any other desired features for all of the species shown.

13. To view neighboring genes in the various species, use the 'Zoom' feature towards the top of the 'Detailed view' box. Click on the larger end of the wedge (or click on the – icon) to display a larger region around the human PTHLH gene and its aligned regions in the other species.

14. At any point, you can access information from the other species by clicking on the feature of interest to bring up a menu of available information. For example, clicking on the **Pthlh** gene

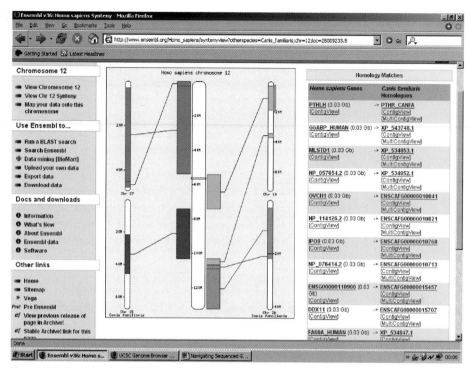

Figure 2. Ensembl viewer showing the synteny viewer.
The PTHLH gene in *Homo sapiens* is on chromosome 12 and shares synteny with dog chromosome 27. The human genes in the region are listed with a reference to the orthologous region in dog, and provide direct access to both the candidate gene and the neighboring genes in both human and dog.

in the *Mus musculus* part of the window and then on the **Gene:ENSMUSG00000048776** item in the menu that pops up will take you to the Ensembl Gene Report for this mouse gene: this is the mouse equivalent of the Gene Report for the human gene, which we saw in step 3 of this protocol. Use your web browser's 'back' button to return to the page displaying the alignments with human.

15. To view a more detailed comparison between the human region and just one other species, go to **View alongside** (upper left-hand part of the screen) and select any of the single species that appear in the pop-up menu. The new window (an Ensembl Human MultiContig View) shows a side-by-side comparison between the genomic regions in human and the second species at three levels of detail: chromosomal ideograms at the top; a 'navigational overview' of the PTHLH region below this; and at the bottom a 'Detailed view'. As before, you can select which features to display in the 'Detailed view' and can also zoom in or out.

Notes

[a]The information and figures given are correct for the NCBI36 assembly current at the time of writing.

[b]Each of the views – the chromosomal ideogram, the overview, and the detailed view – can be closed or opened by clicking on the corresponding '–' or '+' at the top-left corner of the view. A fourth and more detailed 'Basepair view' is closed by default but can be opened at the bottom of the screen by clicking on its '+' icon.

The Ensembl genome browser is very user-friendly with numerous drop-down menus to access further information in a graphical form. If you want to perform your own analyses on the data, there are direct links (such as **Export EMBL file**) in the upper-left part of the screen, which will allow you to export information on the region being examined; or the **Export data** link under 'Use Ensembl to..' lets you specify the region and the type of data to export. Data can also be downloaded from Ensembl using BioMart, which is covered later in this chapter.

Other particularly useful features are listed under 'Other EMBL websites' towards the bottom of the Ensembl front page:

- **Vega** (Vertebrate Genome Annotation) (11). This is a central repository for high-quality, manually annotated data (Ensembl is a fully automated pipeline based on gene prediction programs). It is currently available for four species: *Homo sapiens*, *Mus musculus*, *Danio rerio*, and *Canis familiaris*), so if you are interested in these organisms, it will be worth accessing Vega, at least alongside Ensembl.
- Ensembl Pre! This provides access to recent data that has yet to be entered into Ensembl (very useful for draft genomes or on-going sequencing projects).
- Archive! This is particularly useful for tracing back information in previous releases, if you have data that you accessed previously. Archive versions extending back over a 2 year period are available.

Finally, if you want to explore other genome browsers, there are direct links to both the NCBI and University of California Santa Cruz (UCSC) genome browsers from Ensembl. For example, the Ensembl Contig View, AlignSlice View, and Human MultiContig View pages (seen in *Protocol 2*) all have links in the upper-left part of the screen that will take you to the equivalent information in either of these two alternative browsers.

2.3 Integr8 – an Ensembl lookalike for microbes

Integr8 is a portal maintained by the European Bioinformatics Institute (EBI) for access to information related to completed genomes and their proteomes (12). Integr8 does include some vertebrate data, but its strength is in the application of the Ensembl browser to microbial genomes. It currently does not have an up-to-date listing of all of the organisms that are in Ensembl.

Protocol 3

Using Integr8 to search for dihydroorotase genes in *Acinetobacter*

1. Go to http://www.ebi.ac.uk/[2,3]. Click on the **Databases** tab towards the top left of the screen, then select **Database browsing** and then **Integr8** from the menus that pop up.

2. You first need to specify the species. This can be done using the **Browse Species** menu (left side of screen) and selecting **Acinetobacter sp.** from the resulting list, or by specifying 'Acinetobacter sp.' in the **Search for species** box and clicking the adjacent **Go!** button.

3. To find all genes with 'dihydroorotase' in their name, put 'dihydroorotase' in the **Search for gene/protein** box. The pull-down menu to the right will be set by default to *Acinetobacter* sp. (as we have just selected this species) – leave it set to this default and click the right-most of the **Go!** buttons.

4. This produces 2 matches[a], one of which is 'putative dihydroorotase'. There are then two ways of looking at the data for this gene: using 'Integr8or' or 'Genome Reviews'.

5. To examine the gene using 'Integr8or', click on the **i8** button to the left of 'putative dihydroorotase'. This displays the information on this gene in a very straightforward way via a series of tabs (**Gene**, **Results**, **Context**, and **History**).

6. To identify orthologs and paralogs in other species, click on the **Protein** tab and then on the **Orthologues** and **Paralogues** buttons (next to the 'Homology:' heading underneath the **Gene** tab). This will bring up a table of genes in various species, each of which can be examined in more detail by clicking on its name (on the left of the table).

7. To the right of the table of orthologs or paralogs is a column headed 'Select', with a dot for each of the listed genes. Click on several of these dots (they will turn into 'ticks' when clicked) to select specific genes from the list.

8. Click on the **Compare** button at the top of the column. A table will appear showing each of the selected orthologs (or paralogs) and the genes lying adjacent to it in the respective genome. Clicking on any of the gene names will take you to a new 'Integr8or' page for that gene.

9. Use the 'back' button on your browser to return to the list of *Acetinobacter* genes in step 4.

10. To examine the *Acetinobacter* putative dihydroorotase gene using 'Genome Reviews', click on the **GR** button to the left of the gene name. You will be taken to an Ensembl-style browser displaying information on this gene and the region around it (see *Fig. 3*). See *Protocol 2* for an introduction to the Ensembl-style browser. Note that not all organisms in Integr8 have a Genome Reviews browser.

Note

[a]Correct at time of writing. Details are, of course, likely to change.

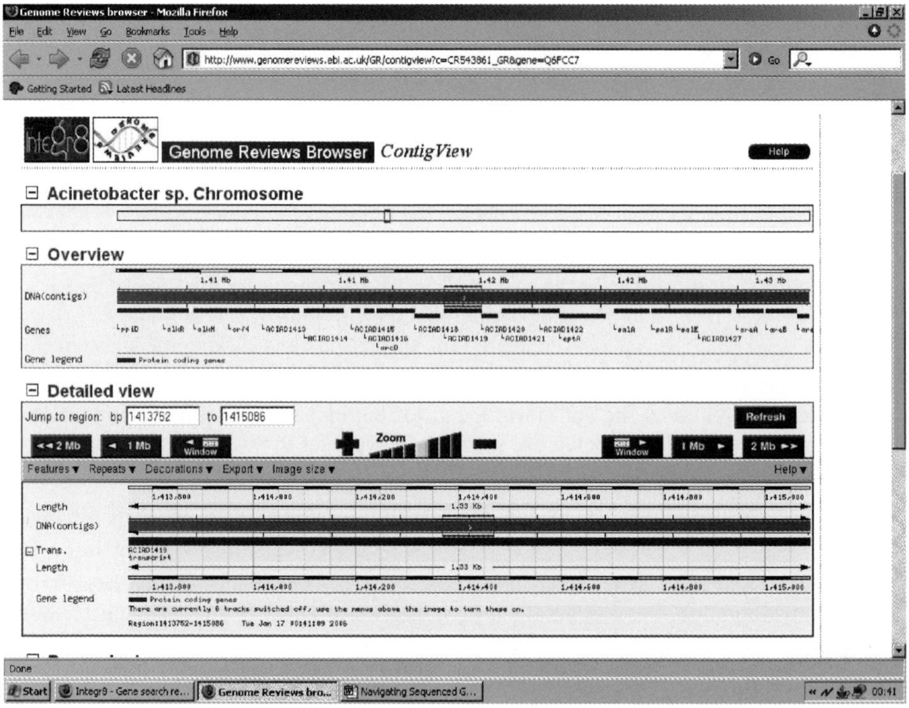

Figure 3. Genome Reviews browser for *Acinetobacter* sp. in the region around the putative dihydrooratase gene.

Integr8 also has some other very useful buttons on the side menu, which provide taxonomy information on your organism, relevant literature, and genome statistics (on amino acid composition, protein length distribution, and triplet usage).

2.4 Other web-based genome browsers

Although we have focused here on Ensembl and Integr8, there are many other web-based genome browsers. All offer access to a different range of data sets, so the best advice is to explore the available sites to find one that suits you and that offers access to the data you need. Below are listed several of the more popular sites, along with brief descriptions.

2.4.1 Genome News Network (GNN)

This online magazine (http://www.genomenewsnetwork.org/[2.4]) covers important developments in genomics research around the world and, under resources, has 'A Quick Guide to Sequenced Genomes'. This provides a brief description of sequenced organisms and a link via 'Abstract' to the relevant Entrez PubMed entry. It also references links to any related science articles by GNN. The site appears to have stopped posting information in 2004; however, it is still a useful source of background information presented in a very accessible format.

2.4.2 Entrez Genomes

This is the NCBI graphical genome viewer (http://www.ncbi.nlm.nih.gov/gquery/gquery.fcgi[2.5]). Entrez Genomes integrates the scientific literature, DNA and protein sequence databases, 3D protein structure and protein domain data, population study datasets, expression data, assemblies of complete genomes, and taxonomic data. There is a comprehensive map viewer, a genome browser for eukaryotic genomes, Plant Genomes Central, microbial and viral genome databases, and gMap, a comparative analysis of microbial genomes. The archaea, bacteria, and eukaryota can all be viewed by either chromosome, plasmid, or organelles. Information leads directly to the relevant NCBI files. NCBI's genome resources are covered in more detail in Chapter 1, and there are extensive links to Ensembl.

2.4.3 UCSC Genome Browser

This browser (http://genome.cse.ucsc.edu/index.html[2.6]) provides access to a comprehensive range of organisms. It has the advantage of linking cDNAs to microarray expression data. Again, there are extensive links to Ensembl.

2.4.4 Joint Genome Institute (JGI)

The JGI (http://www.jgi.doe.gov/[2.7]) combines four national laboratories: Lawrence Berkeley, Lawrence Livermore, Los Alamos, and Oak Ridge; and the Stanford Human Genome Center. They provide information on the genomes they have sequenced, but have a limited number of eukaryotic genomes; for example, there is data for *Homo sapiens* chromosomes 5, 16, and 19 only.

The data is annotated using Vista plots, gene model predictions and BLAST search results. Vista plots (13) are generated from sequence data multiple alignments. If annotation files are present with the segments of DNA chosen, then the Vista file will also show the locations of untranslated regions and exons. The plots show conservation of sequence and similarity over the whole length of the sequence and so are very useful for a quick overview of how similar sequences are between organisms with regard to exons, but also can identify conserved noncoding sequence that may have functional significance.

The JGI's microbial genome browser (IMG) is much more comprehensive and is described in the following section.

2.4.5 Integrated Microbial Genomes (IMG)

This browser (http://img.jgi.doe.gov/cgi-bin/pub/main.cgi[2.8]) (14) provides a framework for the comparative analysis of microbial genomes, many of which have been sequenced by the JGI. Searches include: **Find genes, Find functions,** and **Find organisms.** There is the possibility of browsing across genomes. It is very easy to add sequences from several different genomes to a 'Gene Cart' for further analysis such as DNA and protein alignments, to examine 5′ or 3′ neighborhoods of the gene of interest, and to export sequences. This is an excellent and very easy-to-use tool.

2.4.6 The Institute for Genomic Research (TIGR) Comprehensive Microbial Resource (CMR)

This TIGR-based browser (http://cmr.tigr.org/tigr-scripts/CMR/CmrHomePage.cgi[2.9]) presents data from publicly available complete microbial genomes, many of which have been sequenced by TIGR. It is also possible to BLAST search against TIGR's unfinished data via http://www.tigr.org/db.shtml[2.10]. Access is also provided to a number of parasite and fungal sequence projects.

2.4.7 Gendb

Gendb (http://www.cebitec.uni-bielefeld.de/groups/brf/software/gendb_info/index.html[2.11]) is an annotation system for prokaryotic genomes provided by the Bielefeld University Centre for Biotechnology (CeBiTec). A user log-in is required and there is a limited number of genomes available. However, this site does provide a ready-made annotation system (available via collaboration) for independently sequenced genomes.

2.5 Specialized sites

There are also a great many specialized sites that can provide information on a gene-by-gene rather than genomic basis, and these sites complement the genome browsers. Many of these sites will be discussed in other chapters of this book, but some of particular relevance are listed below.

2.5.1 Gene Cards

This site (http://www.genecards.org/[2.12]) is provided free to non-profit academic institutions and is probably one of the most comprehensive sites with regard to the provision of information. Data include gene name aliases and descriptions, genomic location with protein and transcript data, microarray expression profiles, functional annotation including Gene Ontology (see Chapter 9), orthologs in other species, single-nucleotide polymorphism (SNP) analysis, and research publications. Links are provided to numerous other databases.

2.5.2 Online Mendelian Inheritance in Man (OMIM)

OMIM (http://www.ncbi.nlm.nih.gov/entrez/query.fcgi?db=OMIM[2.13]) is one of the earliest genome resources and is still going strong. This database is a catalog of human genes and genetic disorders authored and edited by Dr Victor A. McKusick and his colleagues at Johns Hopkins and elsewhere, and developed for the web by NCBI. The database contains textual information and references plus links to additional related resources at NCBI and elsewhere.

2.5.3 Entrez Gene

Entrez Gene (http://www.ncbi.nih.gov/entrez/query.fcgi?db=gene[2.14]) has superseded LocusLink and is a searchable database of genes, from RefSeq genomes, and defined by sequence and/or located in the NCBI Map Viewer. This comprehensively links NCBI resources for each gene in a single-page, easy-to-view format.

2.5.4 GeneTests

GeneTests (http://www.genetests.org/[2.15]) is a medical genetics information resource and contains GeneReviews. This is an online publication containing a collection of expert-authored, peer-reviewed descriptions of heritable diseases. It is biased towards clinicians rather than laboratory scientists.

2.6 Downloading data with BioMart

Having seen how to view genomic data, you may wish to download data for your own analysis. Genome browsers will often offer a variety of 'export' options (some of these were mentioned briefly in *Protocol 2*), but these will vary widely depending on the browser in question and on the type of data you are exporting. Here, we will deal with BioMart, a more general tool for downloading genome data.

BioMart is a generic data management system. It can be thought of as a 'shopping tool', allowing you to select and download data for your own analysis. BioMart is not a database *per se* – it is a tool that can be accessed from a variety of databases, providing a standardized way of accessing the respective database's contents. For example, it is accessible from Ensembl (15).

The details that will be displayed in BioMart will of course depend on the database it is accessing, but the basic 'shopping process' consists of three steps that remain the same. First, you choose the dataset (for example, the species and the annotation version) that you want to query. Secondly, BioMart then offers you a range of filters (tailored to the type of data being accessed) to select the data you are after. For example, when using BioMart to access data through Ensembl, you can query for particular regions of the human genome, by chromosome, chromosome bands, base-pair coordinates, or marker location; or you can query for known genes using identifiers such as those from Uniprot, RefSeq, or EntrezGene IDs. You can also query for proteins that belong to particular protein families according to PROFILE IDs, PFAM IDs, InterPro IDs, PRINTS IDs, or PROSITE IDs that have or do not have transmembrane domains or signaling domains. You can even combine the data from two different genomes. Thirdly, after setting the appropriate filters, you can choose the type of information you want to retrieve, such as sequences (including or excluding introns etc.) or features, and the format in which you would like the information downloaded (HTML, CSV, Text, etc.).

Some of the databases that implement BioMart are listed in *Table 1*. The BioMart web site (http://www.biomart.org/[2.16]) also provides a current list of the databases that implement BioMart. *Protocol 4* gives an example of using BioMart to find and download a specific group of genes from Ensembl. Working through this example should enable the reader to perform other types of BioMart queries through Ensembl and, with a little extrapolation, to use BioMart through the other databases given in *Table 1*.

Table 1. Databases providing BioMart access to genome data for various species

Database	Species
Ensembl http://www.ensembl.org/Multi/martview	Various vertebrates and *Apis mellifera, Caenorhabditis elegans, Drosophila melanogaster,* and *Anopheles gambiae*
Gramene http://www.gramene.org/Multi/martview	*Oryza sativa, Zea mays,* and *Arabidopsis thaliana*
Wormbase http://www.wormbase.org/biomart/martview	Caenorhabditis elegans
euGenes http://insects.eugenes.org/BioMart/martview	Various *Drosophila* species
Uniprot http://www.ebi.ac.uk/biomart/martview	Various species, archaea, bacteria, and eukaryota
HapMap http://hapmart.hapmap.org/BioMart/martview	Data collected by the International HapMap Project

Protocol 4

Using BioMart to download transcription factors located on chromosome 9 through Ensembl

1. Go to http://www.ensembl.org/Multi/martview [2.17] (you can also access this page using the **Data mining [BioMart]** link, which appears in the upper-left of most Ensembl pages).

2. On the resulting Martview page, you will be able to build the query. This is done in three stages to specify the dataset you want to search, the criteria ('filters') you want to apply, and the features of the resulting genes you want to see.

Specifying the dataset

3. To specify the dataset you want to search within, on the right-hand part of the screen, there should be menus entitled **Database** and **Dataset** (if not, click on the >>**Dataset** link at upper left of the screen and they should appear). Using these two menus, select **ENSEMBL 42 GENE (SANGER)** and **Homo sapiens genes (NCBI36)** as the database and dataset, respectively[a,b].

Specifying the search criteria

4. Next, click on the >>**Filters** link on the left of the screen. On the right will appear several boxes (**REGION:, GENE:,** etc.), each with a small '+' symbol next to it. These are the filters that will be used to define the type of data (which genes, in this case) you want to download.

5. In general, clicking on '+' next to any of these filters will open a menu within which you can set parameters; check boxes in each menu are used to determine which parameters to take into account.

6. Click on the '+' next to **REGION:** to expand its menu. Below it will appear various options by which you can specify the region (chromosome, base-pair coordinates, band, etc.). From the pull-down menu next to **Chromosome**, select **9**, and ensure that the check box to the left is 'ticked' (see *Fig. 4*).

7. Scroll down to the **Gene** filter, click on the adjacent '+' to expand its menu, and scroll down again to **Gene type**.

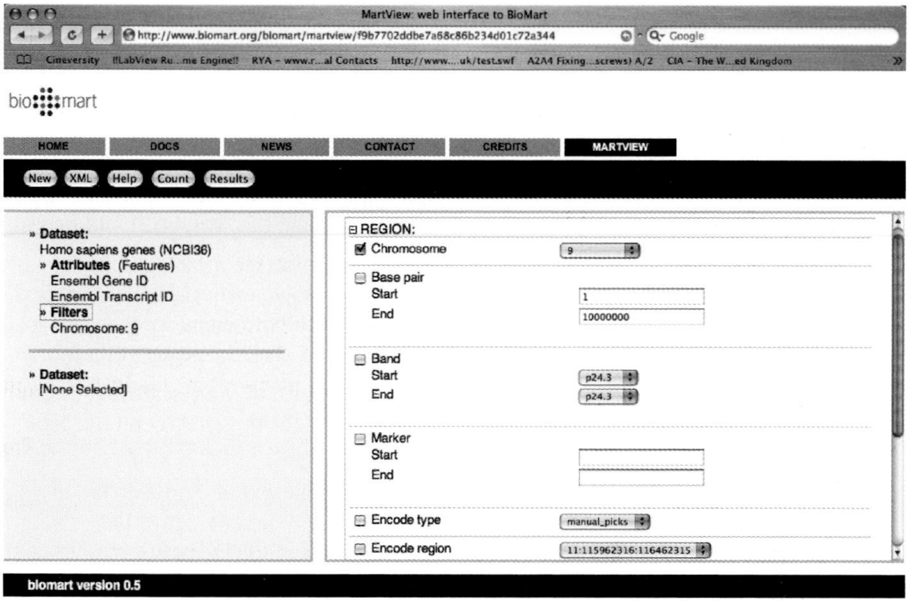

Figure 4. Using BioMart.
In this screenshot, we have chosen to search the dataset NCBI36 (under **Dataset** on the upper left) and have restricted the search to chromosome 9 (set using the menu and check box on the right, and shown under **Filters** on the left). We have not yet specified which **Attributes** to retrieve – the ones shown on the left (**Ensembl Gene ID** and **Ensembl Transcript ID**) are the default values.

8. Click on **Protein_coding** within the **Gene type** menu; the adjacent check box should change to a 'tick' automatically (if not, click to tick it).

9. Still in the **Gene type** menu, ensure that the **Status** menu is set to **Known** (to retrieve only genes of known function) and tick the check box next to it.

10. In the same way, go into the **Gene Ontology** menu (see Chapter 9 for details of Gene Ontology (GO), which is a standardized vocabulary for describing genes). Tick the check boxes next to **Molecular function** and enter 'GO:0003700' into the text field next to it. (GO:0003700 is the GO term for 'transcription factor activity'; if you do not know the relevant GO term, you can use the adjacent **Browse** button to jump to the QuickGO browser, which allows you to browse GO and identify the correct term).

11. Leave the check box next to **Evidence code (Molecular function)** *un*ticked. This means that we will be searching for proteins that have been assigned the GO term 0003700 under 'molecular function', using any type of available evidence.

12. Other menus (**Expression, Protein, SNP**) would allow you to apply further criteria in selecting your data, but we will leave these for now.

Specifying the features to retrieve
13. Click on the **>>Attributes (features)** link on the left. On the right, you will be offered the now-familiar style of menu. Expand the **Gene** submenu (by clicking on the adjacent '+' button)

and, under **Ensembl Attributes**, ensure that only the **Ensembl Gene ID** and **Description** check boxes are ticked.

Previewing and retrieving the results

14. Click on the **>>Dataset** link (the database and dataset will again be displayed on the right) and then on the **Count** button towards the top of the screen. This will display the number of genes matching your criteria and the total number of genes in the dataset, next to the **>>Dataset** link towards the top of the page (at the time of writing, this was **44/31 148 genes**).

15. Now click on the **Results** button. On the right, you will see a table displaying the first few genes retrieved by your query. Above this table, pull-down menus allow you to specify the file format for the output; ensure that the **Export all results to** menu is set to **File** and the **rows as** menu is set to **TSV** (tab-separate value output) and click the adjacent **Go** button.

16. A text file will be downloaded, containing the results of your search. The results of this particular example are given in the Protocol_4 folder for this chapter on the book's web-site as 'BioMart.txt [2.18]'.

17. If you specify **HTML** (in the **rows as** menu), the result will be an HTML file of your results, which can be opened in your browser and which will contain active links for the relevant data. An example of such a file is given in the Protocol_4 folder for this chapter on the book's web-site as 'BioMart.html [2.19]'.

Notes

[a]These were current at the time of writing; there may be more-recent versions by the time you read this.

[b]It is also possible to specify a second dataset to search in combination with the first; this is set using the lower of the two **>>Dataset** links on the left of the screen.

In this example, we started with a 'blank' BioMart query. However, while viewing information in Ensembl (for example, while following *Protocol 2* in this chapter), you will also notice links such as **Export Gene info in region**, with the BioMart logo of colored dots. Clicking on these links effectively completes the first steps of a BioMart query for the gene, region, etc. in question, leaving you only to specify which aspects of the data you want to download and the format for export.

If you have particular queries BioMart does not cater for, you can query the Ensembl database directly by connecting to their MySQL server. You will need to know how to use SQL and to understand the database schema. We cannot go into more detail here, but more information can be found in the Ensembl documentation.

2.7 Browsing genomes 'off line' using stand-alone software

There are alternatives to using a remote web site for genome browsing. ARTEMIS (http://www.sanger.ac.uk/Software/Artemis/ [2.20]) is a stand-alone program that allows you to browse and annotate genomes (16). It can read different formats, including FASTA files, EMBL, and Genbank format, as well as GFF format. These formats contain the sequence data, as well as the annotations of this sequence. The sequence and its features are displayed graphically, in broadly the same way as some of the online genome browsers described above.

ARTEMIS (see *Fig. 5*, also available in the color section) is written in Java (http://java.sun.com/[2.21]) and is therefore available for many different computer systems (Linux, UNIX, Macintosh, and Windows). It is freely available, but you need an installation of Java on your computer (Java is also freely available). For details, see the ARTEMIS web site (http://www.sanger.ac.uk/Software/Artemis/[2.20]).

We recommend that you consult the ARTEMIS manual (available at the same site as the software) to learn about the many available features, as this is beyond the scope of this chapter. For example, ARTEMIS allows you to define your own features, edit and annotate a sequence, and save the results in various standard formats (e.g. Genbank).

Protocol 6 provides a simple example of using ARTEMIS to view an annotated sequence file downloaded from Ensembl.

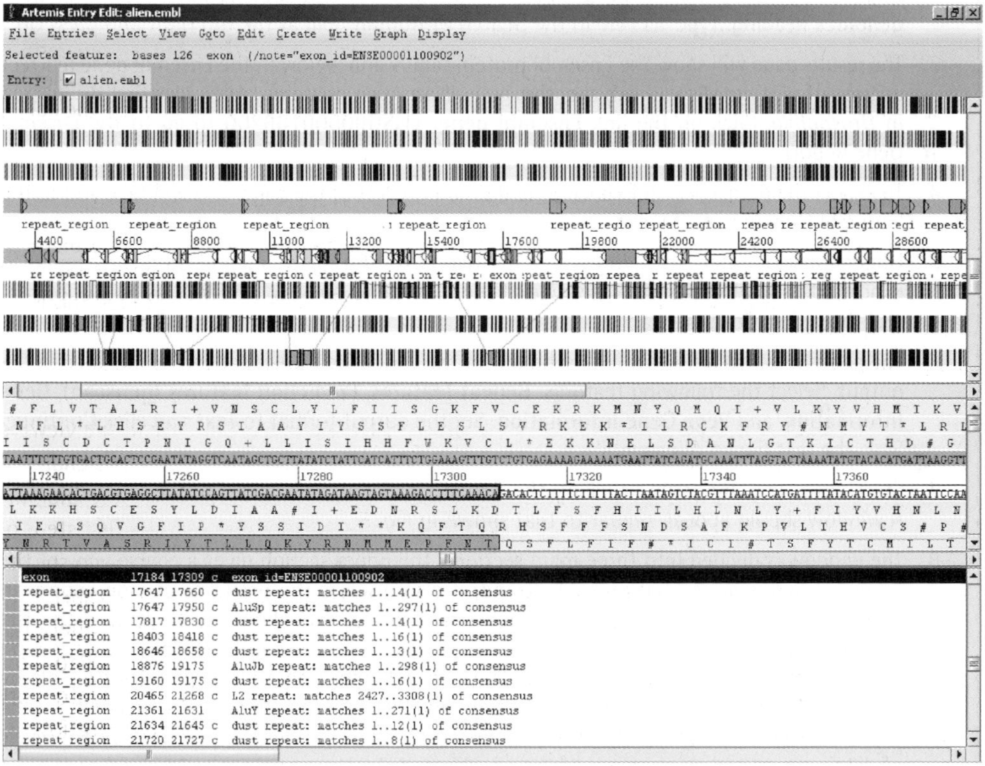

Figure 5. Screenshot of the ARTEMIS sequence viewer and annotation tool (see page xvii for color version).
The main window is divided into several sections. Below the main menu is information about the current selection and the sequences being viewed. Below this (and filling most of the top half of the screen), the 'overview' section shows stop codons in all six reading frames (short vertical black lines) and features on both strands (colored boxes, mainly in blue and yellow); the vertical scroll bar on the right controls the scale (zoom) of this window and the horizontal one scans along the sequence. The 'base view' (just below the overview) shows the sequence of both strands and, above and below these, the translation in all six frames; again, the scroll bars control the zoom and position of this window. The bottom third of the screen shows a list of annotated features. Many other aspects of the sequence can be displayed or hidden (see text).

Protocol 6

Using ARTEMIS to display the human genome sequence surrounding the gene Alien

This protocol assumes that you have installed ARTEMIS on your local computer. ARTEMIS is available from http://www.sanger.ac.uk/Software/Artemis/ [2.20], along with instructions for installing and using it.

1. You first need to download an EMBL file containing the relevant sequence and its annotation. Either download a copy of this file (Alien.embl [2.22]) from the Protocol_6 folder for this chapter at the web site that accompanies this book and proceed to step 7, or recover the data from Ensembl by following steps 2–6.

2. Go to Ensembl via http://www.ensembl.org [2.2] and click on **Homo sapiens** under the **Mammalian genomes** heading. Type 'alien' in the **Search e! Human** field at the top right and press **Go**.

3. The results page will show the Ensembl protein family and, below it, the Ensembl gene (correct at time of writing). As you can see, this gene is an Alien homolog; Alien was first discovered in *Drosophila*. Click on the link **Ensembl gene: ENSG00000166200** to go to the Ensembl Gene Report.

4. To export the data, click on **Export gene dat**a on the left of the page. This will take you to the 'Ensembl Human Export View'.

5. Under **context**, enter '5000' in each of the two boxes (**Bp upstream** and **Bp downstream**) to export 5000 bp of sequence either side of the Alien gene, and choose EMBL as output format and then press **Continue**.

6. The page will show a series of check boxes for the various features that can be included in the output. Select the following features: **Repeat features, Prediction features (genscan), Gene Information**, and **Vega Gene Information**. Set the output format to **Text** and click **Continue**. Save the text file as 'Alien.embl'.

7. Start ARTEMIS (e.g. by double clicking the Artemis.jar icon), and open the file 'Alien.embl [2.22]'. A window should appear, similar to that shown in *Fig. 5*.

8. The window is divided into three main sections. The upper section gives a coarse view of the sequence and is annotated on each strand; it also displays the positions of stop codons in each of the three reading frames on each strand. Scroll bars allow you to scroll along the sequence (horizontal scroll bar) or to zoom in or out (vertical scroll bar). The middle section displays similar information, but is initially set to a higher resolution (again, controllable by its vertical scroll bar) to show the nucleotide sequence and the amino acids encoded in each reading frame. You can use these two sections to look at the same region at different levels of detail, e.g. to see an overview of the exon/intron structure of a gene and the sequence of a particular exon/intron boundary at the same time.

The bottom section lists the annotated features; double-clicking on a feature will bring it to the center of the upper two windows. Many other tools – detailed in the ARTEMIS manual – are available for viewing and editing the sequence and its annotation.

Another option for browsing genomes locally is to install Ensembl on your own computer. This requires some experience with installing software and a knowledge of Perl, MySQL, and Java. You can download the Ensembl data as well as their program code free of charge. This means you can run it on a copy of the public data, run it with your own data, or you can install the whole annotation pipeline. There is some documentation on the Ensembl webpage on how to install the database locally, but installing and running the whole annotation pipeline will require considerable expertise and hardware.

2.8 Linking your own data to a genome browser

The UCSC Genome Browser and Ensembl both allow you to overlay data from other sources, so that you can have additional information displayed in the browser that is not actually part of UCSC's or Ensembl's data, but is provided by other sources (17, 18). This means that you can also feed in your data and have it displayed in the Ensembl or UCSC browser alongside their own data.

There are different ways of achieving this: by uploading your data to the UCSC or Ensembl site; by linking UCSC/Ensembl views to data on your web site; or by setting up a distributed annotation system (DAS) server on your computer. In most cases, the information is stored in text files and can be in various formats (see http://genome.ucsc.edu/goldenPath/help/customTrack.html[2.23]). Unfortunately, the formats differ slightly in the features they support. Check the file formats carefully: some formats require spaces to separate the columns, whilst others require tabs. Please refer to the format descriptions to find out what the best file format for your data might be. Below is shown the example file that we will use in *Protocol 8*, which illustrates some general (although not universal!) points about these formats.

```
#example file
browser position chr19:6901101-7100000
browser hide all
track name=NavigatingGenomesTrack1 description='This is an example of how to ↻
link your data into Ensembl/UCSC' visibility=2 color=255,0,255 useScore=2 ↻
url=http://www.ebi.uniprot.org/uniprot-srv/uniProtView.do?proteinAc=$$
chr19    Navigator    test_element    6903521    6903982    1000    +    .    P61203
chr19    Navigator    test_element    7000000    7001000    400    +    .    P61205
chr19    Navigator    test_element    7010000    7030000    400    +    .    P61205
chr19    Navigator    test_element    7050000    7060000    400    +    .    P61205
chr19    Navigator    test_element    7010000    7090000    200    +    .    P61204
browser dense Test2
track name=NavigatingGenomesTrack2 description='This is an example of how to
show even more data' visibility=1     color=255,0,255 useScore=1 ↻
url=http://www.ebi.uniprot.org/uniprot-srv/uniProtView.do?     proteinAc=$$ ↻
color=0,255,0 visibility=1 .
chr19    Navigator1    test_element1    7010000    7090000    400    -    .    P61204
```

In this example, the first line is a comment (indicated by '#'). The next three lines contain instructions for the browser: the line starting with 'browser' tells the browser which part of the genome to display; the line starting with 'track' states the name of the information track. The 'track' line lets you define some additional information on how to display the information and it also allows you to provide a URL to link to if the user selects the particular element (just as most of the features you have already viewed in Ensembl had links to further information). In this example, we provide links from this track to UniProt entries, by stating:

```
url=http://www.ebi.uniprot.org/uniprot-srv/uniProtView. ↵
    do?proteinAc=$$
```

(we will discuss the '$$' in a moment). Obviously, you can change this to link to point to any web page.

The next five lines contain generic file format (GFF) fields, describing five features that will be displayed in this track. Each line consists of the sequence name (a chromosome or a contig – in this case, 'chr19'), the source (for example, the program that generated this feature – in this case 'Navigator'), the name of this type of feature (such as 'CDS', 'start_codon', or 'exon' – in this case, 'test_element'), the start and end positions of the feature, a score of between 0 and 1000, which determines the level of gray in which this feature is displayed, the strand ('+', '-', or '.' for features to be shown on the plus-strand, minus-strand, or both), the frame (a number between 0 and 2 that represents the reading frame of the first base, or a '.' if the feature is not an exon) and finally the 'group': all lines with the same group (for example, 'P61205') are linked together into a single item, which can be used, for example, if you want to display the linked exons of a single gene. When creating the link for each feature, the genome browser will substitute the group for the '$$' in the generic link, so that in this example each of the five features will link to the respective UniProt entry.

It is possible to display several tracks; for example, you could use one track for each alternative splice variant. The final lines of the example file (starting with 'browser dense Test2') define a second custom track, this time containing only a single feature.

The Ensembl browser behaves slightly differently from the UCSC browser. At the time of writing, it seems that the behavior of the UCSC browser conforms better to the examples given on the instruction pages of the browsers and provides more meaningful error messages. Therefore, we will use a slightly simpler example in Ensembl. In *Protocol 7*, we will use a simplified file format to display basic information in the Ensembl browser. In *Protocol 8*, we will use the more extensive file (shown above) to display slightly more complex data in the UCSC browser.

Protocol 7

Linking your own data to Ensembl

1. Download a copy of the file Linked_dataENS.txt[2.24] from the Protocol_7 folder for this chapter on the book's web site. Open the file in a text editor or word processor and examine its contents. If you experiment by editing this file, be sure to save it as 'text only' after editing[a] and take care with end-of-line and tab characters.

2. Go to http://www.ensembl.org/Homo_sapiens/index.html[2.25].

3. In the **Search** box at the top centre of the screen, type '19: 6900000-7100000' (to view the region from 6.9 to 7.1 Mb on chromosome 19) and then click the **Go** button or press 'Enter' on your keyboard.

4. This brings up the Contig View page for chromosome 19, displaying the specified region. Scroll down if necessary to see the 'Detailed view' panel.

5. Click on the **DAS Sources** menu (at the top of the 'Detailed view' panel) and then click on **Manage sources** at the bottom of the list that appears.

6. A new window (called 'DasconfView') will appear. On the left, under **Manage Sources**, click on the link **Upload your data**.

7. The refreshed page should be headed 'DAS Wizard Step 1 of 3: Data location', and there are fields at the top for your e-mail address and a password of your choice. You need to complete these fields so that Ensembl can contact you at a future date if necessary and so that nobody else can modify your uploaded data.

8. Use the **Choose file** button (or **Browse** button as it will appear in some browsers) to find and upload the file Linked_dataEnsEMBL.txt[2.24]. When this has been done, the filename should appear next to the **Choose file/Browse** button.

9. Click the **Next** button just below this.

10. If everything goes well, you will be taken to a new page headed 'AS Wizard Step 2 of 3: Data appearance', in which case go on to step 12.

11. If this does *not* happen, look for an error message towards the top of the screen (just below where it says 'Please upload your data location'), along the lines of:

    ERROR: Could not upload data due to 'ERROR: Invalid format. Line 1'

    In this case, check that the Linked_dataEnsEMBL.txt[2.24] file has not been corrupted – check especially (using a text editor or Word – but remember to save as a text-only file) that the fields in the file are divided by tab characters (*not* spaces). Try uploading again and, if it still fails, copy and paste the contents of the file into the **Paste your data** window instead of using the **Choose file/Browse** option in step 8.

12. You should now be on the page headed 'AS Wizard Step 2 of 3: Data appearance'. Towards the top of the page are several check boxes (next to **Enable on**; these check boxes determine which views your data will be visible in. The **contigview** box should be checked (if not, click to tick it).

13. Click on the **Next** button and you will be taken to the next page, headed 'DAS Wizard Step 3 of 3: Display configuration'. This page gives you numerous options for the appearance the data you have just uploaded.

14. Under **Name** and **Track label** at the top of the screen, enter any descriptive name you like (for this example, use 'NavigatingGenomes' and 'NavigatingGenomesTrack', respectively). You can leave the other fields blank and the other options at their default settings. Click on the **Finish** button just below these options.

15. You will be taken back to the first step (as in step 6); close this window and return to the Contig View page (as you left it in step 5). Refresh this page using your browser's 'reload' button.

16. Click on the 'DAS Sources' menu to open it (if it is already open, close it and then reopen it). You should see 'NavigatingGenomesTrack' towards the bottom of the list. The check box next to it should be ticked (if not, click to tick it).

17. Now click the **Refresh** button toward the top of the 'Detailed view' panel. The screen will refresh and you will see your added data displayed alongside the other features. Clicking on any one of the features will bring up information from the file you uploaded (see *Fig. 6*).

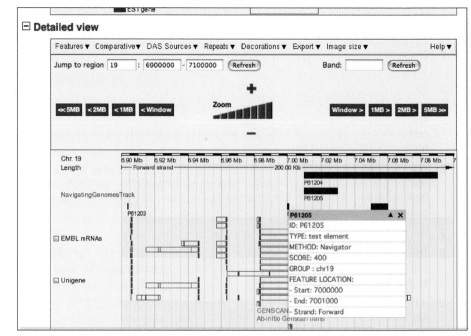

Figure 6. Screenshot showing part of the 'Detailed view' panel of the Contig View page of the Ensembl genome browser (see page xviii for color version).
The data that was uploaded in *Protocol 7* is shown (dark bars just below the chromosome length scale) in the track called 'NavigatingGenomesTrack'. In this shot, the user has clicked on one of the uploaded features (P61205), and the small pop-up window displays information about this feature.

Note

[a]Even this can cause problems: some word processors can change the 'hidden characters' that mark the end of lines, or can substitute spaces for tabs, causing problems later. If in doubt, do not edit or resave the file after you have downloaded it.

By using the other options available through Ensembl, it is possible to upload and present more complex data (including data held on your own or another web site), to provide links (clickable from Ensembl) from your features to other databases such as UniProt, and so on. However, this is beyond the scope of the present chapter. Instead, in the next protocol, we will upload the slightly more complex example file that we discussed earlier into the UCSC browser.

Protocol 8

Linking your own data to the UCSC Genome browser

1. Download a copy of the file Linked_dataUCSC.txt[2.26] from the Protocol_8 folder for this chapter on the book's web site. Open the file in a text editor or word processor and examine its contents. If you experiment by editing this file, be sure to save it as 'text only' after editing and take care with end-of-line and tab characters.

2. Go to http://genome.cse.ucsc.edu/[2.27] and choose **Genomes** from the menu bar at the top.

3. Click the **add custom tracks** button towards the top of the screen.

4. Ensure that the current human genome sequence assembly is selected (menus towards the top of the screen).

5. Click on the button called **Browse** or **Choose file** (the button name will depend on which web browser you are using) just above the first large text field (called 'Paste URLs or data:') and find the file that you saved in step 1. Alternatively, you can copy the contents of the file and paste them into the window or put a URL that points to your file into the window. Press **Submit**.

6. You will be taken to a new page headed 'Manage Custom Tracks'. At the top is a table displaying the tracks you have added so far. There were two tracks defined in the example file, called 'NavigatingGenomesTrack1' and 'NavigatingGenomesTrack2'. (Links and options from this page allow you to delete, view, or edit some tracks, but we will not do this now.)

7. To the right of the table, click the button **go to genome browser**.

8. The UCSC browser should display the appropriate segment of chromosome 19, showing the custom tracks that you have just uploaded (see *Fig. 7*). Try clicking on the newly added features or their names.

9. All other tracks in the genome browser will be disabled, but can be activated by selecting from the numerous pull-down menus underneath the chromosome graphic; click the **refresh** button (underneath the graphic window – not the 'reload' button on your own web browser) after selecting which tracks to activate.

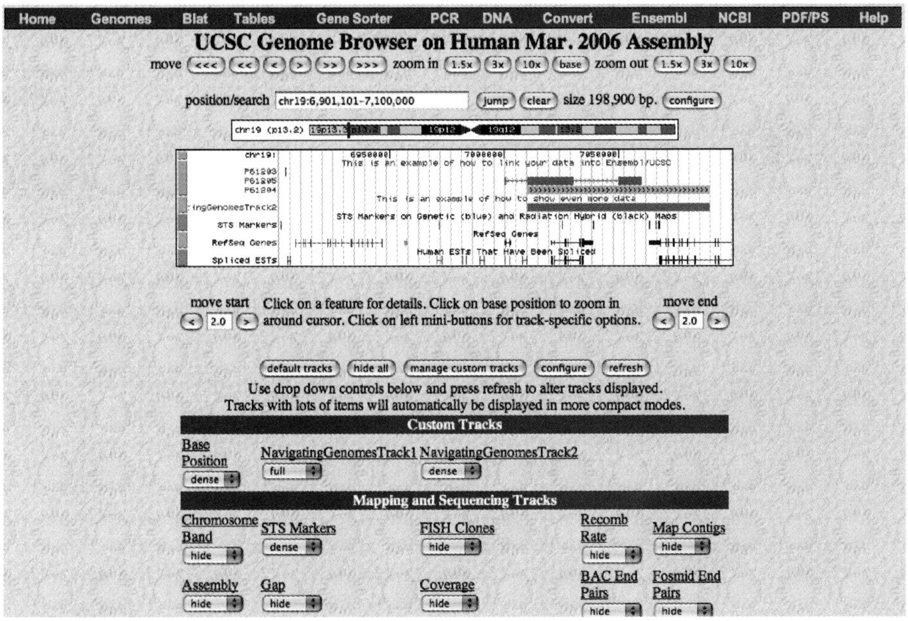

Figure 7. Screenshot showing part of the USCS genome browser.
The data that was uploaded in *Protocol 8* is shown in the graphic window, just beneath the chromosome distance scale. Additional tracks (STS markers, RefSeq genes, and spliced ESTs) have also been activated, using the pull-down menus further down the screen.

Like the Ensembl browser, the UCSC browser offers many opportunities to incorporate your own data, manipulate and display it, and integrate it with other features both within the browser and beyond. Many of these options are beyond the scope of this chapter, but the reader is encouraged to explore and to refer to the online help files.

Acknowledgements

The authors would like to thank all people involved in the many projects presented here, especially the people writing and maintaining the excellent online documentations. T.S. was a British Antarctic Survey/European Bioinformatics Institute/St Edmund's College Research Fellow 2003–2006. This paper was produced by M.S.C. and T.S. within the BIOREACH/BIOFLAME core programs.

3. REFERENCES

1. **Salzberg SL & Yorke JA** (2005) *Bioinformatics*, **21**, 4320–4321.
2. **Hubbard T, Andrews D, Caccamo M, et al.** (2005) *Nucleic Acids Res.* **33**, D447–D453.
★ 3. *EBI 2can Support Portal* http://www.ebi.ac.uk/2can/[2.28]. – *Home page for the EMBL EBI's 2can bioinformatics support portal. Tutorials are offered on a wide range of bioinformatic-related topics from basic biology to detailed protocols.*
★ 4. *Ensembl help pages* http://www.ensembl.org[2.2]. – *The help pages for Ensembl are very readable and helpful, and contain a wealth of information not only on Ensembl but also on related bioinformatic topics.*
★ 5. **Galperin MY** (2007) *Nucleic Acids Res.* **35**, D3–D4. – *Nucleic Acids Research publishes a regularly updated database issue, summarizing major bioinformatic databases. This is a very good source for information on available bioinformatic data resources. Free access is available via* http://nar.oxfordjournals.org/.
★ 6. **FoxJA, McMillan S & Ouelette BF** (2006) *Nucleic Acids Res.* **34**, W3–W5. – *Nucleic Acids Research publishes a regular web server issue, summarizing and providing links to a great many bionformatic web servers. Free access is available via* http://nar.oxfordjournals.org/.
7. **Mullan L** (2004) *Brief. Bioinform.* **5**, 365–369.
8. **Bernal A, Ear U & Kyrpides N** (2001) *Nucleic Acids Res.* **29**, 126–127.
9. **Kyrpides NC** (1999) *Bioinformatics*, **15**, 773–774.
10. **Liolios K, Tavernarakis N, Hugenholtz P & Kyrpides NC** (2006) *Nucleic Acids Res.* **34**, D332–D334.
11. **Ashurst JL, Chen CK, Gilbert JG, et al.** (2005) *Nucleic Acids Res.* **33**, D459–D465.
12. **Kersey P, Bower L, Morris L, et al.** (2005) *Nucleic Acids Res.* **33**, D297–D302.
13. **Frazer KA, Pachter L, Poliakov A, Rubin EM & Dubchak I** (2004) *Nucleic Acids Res.* **32**, W273–W279.
14. **Markowitz VM, Korzeniewski F, Palaniappan K, et al.** (2006) *Nucleic Acids Res.* **34**, D344–D348.
15. **Kasprzyk A, Keefe D, Smedley D, et al.** (2004) *Genome Res.* **14**, 160–169.
16. **Berriman M & Rutherford K** (2003) *Brief. Bioinform.* **4**, 124–132.
17. **Birney E, Andrews D, Caccamo M, et al.** (2006) *Nucleic Acids Res.* **34**, D556–D561.
18. **Hinrichs AS, Karolchik D, Baertsch R, et al.** (2006) *Nucleic Acids Res*, **34**, D590–D598.

CHAPTER 3

Sequence similarity searches

Jaap Heringa and Walter Pirovano

1. INTRODUCTION

1.1 Comparative sequence analysis

Comparative sequence analysis is a common first step in the analysis of sequence-structure–function relationships in protein and nucleotide sequences. To obtain knowledge about the role of a certain unknown protein, comparing the protein's sequence with the many sequences in annotated protein databases often leads to useful suggestions regarding the protein's three-dimensional structure or molecular function. As the prediction of a protein's structure and function on first principles is still a major unsolved problem in molecular biology (see Chapters 8 and 10), the method of indirect inference by comparative sequence techniques has become essential for structural and functional genomics initiatives. Over the last decade, this approach has led to the novel annotation of more sequences than any other individual technology.

Analysis of sequence similarity also underpins many other areas of bioinformatics, including the identification of coding regions in genomic sequence by interspecies comparison (see Chapter 4) and the analysis of evolutionary relationships (see Chapter 12). In short, analysis of the similarities between protein or DNA sequences is a cornerstone of bioinformatics.

Given the fact that sequence databases are growing exponentially, many current research projects are aimed at improving the sensitivity of sequence similarity searching techniques, whilst trying to ensure that the speed of the algorithms is sufficient to scour all of the available sequence data.

1.2 Sequence alignment as a reflection of similarity

Although many properties of nucleotide or protein sequences can be used to derive a similarity score (e.g. nucleotide or amino acid composition, isoelectric point, or molecular weight), the vast majority of sequence similarity calculations rely on an alignment between two sequences from which a similarity score is inferred.

Bioinformatics: *Methods Express* (Paul H. Dear, ed.)
© Scion Publishing Limited, 2007

Ideally, the alignment matches the nucleotide or amino acid sequences from either sequence according to their evolutionary descent from a common ancestor, with conserved residues at matched positions and inserted/deleted fragments intervening at proper sequence positions. Often, however, evolution has led to widely diverged sequences where the ancestral ties have become blurred beyond recognition, leading to biologically incorrect alignment.

Another confounding issue is the fact that an increasing number of cases are identified with nonorthologous displacement, where enzymes carrying out an identical function in different organisms belong to entirely different protein families and thus are not expected to show any sequence similarity. For example, the ornithine decarboxylase spe1 in *Saccharomyces cerevisiae* has a completely different domain structure from – and is not related to – the *Escherichia coli* ornithine decarboxylase isozymes speC and speF (1). Nor are sequence alignment techniques able to trace evolutionary cases of horizontal gene transfer or functional displacement of one gene by another within a genome.

1.3 Similarity versus homology

The term 'homologous sequence' is often used when in fact a sequence should only be described as 'similar' to a given reference sequence (2). Whereas sequence similarity is a quantification of an empirical relationship of sequences expressed using a gradual scale, 'homology' denotes an inference of a common ancestor between the sequences. Sequence similarity is normally used to assess the *likelihood* of homology, but homology itself is a qualitative state: *a pair of sequences is either homologous or not*. As protein tertiary structures are more conserved during evolution than their coding sequences, homologous sequences are assumed to share the same protein fold. Although it is possible in theory that two proteins evolve different structures and functions from a common ancestor, this situation cannot be traced and so such proteins are seen as unrelated. However, numerous cases exist of homologous protein families where subfamilies with the same fold have evolved distinct molecular functions. The term homology is often used in practice when two sequences have the same structure or function, although in the case of two sequences sharing a common function this ignores the possibility that the sequences are analogs resulting from convergent evolution, now often referred to as nonorthologous displacement.

Unfortunately, it is not straightforward to infer homology from similarity, as enormous differences exist between sequence similarities within homologous families. Many protein families of common descent comprise members that share pairwise sequence similarities that are only slightly higher than those observed between unrelated proteins. This region of uncertainty has been characterized to lie in the range of 15–25% sequence identity (3) (see below) and is commonly referred to as the 'twilight zone' There are even some known examples of homologous proteins with sequence similarities below the randomly expected level given their amino acid composition (4). As a consequence, it is impossible to prove using sequence similarity that two sequences are not homologous.

The similarity score for two sequences can be calculated from their alignment (see below), such that it depends on the actual scoring matrix and gap penalties used. It has also been calculated as a fraction of a maximal score possible for two sequences using a normalized scoring matrix and by normalizing the raw alignment score by the length of the shorter sequence (5).

1.4 Techniques for pairwise alignment

1.4.1 The dynamic programming algorithm

Protein sequences mutate to varying degrees of divergence through evolution. In order to identify homologous proteins and reveal important similarities, a range of sequence alignment methods are commonly used (for a recent overview, see 6). These methods rely mainly on approximated evolutionary models that aim to reflect as accurately as possible the evolutionary paths that connect two or more protein sequences.

Many methods for the calculation of sequence alignments have been developed, of which implementations of the dynamic programming algorithm (7, 8) are considered the standard in yielding the most biologically relevant alignments. (For three or more sequences, these methods apply the progressive strategy (9), where sequences are hierarchically aligned in pairs according to a pre-generated tree, based on their sequence similarity; see Chapter 11 for a discussion of multiple sequence alignment).

The dynamic programming algorithm (7) requires a *scoring matrix*, which is an evolutionary model expressed in the form of a symmetrical 4×4 exchange matrix for nucleotide sequences or a 20×20 matrix for amino acids: each matrix cell approximates the evolutionary propensity for the mutation of one nucleotide or amino acid type into another, including self-conservation. For this purpose, it is common to use pre-determined substitution scores (e.g. the scores from the BLOSUM (10) and PAM (11) series and more recently the JTT (12), GONNET (13), VT (14), and VTML (15) series) that have been derived using a specific set of 'true' alignments. However, these 'standard' substitution scores reflect a standardized evolutionary model and introduce inconsistencies when applied to nonstandard cases (16). Although this does not impact too severely on alignments between closely related sequences, sequences in the so-called 'twilight zone' (<30% sequence identity) are extremely difficult to align (3), partly for this reason. This is because the evolutionary scenario relating them becomes virtually undetectable against the 'noise' introduced by the extent of mutational change that has occurred (17).

The dynamic programming algorithm also relies on the specification of *gap penalties*, which model the relative probabilities for the occurrence of insertion/deletion events during evolution. In most available methods, a penalty score is applied for creating (opening) a gap, and a further penalty score is added for each extension of the gap (*affine* gap penalties), so that the chance for an insertion/deletion depends linearly upon the length of the associated fragment. Given an exchange matrix and gap penalty values (which together are commonly called the *scoring scheme*), the dynamic programming algorithm is guaranteed to produce the highest scoring alignment of any pair of sequences, the *optimal alignment*.

1.4.2 Global versus local alignments

Two types of alignment are generally distinguished: global and local alignment. *Global alignment* (7) denotes an alignment over the full length of both sequences, which is an appropriate strategy to follow when two sequences are similar or have roughly the same length. However, some sequences may show similarity limited to a motif or a domain only, whilst the remaining sequence stretches may be essentially unrelated. In such cases, global alignment may well misalign the related fragments, as these become overshadowed by the unrelated sequence portions that the global method attempts to align, possibly leading to a score that would not allow the recognition of any similarity. If not much knowledge about the relationship of two sequences is available, it is usually better to align selected fragments of either sequence. This can be done using the *local alignment* technique (8). The first method for local alignment, often referred to as the Smith–Waterman algorithm (8), is in fact a minor modification of the dynamic programming algorithm for global alignment. The algorithm selects the best scoring subsequence from each sequence and provides their alignment, thereby disregarding the remaining sequence fragments. Later elaborations of the algorithm include methods to generate a number of suboptimal local alignments in addition to the optimal pairwise alignment (18).

1.5 Alignment scores as a measure of similarity

In order to optimize alignments and determine the degree of similarity they reflect, it is obviously necessary to have a measure by which to score an alignment. As the dynamic programming algorithm essentially models the alignment of two sequences as a Markov process, where the amino acid (or nucleotide) matches are considered independent, the product of the probabilities for each match within an alignment should be taken. As many of the scoring matrices contain exchange propensities converted to logarithmic values (*log odds*), the score for any alignment can be calculated by summing the log-odd values corresponding to matched residues minus appropriate gap penalties:

$$S_{a,b} = \sum_l S(a_i, b_j) - \sum_k N_k \cdot gp(k)$$

where the first summation is over the exchange values associated with l matched residues and the second over each group of gaps of length k, with N_k the number of gaps of length k and $gp(k)$ the associated gap penalty. In case affine gap penalties are used (see above), $gp(k) = pi + k \cdot pe$, where pi and pe are the penalties for gap initialization and extension, respectively. In other words, the alignment score is composed of a term reflecting the similarities between the aligned residues at each point in the alignment, minus a term reflecting the number and sizes of gaps that were needed to make the alignment. A consequence of the widely used affine gap penalty scheme is that the long gaps required, for example to span an inserted domain B in aligning a two-domain sequence AC (where A and C represent domains) with a three-domain sequence ABC, are often too costly so that such sequences become misaligned.

1.6 Sequence identity as a measure of similarity

In addition to similarity scores calculated as above, a measure of the *sequence identity* between the two sequences in an alignment is valuable because it is simple and also because it gives a good rule-of-thumb indication of the likely structural and functional relationship between the aligned sequences.

Sequence identity is normally expressed as the percentage of identical residues found in a given alignment, normalized using either the length of the alignment or the length of the shorter sequence. This measurement does not depend on an exchange matrix or on gap penalties and therefore is not directly biased by assumptions about the underlying evolutionary model. However, the alignment itself will almost always have been constructed in the first place using a dynamic programming algorithm, which depends on an exchange matrix and gap penalty values, so sequence identity cannot be regarded as independent from sequence similarity.

Using sequence identity as a measure, Sander and Schneider (19) estimated that if two protein sequences are longer than 80 residues, they could relatively safely be assumed to be homologous whenever their sequence identity is 25% or more. Another commonly used notion is that if two sequences share more than 50% sequence identity, their enzymatic function will be the same (20). Contrary to this notion, however, it has been estimated that 70% of pair fragments above 50% sequence identity might not have a completely identical function (20). An example is *Bacillus subtilis* exodeoxyribonuclease and rat DNA lyase, which share 57% identity over 122 alignment positions, yet fulfill different functions (DNA degradation and repair, respectively). Despite its popularity and use in empirical rules as above, the use of sequence identity percentages is not optimal for homology searches (5). As a result, no major sequence comparison methods employ sequence identity scores in deriving statistical significance estimates.

1.7 Statistics of alignment similarity scores

Sequence alignment methods can always produce alignment with an associated similarity score, even in the case of absence of any biological relationship. Although similarity scores of unrelated sequences are essentially random, they can behave like 'real' scores and, for example, like the latter are correlated with the length of the sequences compared.

For this reason, and particularly in the context of database searching, it is important to know what scores can be expected by chance and how scores that deviate from random expectation should be assessed. If, armed with this knowledge, a similarity between two sequences is deemed to be statistically significant, this provides confidence in inferring a biological relationship. Because of the complexities of protein evolution and distant relationships observed in nature, any statistical scheme will inevitably lead to situations where a sequence is assessed as unrelated whilst it is in fact homologous (*false negative*), or the inverse, where a sequence is deemed homologous whilst it is in fact biologically unrelated (*false positive*). A frequent cause of false positives – and hence of erroneous transfer of annotation – is based on similarity found over relatively short sequence regions, or similarity based on different domains in multi-domain structures (20).

1.8 Protein domains

Many protein families have diverged from common ancestors by evolving different combinations and associations of domains (21–23). Domains are characterized as semi-independent three-dimensional units in proteins, often with a particular function, observed to be genetically mobile and frequently moving within and between biological systems through mechanisms of gene or exon shuffling. An understanding of the domain organization of a protein sequence is crucial for structural and functional genomics initiatives and the reader is referred to Chapter 8 for a discussion of protein architecture and domains.

The correct partitioning of a protein into its putative domains is especially important in the comparative analysis of entire genome sequences. Consideration of domain architecture will shed light on the evolution, structure, and function of a protein family. For example, the 'Rosetta Stone' genome analysis method (24) exploits the fact that a multi-domain protein in one organism may be present as separate (and hence, presumably, interacting) proteins in another organism. It is clear that such analysis requires accurate sequence comparison tools at the level of the domain rather than of the whole protein.

Domain annotation of a protein sequence in the absence of structural information has proved to be a difficult problem. For example, the method of Wheelan *et al.* (25) is based on the fact that domains have a distinct size distribution, averaging at 100 residues. Accurate predictions are limited to two-domain proteins with less than 300 residues. George and Heringa (26) improved the delineation of protein domain boundaries to 52% using a consistency-based protocol over sets of protein *ab initio* three-dimensional model structures generated using distance geometry. Currently, most annotated domain databases are based on inferring domains by sequence similarity searches (27–32). A number of these search techniques will be discussed in the next section.

2. METHODS AND APPROACHES

A typical application to infer knowledge for a given query sequence is to compare it with all sequences in an annotated sequence database. Unfortunately, the dynamic programming algorithm (see above) is too slow for repeated searches over large databases and may take many hours for a single query sequence on a standard workstation. However, for any biologist who has a new protein sequence of unknown functionality, comparison with all known and annotated sequences is paramount.

Therefore, fast routines have been devised that enable database searches on even small computers with only a small loss of sensitivity compared with searches using full dynamic programming. With the recent advent of parallel multi-processor computers at central sites, researchers can routinely perform multiple sequence searches over complete sequence databases. However, for large-scale application of the dynamic programming technique, the computational requirements are still prohibitive. For example, consider the task of searching the Swiss-Prot database against a query sequence of 400 amino acids. As release 50.0 of UniProtKB/Swiss-Prot contains 222 289 sequence entries, comprising 81 585 146 amino acids, finding local alignments via dynamic programming over this database would entail about 10^{10} matrix operations. Given the fact that many servers routinely handle thousands of such queries a day (over 50 000 per day in the case of the NCBI server), it is clear that the application of dynamic programming would lead to unfeasible waiting times.

Although some special hardware has been designed to accelerate the dynamic programming algorithm, the solution has depended largely on the development of several heuristic algorithms that represent shortcuts to speed up the basic alignment procedure. These include the currently most widely used heuristic method for scouring sequence databases for homologies, PSI-BLAST (33), an extension of the BLAST technology (34), and FASTA (35), which is another commonly used heuristic method for fast sequence comparison. At the same time, advances in computer hardware have made it possible to use some more computationally intense approaches such as the hidden Markov modeling-based tools SAM-T99 (36), SAM-T2K (37), and HMMER2 (38).

2.1 Should one compare protein or nucleotide sequences?

As long as we are considering sequences between encoded proteins, the actual pairwise comparison between two sequences can take place at the nucleotide or peptide level. However, the most effective way to compare sequences is at the protein level (39), which requires that nucleotide sequences must first be translated in all six reading frames followed by comparison with each of these conceptual protein sequences.

Although mutation, insertion, and deletion events take place at the DNA level, there are several reasons why comparing protein sequences can reveal more distant relationships:

1. Many mutations within DNA are synonymous, which means that they do not lead to a change in the corresponding amino acids. As a result of the fact that most evolutionary selection pressure is exerted on protein sequences, synonymous mutations can lead to an overestimation of the sequence divergence if compared at the DNA level.
2. Evolutionary relationships can be expressed more finely using a 20×20

amino acid exchange table than by using exchange values among four nucleotides, leading to a significant increase in statistical subtlety for protein sequences. Amino acid substitution matrices incorporate subtle differences in physicochemical properties among the 20 residue types, rendering protein sequences more informative than nucleotide sequences.

3. DNA sequences contain noncoding regions, which should be avoided in homology searches. Note that the latter is still an issue when using DNA translated into protein sequences through a codon table. However, a complication arises when using translated DNA sequences to search at the protein level because frame shifts can occur, leading to stretches of incorrect amino acids in the wrongly transcribed product and possible elongation of sequences due to missed stop codons. On the other hand, frame shifts typically result in stretches of highly unlikely and distant amino acids, which can be used as a signal to trace their occurrence.

2.2 Curated and annotated sequence databases

The success of sequence similarity searches depends crucially on the quality and coverage of the sequence database used. Although the amount of raw sequence data is increasing rapidly, and although modern sequencing techniques achieve a very high accuracy, the utility of this data depends crucially upon its annotation. Incorrect annotation of database sequences can distort similarity searches (for example, when the location or structure of predicted genes in the database sequence is incorrect), or can lead to false inferences when genuine similarities are found but the database sequence has been annotated with an incorrect function.

As inferring and experimentally validating the annotations represents a bottleneck, there is a rapidly widening gap between sequence and annotation data. This is reflected by the fact that many sequences have 'unknown' as their functional annotation, whilst an increasing number of sequences, especially those originating from bacterial genomes, have annotations such as 'conserved hypothetical'. Conserved hypothetical open reading frames have homologs, usually in other organisms (which at least gives reassurance that the open reading frame truly is a gene – see Chapter 4), but none of these homologs have known functions.

Although many new protein structures are now being determined using X-ray crystallography, nuclear magnetic resonance spectroscopy, and cryoelectron microscopy, without direct experimental evidence there is considerable difficulty in assigning functions to proteins from their structures. This can even be the case for homologs of well-characterized proteins because of the recruitment of similar proteins for divergent functions. Computational prediction methods can aid to some extent, but for reliable annotation, manual curation is often essential.

Widely used annotated databanks for homology searches include the annotated EMBL, GenBank, and DDBJ for nucleotide sequences, whilst for protein sequences the Swiss-Prot, Protein Information Resource (PIR), TrEMBL, GenPept, NR-NCBI, and NR-ExPasy databases are popular. Also the genome survey sequence (GSS), expressed sequence tag (EST), sequence-tagged site (STS) and high-throughput genomic sequence nucleotide databases can be scoured to find homologies, gain insight in expression data, or locate a gene on the genome map. The NR-NCBI database is compiled by the National Center for Biotechnology Information (NCBI) as a nonredundant (NR) protein sequence database for BLAST searches. It contains a total of about one million nonidentical sequences from GenBank CDS (coding sequence) translations, Protein Data Bank (PDB), Swiss-Prot, PIR, and Protein Research Foundation (PRF).

2.3 Heuristic sequence similarity searching methods

Both the FASTA and BLAST suite of programs feature a quick step for initial filtering of the database sequences, followed by a second slower step to scrutinize the sequences and compile the final alignments between the query and each of the database sequences. If the initial filtering step is too strict, there is a biological risk: homologous sequences will be discarded before the more detailed analysis and are lost (false negatives). If the initial filtering step is too permissive, however, there is a computational penalty because too many unrelated sequences are passed through to the slower subsequent step. In both the FASTA method and a recent implementation of the BLAST algorithm, the slow step incorporates the dynamic programming algorithm to compile a local alignment.

2.3.1 FASTA

In the early years of sequence database searching, the heuristic method FASTA (35) was the most widely used technique. The FASTA program compares a given query sequence with a library of sequences and calculates for each pair the highest-scoring local alignment. The speed of the algorithm is obtained by delaying application of the dynamic programming technique to the moment where the most similar segments are already identified by faster and less-sensitive techniques. To accomplish this, the FASTA routine operates in four steps of which the first two represent a quick filter to eliminate sequences that have no fragments scoring beyond a specified threshold value. Sequence fragments that score beyond a given threshold value after the first two steps are combined and realigned in the last two steps.

The four basic steps of FASTA are as follows:

1. The first step searches for identical 'words' (short segments of sequence) of a user-specified length ('*ktup*') occurring in the query sequence and the target sequence(s). For each target sequence, the ten regions with the highest density of ungapped common words are determined. The technique is based on that of Wilbur and Lipman (40, 41) and, for not-too-distant sequences (>35% residue identity), little sensitivity is lost whilst speed is greatly increased. The search is performed by 'hashing techniques', where a look-up table is constructed for all words in the query sequence and is then used to compare all encountered

identical words in the database sequence(s). Generally, for proteins, a word length of two amino acids is sufficient ($ktup$=2), whilst for nucleotide sequences $ktup$=6 is the default word length. Searching with higher $ktup$ values increases the speed but also the risk that similar regions are missed.

2. In the second step, these ten regions are rescored using the Dayhoff PAM-250 residue exchange matrix (42).

3. In the third step, a threshold value is applied to filter the ten regions: sequences with none of the ten regions scoring beyond the threshold are effectively discarded at this point; regions scoring higher than the threshold value and being sufficiently near to each other in the sequence are joined, now allowing gaps. The highest-scoring region of these new fragments is retained.

4. The fourth and final step performs a full dynamic programming alignment over the region yielded in the preceding step, which is widened by 32 residues on either side (43).

In early FASTA versions, the best-scoring regions resulting from steps 2 and 3 above were reported as *init1* and *initn* in the FASTA output, respectively, whilst the final alignment score (step 4) was written under *opt*. Modern implementations of FASTA, however, only report an *E* value for each of the database sequence fragments aligned with the query as a measure of their statistical significance as putative homologs (see section 2.4).

In *Protocol 1*, we will give an example of the use of FASTA, focusing on a subunit of a large enzymatic complex called cytochrome c oxidase. This complex is found both in bacteria and mitochondria where it catalyzes electron transfer through the last part of the respiratory chain. The starting point will be the mouse (*Mus musculus*) subunit IV of the mouse cytochrome c oxidase complex.

Protocol 1

A typical search using FASTA

1. First, retrieve the query sequence that we will be using for this example. Go to the NCBI web site at http://www.ncbi.nlm.nih.gov[3.1]. From the pull-down menu headed **Search** at the top left, select **Protein** and type 'NP_034071' in the adjacent text box. Click **Go** or press 'Enter' on your keyboard; we now see a link to the corresponding entry.

2. Click on the link (**NP_034071**) at the top of the entry and then, from the **Display** menu (left, towards the top), select **FASTA** to display the protein sequence in the FASTA format. Copy the entire entry (from the header line starting '>gi|6753498' to the end of the protein sequence, ending '...DKNEWKK').

3. Paste the text into a new document in Word or another text processor and save the file as 'NP_034071.fa' on your computer. It is important to save it as 'text only'.

4. There are several other ways to obtain the same sequence from NCBI or from other databases – for example, see Chapters 1 and 2 for general information on retrieving sequences from databases. Also, a copy of the file NP_034071.fa can be downloaded from the Protocol_1 folder for this chapter on the book's web site.

5. For running the actual FASTA routine, go to the home page of the method at http://www.ebi.ac.uk/ fasta33/[3.2]. Note that the FASTA method should not be confused with the FASTA sequence format mentioned in the preceding step.

6. This page offers many variants of FASTA, but we will use FASTA3 – make sure that this is selected (highlighted) in the **Program** menu.

7. There are numerous options on the page, but most of them should be left at their default values. You can (for example) choose to receive the results by e-mail rather than interactively. You can also specify which protein sequence databases to search against (menu at upper right); you can select multiple databases by shift-clicking or by other key/mouse combinations (depending on your computer and browser) but, for this example, leave the **Databases** menu set to **UniProt**.

8. The other parameters define the details of the search. The values for the gap penalty (both to open a gap and to extend it by one residue), the **ktup** value (the size of the 'words' that are used in the early stages of the search – see above), and the **substitution matrix** that is used to evaluate the similarity between amino acids can all be altered, but for now, leave them at their default values.

9. The expectation upper and lower values (E values) can also be altered. The E value is a measure of the statistical significance of a hit (see below) – higher values correspond to lower significance. The default upper value of 10 ensures that we will find even fairly distantly related proteins. The default lower value is effectively zero – so that we will also find extremely closely related (or identical) proteins. Leave these settings at their defaults.

10. It is also possible to restrict the search to a part of the query sequence (using **Sequence range**), or to compare the query only against database proteins that have a certain range of sizes (**Database range**), but we will leave these at their default settings, so that we search all of our query sequence against all proteins in the database.

11. Under **Scores & Alignments**, set both **Scores** and **Align** to 100 (the default values are 50) – this ensures that we will retrieve up to 100 matches.

12. Further help can be obtained by clicking on any of the colored menu titles.

13. Either copy and paste the complete contents of NP_034071.fa into the large text window lower down the screen, or use the **Choose file** (or **Browse**) button to choose this file. (Note that many other sequence formats can also be used.)

14. Click **Run Fasta3** and wait for your job to be processed (this should take only a minute or so).

15. When your results page appears[a], it should look similar to *Fig. 1*. (Of course, the UniProt database that was searched is updated frequently. It is therefore quite likely that some of the 'hits' will change by the time you read this. At the time of writing, this search produced 81 hits.)

16. The upper part of the screen (**Submission parameters**) simply summarizes the settings you have used and some aspects of your query sequence.

17. The lower part of the screen is a table listing the hits, starting with the most significant (lowest E value – see section 2.4). (Clicking on any of the other headings in the table will cause the results to be resorted by that parameter; if you try this, click on E() afterwards, to again sort the list by E value.)

18. For each hit, there is a link to UniProt (under **DB:ID**). Also reported are the length of the protein, the percentage of residues in the alignment that were either identical or similar to those in the query sequence, and the length of the overlap between the two sequences in the alignment.

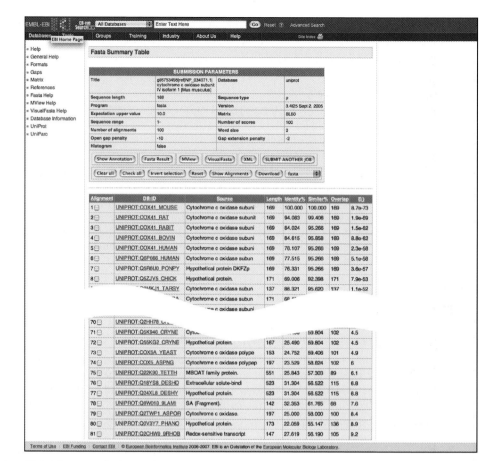

Figure 1. FASTA **output.**
Below the summary table (which gives details of the search parameters), 81 hits are listed in order of increasing *E* value (decreasing significance).

19. Note that the top sequences have very low *E* values and can therefore be trusted to be homologous to the query. (The top match, in fact, is identical to the query sequence). However, at the bottom of the list we also find some sequences that apparently are not related to the query (e.g. UNIPROT:Q2CHW9_9RHOB showing an *E* value of 9). Although there might well be genuinely homologous sequences having unfavorable *E* values (e.g. UNIPROT:Q2TWP1_ASPOR with an *E* value of 8.1), users should generally be cautious of *E* values above 0.001, as at this score level, false positives can arise.

20. Click on **Show aligmnents** (underneath the **Submission Parameters**). The display will now show alignment details and the alignment itself for each of the hits. Each alignment shows the aligned parts of the two proteins plus (if the alignment covers only part of the sequences) some flanking sequence. A ':' indicates that the amino acids in the query sequence and the hit are identical, and a '.' that they are similar (for example, lysine and arginine).

21. Click on **Summary table** to return to the previous page.

22. Click on **MView**. The hits are now displayed in color-coded form: residues identical to the query sequence in the respective alignment are colored, whilst nonidentical residues are gray. (Note that the display is in 'blocks' – the first 80 residues for each aligned protein, then the

next 80, etc.). Also shown are consensus sequences at different levels of stringency, although this is only meaningful when all of the proteins are well aligned.

23. Click on **Return to result** to return to the results table.

24. There are many options for viewing and downloading the results, or for selecting only some of the hits (using the check boxes under **Alignment** on the left of the table) to examine further. In particular, the **VisualFasta** option gives a quick graphical indication of the strength and extent of each of the alignments.

Note

[a]Note that results are stored for about 24 h. Thus, if you copy the web address of the results page, you can return to it at a later point.

Protocol 2 suggests a variant on the previous search, this time examining only more distantly related proteins.

Protocol 2

A FASTA search for distant homologs

1. Repeat the FASTA search from *Protocol 1*, but this time set the **Expectation upper value** and **Expectation lower value** to 20 and 0.001, respectively. This will find only those sequences with a low similarity (down to an *E* value of 20, which is of very low significance) and will exclude the most similar sequences (with *E* values below 0.001). At the time of writing, this search produced 24 hits.

2. If you look at these hits using the **Mview** option, you will see that the identity with the query sequence is generally very sparse, and that the identical residues are widely scattered and tend to be at different places in the different hits. This suggests that many of these hits are spurious.

3. There will certainly be some true homologs among these less-significant hits, but it is difficult to spot these purely from the sequence identities in Mview. Examining the alignments gives a little more information; a true but distant homolog might be expected to have some similarity across most of the protein length (rather than good similarity on only one or a few small areas). However, more confident identification of true distant homologs is not possible using this simple search strategy.

2.3.2 BLAST

Since its inception in 1990, the BLAST (basic local alignment search tool) program has quickly gained a dominant position, and the original publication of the technique (34) is the most cited paper in molecular biology to date. BLAST is a speed-optimized technique that maintains significant sensitivity through the combination of a fast and subsequent slow algorithmic step.

The BLAST suite includes a number of variants to allow all possible combinations of comparisons between nucleotide or protein sequences. In particular, nucleotide sequences can be translated in all six possible reading frames for comparison either with protein sequences or with other, similarly translated nucleotide sequences (see *Table 1*).

Table 1. BLAST **variants**

Note the distinction between BLASTN (which compares query and database nucleotide sequences directly with each other) and TBLASTX (which first translates the query and database sequences in all possible frames and then compares the resulting protein sequences).

Program	Query sequence	Database	Notes
BLASTN	Nucleotide	Nucleotide	Direct comparison between nucleotide sequences
BLASTP	Protein	Protein	Direct comparison between protein sequences
BLASTX	Nucleotide	Protein	All six translations of the query sequence are compared with the protein database
TBLASTN	Protein	Nucleotide	The query protein is compared against all six translations of each sequence in the nucleotide database
TBLASTX	Nucleotide	Nucleotide	All six translations of the query sequence are compared against all six translations of each sequence in the nucleotide database

The basic idea behind the initial quick filtering step in BLAST is the generation of all consecutive tripeptides in a given protein query sequence, or 11-nucleotide words when searching with a DNA sequence. For each of the words, a table is constructed of words deemed to be 'similar', where the number of similar tripeptides corresponds to only a fraction of the 20^3 possible tripeptides, or 4^{11} possible nucleotide 11mers. The BLAST program uses the tables of similar words to quickly scan a database of protein or nucleotide sequences for ungapped regions showing high similarity; each time, a database word is accepted whenever it occurs in the table for the query word considered.

In this respect, BLAST differs significantly from FASTA, in that it can consider similar 'words' in the early stages of the search whereas FASTA considers identical words. Similar regions found by BLAST between the query and database sequences scoring beyond a given threshold, T, are referred to as high-scoring sequence pairs (HSPs) and are retained for further processing. To score these regions, BLAST employs the BLOSUM62 amino acid exchange matrix (10), such that the existence of HSPs scoring higher than T signifies pairwise similarity beyond random probability, which is taken as a signal that the database sequence considered is related. The computational strategy involved behind the quick initial step in the BLAST method is based on deterministic finite automata, which allows very quick searching of the similar-words table associated with each query word.

The original BLAST method features a slow algorithmic step that tries to refine the database hits by extending each HSP in either direction in an attempt to generate a longer alignment with a higher score than the nonextended region. During extension, the alignment score is temporarily allowed to drop but not more than a pre-set drop threshold of S, which is set to 20 for protein sequences and 22 for DNA sequences, before the score picks up again to arrive at a higher value. The maximum scoring pairs (MSPs) resulting from the word extensions are presented as the final result in the original BLAST version.

As the original BLAST program spends more than 90% of its time on extending words, a key improvement of the BLAST method (33) has been to extend words only when there are two hits on the same diagonal within a given distance, A, of each other (in other words, where there is a chance of the extension running into further regions of similarity). As this would lead to far fewer words being extended, sensitivity is maintained by lowering the value of T for finding the initial HSPs. With a lower value of T, far more single hits are produced, but only a minority has an associated second hit nearby on the same diagonal. In the more recent version of BLAST, word extension is done using dynamic programming, leading to gapped alignments. The updated technique, referred to as gapped BLAST (33), is therefore more similar in spirit to the earlier FASTA program (35) than to the earlier BLAST method (34). It is also slightly faster than the earlier BLAST method, as extension by dynamic programming is only triggered when the aforementioned two-hit extension has a sufficiently large score. If this is the case, the highest scoring segment of length 11 along the region covered by the two-hit extension is taken as the seed. Dynamic programming is then initiated in the forward and backward directions from the central pair in the 11-long HSP. Gapped extension proceeds as long as the score remains above a given threshold, whilst the score is temporarily allowed to drop below the threshold as long as it takes off again and rises above the threshold value. The ends of the alignment are finally pruned to yield the best local alignment given the 11-residue seed, and this alignment is reported to the user.

A difference between FASTA and BLAST is that FASTA uses the BLOSUM50 substitution matrix when calculating similarities between protein sequences, whilst BLAST uses BLOSUM62 (although this can be changed in both programs). BLOSUM62 is a 'harder' matrix (i.e. overall it tends to report a lower similarity between any two nonidentical amino acids than BLOSUM50), which is amenable to less-divergent sequence comparisons. Another difference is that the BLAST server can return a maximum of 20000 hit sequence descriptions and alignments, whilst the FASTA server (*Protocol 1*) is limited to a maximum of 100.

Protocol 3 takes the user through a basic BLAST search, again using subunit IV of the mouse cytochrome c oxidase complex from *M. musculus* as an example.

Protocol 3

A typical search using BLAST

1. Open the BLAST homepage at http://www.ncbi.nlm.nih.gov/blast/[3.3].

2. Click on **protein blast**.

3. Paste the cytochrome c oxidase sequence from *M. musculus* in FASTA format (see *Protocol 1*) into the **Enter Query Sequence** box.

4. Below we can see that the choice of available databases is different from that available from the FASTA search page (*Protocol 1*), although most are major protein databases that will have essentially the same content. Leave **Database** at its default of **nr** – this encompasses all nonredundant translations from GenBank and several other major databases (click on the **Database** link, or indeed on any of the highlighted headings, for a 'Help' page giving more details).

5. Leave all other settings at their defaults for now and click **BLAST** to start the search using standard BLASTP (protein–protein BLAST).

6. Typically within a few seconds the results of the BLASTP run will be displayed (see *Fig. 2*).

7. At the top of the results page (after the references), a graphic similar to FASTA's Mview (see *Protocol 1*) represents the extent and significance of hits against the query sequence. Moving the mouse over any of the colored bars will cause details of that alignment to appear in the small text window above the graphic.

8. Below the graphic, the hits are listed in order of increasing *E* value (decreasing significance). The hit sequences found for the cytochrome c oxidase query are in line with those retrieved by the FASTA method (*Protocol 1*), but the different nomenclatures do not make this obvious.

9. The default threshold for the *E* value (see section 2.4) is 0, but can be adjusted by the user. As with the FASTA method, sequences with *E* values beyond 0.00 should be treated with caution as they may well be unrelated (e.g. sequence `gi|118401825|ref|XP_001033232.1|`, which has an *E* value of 2.6).

10. Links on the left of each of the named hits link to the Entrez protein database entry for the protein. **U** and **G** symbols to the right of some hits link to the entries in UniGene or Entrez Gene. From all of these linked databases, there are numerous links to other resources for that protein.

11. Below the list of hits, each alignment is shown in detail. (You can jump directly to one of the alignments by clicking on the link under **Score** in the list of hits, or by clicking on the colored bar in the graphic window.) Note that redundant entries are listed here (although only one of them is given in the graphic window and in the list of hits).

12. On the query page (step 3), there are numerous options to alter the search parameters or limit the search. In particular, under **Algorithm parameters** (below), the *E* value threshold (**Expect threshold**), word size (equivalent to *ktup* in FASTA), and the gap penalties (**Gap costs**) can be adjusted, as can more-complex parameters.

13. For example, the **Organism** box (under **Choose Search set**) can be used to limit the search to specific organisms or groups of organisms. (Try repeating the search with this option set to **Custom . . .** combined with **Sus scrofa**; the search should return only a few hits, all from pig.) More complex limits can be imposed by typing an Entrez query into the adjacent text box; see Chapter 1 for more information on Entrez search terms.

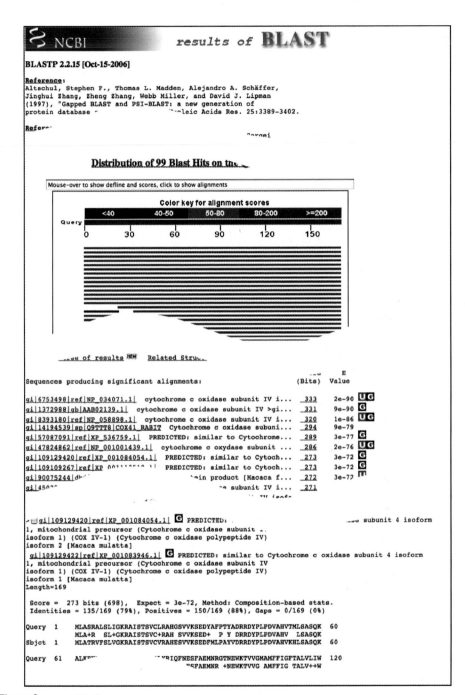

Figure 2. BLAST **output.**
Only sections of the full output page are shown. Beneath the references at the top of the page, a graphic shows the distribution and strength of hits against the query sequence, as a series of colored bars. Below this, a table lists the hits in order of increasing *E* value (decreasing significance), with links to other databases. Below this, each alignment is shown in detail (duplicate entries, such as gi109129420 and gi109129422, are shown as a single alignment).

2.3.3 BLAT

A recent adaptation of the BLAST routine is the BLAT (BLAST-like alignment tool; 44) program. BLAT performs rapid mRNA/DNA and cross-species protein alignments. It is more accurate and about 500 times faster than other tools such as BLAST when used for mRNA/DNA alignments, and 50 times faster for protein alignments at sensitivity settings typically used when comparing vertebrate sequences. When BLAT is applied to DNA sequences, it builds an index of the entire genome in memory. The index consists of all nonoverlapping 11mers except for those heavily involved in repeats. As the total index amounts to less than a gigabyte, it can be kept in RAM for quick access, allowing BLAT to perform very quick searches on a standard Linux box. The index is used to delineate areas that are likely to be homologous, which are then loaded into memory for further detailed alignment. Protein BLAT works in a similar manner, except with 4mers rather than 11mers. The protein index takes a little more than 2 gigabytes, which is also feasible on modern workstations.

The standard implementation of BLAT quickly finds sequences of ≥95% similarity and length 40 bases or more when applied to DNA. It may miss more divergent regions of longer length or more similar ones of shorter length. None the less, BLAT is guaranteed to detect sequence matches down to 33 bases and sometimes detects identical regions as short as 20 bases. When applied to search protein sequences, BLAT finds sequences of ≥80% sequence identity and of length 20 amino acids or more. In practice, DNA BLAT works well on primates and protein BLAT on land vertebrates.

2.4 Statistical significance of search results – *E* values

The BLAST method is based on an exhaustive statistical analysis of ungapped alignments (45) and provides a rigorous statistical framework, based on the extreme value theorem, to estimate the statistical significance of putative homologs.

The *E* (or expectation) value indicates the expected number of sequences with an alignment score equal to or greater than that of the alignment considered, taking into account factors such as the size of database being searched and the composition of the query sequence. For example, if an alignment has an *E* value of 1e-9 (10^{-9}), this means that a match with that score (or better) would only be expected to occur by chance (i.e. in the absence of true homology) in the database with a probability of 1 in a billion and is thus highly significant. Conversely, if a hit has an *E* value of 3.0, this means that one might expect about three equally similar sequences to be found in the database by chance alone – clearly, therefore, such a hit is not necessarily a homologous sequence.

The original BLAST program could detect only local alignments without gaps and therefore might miss some significant similarities. The more recent version of the BLAST algorithm, gapped BLAST, is able to insert gaps in the alignments, leading to increased sensitivity (33). The original statistical framework for ungapped alignments is used to assess the significance of the gapped alignments as well, although no mathematical proof for this is yet available (46). However, computer simulations have indicated that the theory probably applies to gapped alignments as well, so that its application to general pairwise alignment is not likely to introduce error. Therefore, *E* values are roughly comparable across the various search tools. A number of state-of-the-art homology search techniques adopt the Karlin–Altschul statistical framework and routinely calculate *P* or *E* values for each query–database pairwise sequence comparison (see next section).

2.4.1 Statistics of local alignments without gaps

An important contribution for fast sequence database searching has been the realization (45, 47, 48) that local similarity scores of ungapped alignments follow the *extreme value distribution* (EVD) (49). This distribution is unimodal but not symmetrical like the normal distribution, because the right-hand tail at high-scoring values falls off more gradually than the lower tail, reflecting the fact that the best local alignment is associated with a score that is the maximum out of a great number of independent alignments (see *Fig. 3*).

Following the EVD, the probability (P) of a score *S* being larger than a given value *x* can be calculated as:

$$P(S \geq x) = 1 - \exp(-e^{-\lambda(x-\mu)})$$

where $\mu = (\ln Kmn)/\lambda$ and *K* is a constant that can be estimated from the background amino acid distribution and scoring matrix (for a collection of values for λ and *K*

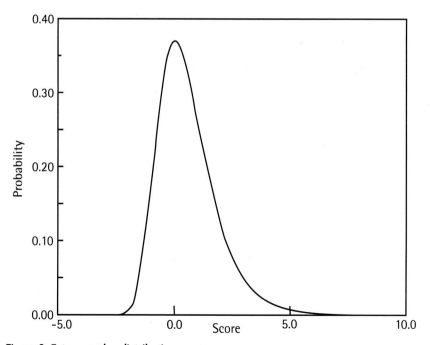

Figure 3. Extreme value distribution.
Shown is the probability density function for the extreme value distribution (EDV) resulting from parameter values $\mu = 0$ and $\lambda = 1$.

over a set of widely used scoring matrices, see 50). Using the equation for μ, the probability for S becomes:

$$P(S \geq x) = 1 - \exp(-Kmne^{-\lambda x})$$

In practice, the probability $P(S \geq x)$ is estimated using the approximation:

$$1 - \exp(-e^{-x}) \approx e^{-x}$$

which is valid for large values of x. This leads to a simplification of the equation for $P(S \geq x)$:

$$P(S \geq x) \approx e^{-\lambda(x-\mu)} = Kmne^{-\lambda x}$$

The lower the probability for a given threshold value x, the more significant the score S.

In spite of the usefulness of the above statistical estimates in recognizing sequence similarity, it should be noted that they do not judge the distribution of similarity along the sequences, which is a crucial aspect in assessing homology. For example, a statistically significant alignment score can correspond to a single domain in a multi-domain protein sequence or to a single motif within a domain, thereby still conferring an incomplete biological picture.

2.4.2 Statistics of local alignments with gaps

Although similarities between sequences can be detected reasonably well using methods that do not allow insertions/deletions in aligned sequences, it is clear that insertion/deletion events play a major role in divergent sequences. This means that accommodating gaps within alignments of distantly related sequences is important for obtaining an accurate measure of similarity. Unfortunately, a rigorous statistical framework as obtained for gapless local alignments has not been conceived for local alignments with gaps. However, although it has not been proven analytically that the distribution of S for gapped alignments can be approximated with the EVD, there is accumulated evidence that this is the case: for example, for various scoring matrices, gapped alignment similarities have been observed to grow exponentially with the sequence lengths (51). Other empirical studies have shown it to be likely that the distribution of local gapped similarities follows the EVD (52, 53), although an appropriate downward correction for the effective sequence length has been recommended (50). The distribution of empirical similarity values can be obtained from unrelated biological sequences (54). Fitting of the EVD parameters λ and K (see above) can be performed using a linear regression technique (54), although the technique is not robust against outliers, which can have a marked influence. Maximum likelihood estimation (55, 56) has been shown to be superior for EDV parameter fitting and, for example, is the method used to parameterize the gapped BLAST method (33). However, when low gap penalties are used to generate the alignments, the similarity scores can lose their local character and assume more global behavior, such that the EVD-based probability estimates are no longer valid (51).

2.4.3 Statistics of database searches

In order to be useful in sequence database searches, the above framework for comparing a pair of random sequences should be adapted to multiple pairwise comparisons. Here, it becomes important to establish the probability for a given query sequence to have a significant similarity with at least one of the database sequences. The P value is the probability of seeing at least one unrelated score S greater than or equal to a given score x in a database search over n sequences. This probability has been demonstrated to follow a Poisson distribution (53):

$$P(x, n) = 1 - e^{-n \cdot P(S \geq x)}$$

where n is the number of sequences in the database. In addition to the P value, some database search methods employ the E value of the Poisson distribution, which is defined as the expected number of nonhomologous sequences with a score greater than or equal to a score x in a database of n sequences:

$$E(x, n) = n \cdot P(S \geq x)$$

For example, if the E value of a matched database sequence segment is 0.01, then the expected number of random hits with score $S \geq x$ is 0.01, which means that this E value is expected by chance only once in 100 independent searches over the database. However, if the E value of a hit is 5, then five fortuitous hits with $S \geq x$ are expected within a single database search, which renders the hit not significant. Database searching is commonly performed using an E value of between 0.1 and 0.001. Low E values decrease the number of false positives in a database search, but increase the number of false negatives such that the sensitivity (see below) of the search is lowered.

In addition to P or E values, a number of sequence similarity searching routines provide an additional normalized alignment score based on the raw alignment score, S. This score, called the *bit score*, is defined as:

$$B = (\lambda S - \ln K)/\ln 2$$

where S is the raw alignment score and λ and K are the aforementioned statistical parameters of the scoring system (50). The bit score, B, is a linear transformation of the raw score and has a standard set of units – the higher the score, the more significant the alignment. As bit scores are normalized with respect to the scoring system, they can be used to compare alignment scores from different searches based on different scoring schemes, which is not warranted using raw alignment scores.

2.5 Fast Smith–Waterman local alignment searches

Collins and Coulson (57) devised a parallel computer protocol to perform database searches based on an implementation of the full Smith–Waterman (8) local alignment technique. They implemented their MPSRCH protocol (58) on massively parallel computers with single-instruction multiple data-type processors. Following Collins and Coulsen (57), a number of implementations that enable fast Smith–Waterman-based local searches have emerged. One of the central computer sites where such programs are running as web servers is the European Bioinformatics Institute outstation of the European Molecular Biology Laboratory. Available are MPSRCH (http://www.ebi.ac.uk/MPsrch/index.html[3,4]) and a fast implementation of the true Smith–Waterman algorithm (SCANPS) (59), both allowing users to perform database queries via the Internet. The output of a query is a list of top-scoring local alignments (one per protein) where statistical significance measures are also given based on the mean value and standard deviation of the distribution of scores over the entire database (57). The speed of the techniques allows several PAM exchange weight matrices (based on different evolutionary distances) to be used in searching the databanks with the same query sequence.

2.6 Profile searching

A natural extension of sequence database searching is provided by methods that use information over an entire sequence alignment of a certain protein family to find additional related family members. The earliest conceptually clear technique of this kind of sequence searching was called profile analysis (60), which combines a full representation of a sequence alignment with a sensitive searching algorithm. The procedure takes as input a multiple alignment of n sequences. First, a profile is constructed from the alignment, i.e. an alignment-specific scoring table, which comprises the likelihood of each residue type occurring in each position of the multiple alignment. A typical profile has $L \times (20 + 2)$ elements, where L is the total length of the alignment and 20 rows are reserved for the number of amino acid types, whilst the last two rows are often reserved for affine gap penalties (see above). Gribskov et al. (60) used a single extra column in the profile to describe the local weight for both the gap opening and the gap extension penalty. For gapless alignment positions, the weight is the maximal value, whereas for positions with insertions/deletions, the weighting factor is lowered according to the maximum length of the gap crossing a given alignment position. The advantage of positional gap weights is that multiple alignment regions with gaps (loop regions) will be assigned lowered gap penalties and hence will be more likely than core regions to attract gaps in a target sequence during profile searching, consistent with structural considerations. However, the implementation by Gribskov et al. does not take the frequency of gaps at each alignment position into account for the estimation of gap opening and/or extension penalties. Many alternative profile implementations therefore reserve the two last columns of the profile for positional gap opening (P_{open}) and gap extension (P_{extend}) penalties, which can be individually determined using protocols that take the above considerations into account.

In the approach of Gribskov et al., a profile calculated as described above is aligned with the databank sequence by means of the Smith–Waterman dynamic programming procedure (8). For each database sequence, the alignment score corresponding to the best local alignment quantifies the degree of similarity of this sequence with the probe profile. The scores are then corrected for sequence length, represented in the form of Z scores, and ranked to create the final list of databank search hits. Top-scoring sequences with scores above some threshold level are then likely to be related to the multiply aligned sequences used to build the profile. In addition to aligning a single sequence to a profile, it is also possible to align two profiles. In this case, two matched profile positions receive a score by summing the products of the corresponding propensities from the two profiles over the 20 residue types.

A number of improvements have been effected since the early Gribskov et al. (60) approach. A subclass of profiles for ungapped alignments is referred to as position-specific scoring matrices, a term developed by the BLAST team (33). An approach adopted in many profile-based methods is to implement a more probabilistic and informational scheme based on log likelihoods and normalization using expected residue compositions, which has been shown to lead to more sensitive comparisons than the classic Gribskov et al. approach. A common problem in this approach is the occurrence of zero values at some positions in the matrix. This not only leads to divide-by-zero problems in the analysis, but also fails to recognize the potential for diversity at some sites, which might be seen if a large enough set of sequences was available. A common way to deal with the under-representation of nucleotides or amino acids at alignment positions is the application of *pseudo-counts* (61–64). Pseudo-counts effectively extrapolate the number of amino acids at each alignment position by adding extra artificial residue counts to the profile, based for example on a known residue composition observed in the database. This generally enhances the predicted power of the profile.

In 1994, Baldi et al. (65) and Krogh et al. (66) pioneered the use of hidden Markov models (HMMs) to represent an aligned block of sequences. A distinct advantage of HMMs over traditional profiles is that an HMM incorporates a richer probabilistic description of insertions and deletions probabilities. Whereas in profiles there is just a single gap penalty for each position, such that the introduction of a gap in either the profile or the sequence aligned against the profile leads to the same penalty, in an HMM these two events can be modeled with different probabilities. An extensive library of HMMs for protein domains is deposited in the Pfam database (30). Profile searching using HMMs is currently one of the most sensitive search techniques.

Bucher et al. (67) unified the profile, motif, and HMM approaches through extension of the profile definition with regular expression-like patterns, weight matrices, and HMMs. They proved that their generalized profiles are equivalent to certain types of HMM. The generalized profiles have been used to extend the PROSITE protein motif database (67, 68), which in its basic form is a library of regular expressions. The profile syntax enables the emulation of most common motif search techniques, such as direct searching for PROSITE patterns, searching for patterns without gaps (69), searching using the profile definition of Gribskov et al. (60), flexible pattern searches (70), searching using HMMs (66), and domain and fragment searches using the HMMER method (71).

2.6.1 PSI-BLAST – iterative searching

The most widely used algorithm of the BLAST suite is position-specific iterated BLAST (PSI-BLAST) (33), which exploits increased sensitivity offered by multiple alignments and derived profiles in an iterative fashion.

The program initially operates on a single query sequence by performing a conventional gapped BLAST search (see above). Low-complexity regions of the query sequence (such as coiled-coil or transmembrane regions (71), which are widespread) are filtered and masked out, as they are not likely to convey a specific signal for recognizing a particular protein family. The PSI-BLAST program searches the database with the filtered query sequence, selects the significant hits, and then aligns these with each other.

From the aligned 'first-generation' hits, PSI-BLAST then creates a profile that reflects regions conserved across the aligned proteins. This profile is known as a position-specific scoring matrix (PSSM): in effect, it is a way of giving extra weight to those parts of the protein that are most characteristic of all members of the family. The PSSM is then used to rescan the database in a second round aimed at finding new sequences. These new sequences will be ones that had too little similarity to the original query sequence to be found in the first round, but that do have significant similarity when the PSSM is used to concentrate on the conserved motifs.

This process can be repeated and, at each iteration, the 'profile' (PSSM) of the group of proteins is refined, making the next round more sensitive to more distantly related family members. This continues until the user decides to stop or until the search has converged (i.e. further iterations produce no additional hits). The main web server for PSI-BLAST is located at http://www.ncbi.nlm.nih.gov/blast/[3.3], but a large number of mirror sites exist.

The web server enables the user to specify at each iteration round which sequences should be included in the profile, whilst by default all sequences are included that score beyond a user-set E value. However, the user needs to ask for each subsequent iteration. An alternative to the PSI-BLAST web server is a stand-alone version of the program, downloadable from the aforementioned URL, which allows the user to specify beforehand the desired number of iterations. An important limitation of the original PSI-BLAST program is that the statistics do not take compositional biases into account. This was addressed in an update of the PSI-BLAST technique (72). Biased amino acid compositions could confuse the original algorithm and lead to a build-up of errors in subsequent iterations. This is significant for cross-genome comparisons, as many large overall compositional differences exist among individual genomes. Another matter of debate is the way in which the PSSM is generated. It is clear that erroneous alignments are likely to drive the engine into inclusion of false positives.

In *Protocol 4*, we will use PSI-BLAST in an iterative search, again starting from the mouse cytochrome c oxidase (subunit IV).

Protocol 4

A search for divergent family members using PSI-BLAST

1. Open the BLAST home page at http://www.ncbi.nlm.nih.gov/blast/ [3.3].

2. Click on **protein blast** and then select **PSI-BLAST** (below under **Program selection**).

3. Paste the cytochrome c oxidase sequence from *M. musculus* in FASTA format (see *Protocol 1*) into the **Enter Query Sequence** box. Leave all other settings at their default values, click **BLAST**, and then retrieve your results as in *Protocol 3*.

4. The output of PSI-BLAST is quite similar to that of normal BLASTP except that the *E* value cut-off point is now set to 0.00. A dividing line in the hit list clearly shows which sequences fall below this new *E* value cut-off point. This cut-off value is chosen to ensure that only reasonably significant hits to the query protein are used in building the profile for the next iteration.

5. All of the hits above the cut-off line will be marked 'NEW', as they were found in this iteration of the search.

6. Now click on the button **Run PSI-Blast iteration 2** below the list of 'NEW' hits. The program will run again, now using a query profile (which holds information for all selected sequences above the cut-off value in the preceding iteration) instead of the original query sequence.

7. The result (see *Fig. 4*, also available in the color section) shows a number of newly identified putative homologs (indicated by 'NEW' to the left of the hit; those hits from the previous round are marked with a green dot).

8. Note that the *E* values for some of the earlier sequences (those found in the first round) are much lower (more significant) than before. This is because in the refined search these sequences were found to conform more closely to the 'profile' of the family (the PSSM), even if they had only modest similarity to the original query sequence. For instance, 'hypothetical protein C1G_08304' had an *E* value of 0.8 in the original search (and was below the cut-off line), but now has an *E* value of 6e-9 in this second iteration: it matches the family profile well, even though it matched the original query protein only loosely.

9. In some cases, proteins found in the first search will be rejected (falling below the cut-off line) in later iterations. This normally indicates that the protein matched the query protein overall to a reasonable degree, but did not conform well to the specific profile of the family of related proteins.

10. You can continue this process, running further iterations. The advantage is that more distant homologs can be found, but there is also a higher chance that more unrelated sequences will end up above the *E* cut-off value. Typically, PSI-BLAST runs are iterated three to ten times. In this example (and searching the database as at the time of writing), no new sequences were found after the seventh iteration.

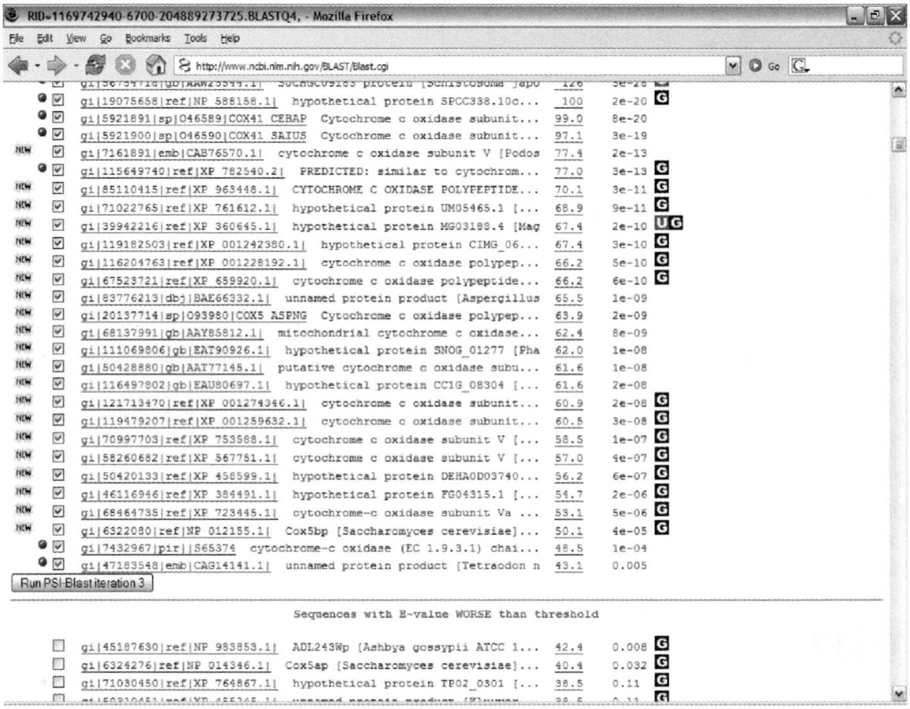

Figure 4. PSI-BLAST output (see page xix for color version).
Part of the results after the second PSI-BLAST iteration are shown. The output format is essentially the same as for BLASTP (see *Fig. 2*), but new family members found in this iteration are indicated by 'NEW'. Family members found in the previous round are indicated by green dots. Sequences shown below the horizontal dividing line have too high an *E* value; only those above the line will be used in compiling the PSSM (or 'profile') of the protein family for the next iteration of the search.

2.6.2 Profile–profile alignment

The previous sections described how a query multiple sequence alignment, based on a given query sequence and a number of putative homologs, can confer an enhanced signal for recognizing distantly evolved members of a given family. Over the last few years, further improvements to the alignment of distant sequences have been achieved using several approaches. As a first improvement, the evolutionary model describing the relationship of a set of sequences can be readjusted to fit the sequence set rather than using a pre-set generic model incorporated in a single-residue exchange matrix such as the PAM or BLOSUM series. Recently, Yu *et al.* (16) showed that the use of organism-specific or alignment-set-specific background frequencies for contextual readjustment of the standard amino acid exchange weights provides a more sensitive and biologically accurate way of aligning sequences. Alternatively, structural or homologous sequence information can be incorporated into the alignment process to help identify the distant relationships between sequences. The benefits of using related sequence information has been shown in numerous profile–profile alignment methods that apply different profile-scoring schemes (73–89).

Many of these scoring schemes have been assessed in recent comparison studies and have shown little significant difference in their respective performances (90, 91). However, most of the profile–profile alignment approaches to date have been used mainly for sequence database searching (local pairwise alignment), where a popular application has been to use profile–profile comparisons for aligning a profile derived from a query multiple alignment with a number of profiles describing a collection of different protein families.

A direct application of the profile–profile alignment technique is implemented in PRALINE-PSI (92), a multiple alignment technique that relies on constructing a profile for each of the query sequences using the PSI-BLAST method. Pre-alignment profiles (pre-profiles) are generated using each sequence in a set as a PSI-BLAST (33, 46) query. The resulting PSI-BLAST local alignments are filtered for redundancy and converted to PRALINE pre-profiles, which replace the single sequence input that would otherwise be used for the alignment (see 93–95 for further details). The increased sensitivity of the PRALINE-PSI method in detecting similarities becomes most evident in aligning distant sequence pairs (or sequence–profile and profile–profile pairs in multiple sequence alignment).

3. TROUBLESHOOTING

3.1 Iterative homology searching problems

Iterative sequence search methods such as PSI-BLAST can be a powerful way of finding distant homologies, but often fail when querying a multi-domain protein or a protein with regions of compositional bias. For example, common conserved protein domains such as the tyrosine kinase domain can obscure weak but relevant matches to other domain types (96), whilst sequences containing low-complexity regions, such as coiled coils and transmembrane regions, can cause an explosion of the search rather than convergence due to the absence of any strong sequence signals. Conversely, some searches may lead to premature convergence; this occurs when the PSSM is too strict, only allowing matches to very similar proteins, i.e. sequences with the same domain organization as the query are detected but no homologs with different domain combinations.

An additional problem with iterative searches is 'matrix migration' (also referred to as 'profile wander'), which occurs when the search strategy is too permissive so that information from false-positive sequences is included in the profile, resulting in the possible loss of truly homologous sequences found in earlier rounds. A further loss of information can be incurred with PSI-BLAST, as PSI-BLAST PSSMs are trimmed to use only the highest-scoring region in a search, ignoring less-conserved regions.

The alternative database search method QUEST (97) alleviates these problems by using an independent multiple-alignment program to generate a true multiple sequence alignment between iterations, and not a 'master–slave' alignment, thereby improving the quality of the PSSM. The QUEST method also removes any sequences that are deemed to be too divergent as a reliable family member in order not to 'pollute' the PSSM, which leads to increased search capabilities.

3.2 Post-processing of homology searches

A few methods exist to predict domain boundaries through post-processing BLAST searches. The BALLAST method can be used to visualize conservation profiles for a query sequence based on sequence searching (98, 99), although the method does not delineate domain boundaries. Another technique is the PASS (prediction of autonomous folding units based on sequence similarities) method, which uses a simple and noniterative method of domain delineation based on the stacking of sequences from a gapped BLAST search onto the query sequence (100). Regions along a query sequence often have a varying number of matching sequences from the BLAST data leading to abrupt increases and decreases in sequence numbers along the query. The PASS method is based on a single BLAST run and does not use iteration to include information from distant homologs. Furthermore, the current release of the PRODOM domain database (29) is created using the MKDOM2 method (101), which performs PSI-BLAST searches starting with the smallest sequence in the database as a query, which is supposed to represent a single domain. All domain sequences identified are removed from the database, after which the process is iterated with the remaining subsequences and terminated when the database becomes empty. The MKDOM2 method is an iterative protocol but does not address the aforementioned problems connected to PSSM-based iterative searches.

The DOMAINATION method (102) assigns domain boundaries by applying PSI-BLAST in a repetitive fashion. The distribution of the aligned positions of N and C termini from PSI-BLAST local sequence alignments is used to identify potential domain boundaries. DOMAINATION incorporates an iterative strategy for chopping and joining domains and domain segments based on the loss and gain of domains. This allows the recognition of both continuous and discontinuous domains. For each domain inferred from the corresponding PSI-BLAST local alignments, profiles are created by filtering redundant sequences and subsequent multiple sequence alignments. Each profile filtered in this way is then used in further iterative database searches using PSI-BLAST. All profiles are required to contain the original query sequence at each iteration of PSI-BLAST to avoid profile wander, but parameters are set to ensure that the profiles are divergent enough to capture distant sequence fragments. The whole process of iterative PSI-BLAST searches is repeated until domain assignment ends and no new homologs are found anymore. DOMAINATION can successfully assign domain boundaries within a given query sequence, whilst the added information gleaned from the putative domains delineated during the parallel and iterated searches leads to a search performance enhanced by 15 percentage points over a wide range of E values compared with stand-alone PSI-BLAST searches.

3.3 Evaluating sequence database searches

A few useful measures are commonly used to gauge the accuracy of sequence database search methods over an annotated nonredundant database. The *sensitivity* of a search is defined as TP/(TP + FN), where TP is the number of true positives and FN the number of false negatives, which reflects the fraction of true hits found relative to the total number of sequences in the database that are homologous to the query. The sensitivity reflects the extent to which the method is able to identify distantly related sequences. In many studies, this measure is also referred to as *coverage*. The *specificity* (or *selectivity*) is defined as TN/(FP + TN), where FP is the number of false positives, which denotes the fraction of entries correctly excluded as hits and hence measures the avoidance of unrelated hits. Yet another widely used measure is the *positive predictive value*, defined as TP/(TP + FP), which measures the proportion of true homologs within all sequences designated by the search tool as related. In practical database searches, there is a trade-off between sensitivity and specificity: the more the *P* or *E* values are relaxed to allow more distantly related sequences to be found, the more likely it becomes that chance hits infiltrate the search. Moreover, even if a statistically highly significant similarity is encountered, problems remain. For example, if high similarity is found over only a portion of the sequences, the sequences may each contain multiple domains and share a single homologous domain only (see above), so that only an aspect of the overall function might be inferred. In iterative homology searches, protein sequences containing more than one structural domain can be problematic in that they cause the search to terminate prematurely or lead to an 'explosion' of common domains (102). For example, the occurrence in the query sequence of a common and conserved protein domain such as the tyrosine kinase domain, which is then hit many times in the database, can obscure weaker but also relevant matches to other domain types (102), particularly when the *E* value is set to include only strong hits. Conversely, when multi-domain sequences with the same sequential order of domains as in the query sequence are found initially during an iterative search, homologs with different domain combinations might well be missed due to early convergence of the search. To reduce the chance of including spurious hits, some database search engines, such as PSI-BLAST (33), scan query sequences for the presence of so-called low-complexity regions. These are then excluded from the alignment to limit the inclusion of false-positive hits due to database sequence matches with these regions. However, the occurrence of database sequences with low-complexity regions can still cause an explosion of false positives in iterative homology searches (102). Despite recent improvements in search techniques, complications such as the above illustrate that automatic biological evaluation of homology searches in genomic pipelines remains elusive for biologically intricate relationships.

4. REFERENCES

1. **Teichmann SA, Rison SC, Thornton JM, Riley M, Gough J & Chothia C** (2001) *J. Mol. Biol.* **311**, 693–708.
2. **May AC** (2001) *Nature*, **413**, 453.
3. **Doolittle RF** (1981) *Science*, **214**, 149–159.
4. **Pascarella S & Argos P** (1992) *Protein Eng.* **5**, 121–137.
★ 5. **Abagyan RA & Batalov S** (1997) *J. Mol. Biol.* **273**, 355–368. – *An important and comprehensive study of the relationship between sequence similarity and homology.*

Various sequence similarity measures are evaluated and a protocol for calculating EVD is presented.

6. **Simossis VA, Kleinjung J & Heringa J** (2003) In *Current Protocols in Bioinformatics*, pp. 3.7.1–3.7.25. Edited by AD Baxevanis. John Wiley, New York.

7. **Needleman SB & Wunsch CD** (1970) *J. Mol. Biol.* **48**, 443–453.

8. **Smith TF & Waterman MS** (1981) *J. Mol. Biol.* **147**, 195–197.

9. **Feng DF & Doolittle RF** (1987) *J. Mol. Evol.* **25**, 351–360.

★ 10. **HenikoffS & Henikoff JG** (1992) *Proc. Natl. Acad. Sci. U.S.A.* **89**, 10915–10919. *– In this paper, the widely used BLOSUM series of amino acid exchange matrices is introduced.*

11. **Barker WC, Ketcham LK & Dayhoff MO** (1978) *J. Mol. Evol.* **10**, 265–281.

12. **Jones DT, Taylor WR & Thornton JM** (1992) *Comput. Appl. Biosci.* **8**, 275–282.

13. **Gonnet GH, Cohen MA & Benner SA** (1992) *Science*, **256**, 1443–1445.

14. **Muller T & Vingron M** (2000) *J. Comput. Biol.* **7**, 761–776.

15. **Muller T, Spang R & Vingron M** (2002) *Mol. Biol. Evol.* **19**, 8–13.

16. **Yu YK, Wootton JC & Altschul SF** (2003) *Proc. Natl. Acad. Sci. U.S.A.* **100**, 15688–15693.

17. **Rost B** (1999) *Protein Eng.* **12**, 85–94.

18. **Waterman MS & Eggert M** (1987) *J. Mol. Biol.* **197**, 723–728.

19. **Sander C & Schneider R** (1991) *Proteins*, **9**, 56–68.

20. **Rost B** (2002) *J. Mol. Biol.* **318**, 595–608.

21. **Bork P** (1991) *FEBS Lett.* **286**, 47–54.

22. **Doolittle RF** (1995) *Annu. Rev. Biochem.* **64**, 287–314.

23. **Heringa J & Taylor WR** (1997) *Curr. Opin. Struct. Biol.* **7**, 416–421.

24. **Marcotte EM, Pellegrini M, Thompson MJ, Yeates TO & Eisenberg D** (1999) *Nature*, **402**, 83–86.

25. **Wheelan SJ, Marchler–Bauer A & Bryant SH** (2000) *Bioinformatics*, **16**, 613–618.

26. **George RA & Heringa J** (2002) *J. Mol. Biol.* **316**, 839–851.

27. **Gracy J & Argos P** (1998) *Bioinformatics*, **14**, 164–173.

28. **Gracy J & Argos P** (1998) *Trends Biochem. Sci.* **23**, 495–497.

29. **Corpet F, Servant F, Gouzy J & Kahn D** (2000) *Nucleic Acids Res.* **28**, 267–269.

30. **Bateman A, Coin L, Durbin R, et al.** (2004) *Nucleic Acids Res.* **32**, D138–D141.

31. **Ponting CP, Schultz J, Milpetz F & Bork P** (1999) *Nucleic Acids Res.* **27**, 229–232.

32. **Gracy J & Argos P** (1998) *Bioinformatics*, **14**, 174–187.

★ 33. **AltschulSF, Madden TL, Schaffer AA, et al.** (1997) *Nucleic Acids Res.* **25**, 3389–3402. *– This paper introduces the homology search tool PSI-BLAST.*

34. **Altschul SF, Gish W, Miller W, Myers EW & Lipman DJ** (1990) *J. Mol. Biol.* **215**, 403–410.

35. **Pearson WR & Lipman DJ** (1988) *Proc. Natl. Acad. Sci. U.S.A.* **85**, 2444–2448.

36. **Karplus K, Barrett C & Hughey R** (1998) *Bioinformatics*, **14**, 846–856.

37. **Karplus K, Karchin R, Barrett C, et al.** (2001) *Proteins* (Suppl. 5), 86–91.

38. **Eddy SR** (1998) *Bioinformatics*, **14**, 755–763.

39. **Pearson WR** (1996) *Methods Enzymol.* **266**, 227–258.

40. **Wilbur WJ & Lipman DJ** (1983) *Proc. Natl. Acad. Sci. U.S.A.* **80**, 726–730.

41. **Wilbur WJ & Lipman DJ** (1984) *SIAM J. Appl. Math.* **44**, 557–567.

42. **Dayhoff MO, Barker WC & Hunt LT** (1983) *Methods Enzymol.* **91**, 524–545.

43. **Chao KM, Pearson WR & Miller W** (1992) *Comp. Appl. Biosci.* **8**, 481–487.

44. **Kent WJ** (2002) *Genome Res*, **12**, 656–664.

45. **Karlin S & Altschul SF** (1990) *Proc. Natl. Acad. Sci. U.S.A.* **87**, 2264–2268.

46. **Altschul SF & Koonin EV** (1998) *Trends Biochem. Sci.* **23**, 444–447.

47. **Dembo A & Karlin S** (1991) *Ann. Prob.* 1737.

48. **Dembo A, Karlin S & Zeitouni O** (1994) *Ann. Prob.* **22**, 2022.

49. **Gumbel EJ** (1958) *Statistics of Extremes*. Columbia University Press, New York.

★ 50. **AltschulSF & Gish W** (1996) *Methods Enzymol.* **266**, 460–480. *– A seminal overview of the statistical framework incorporated in the BLAST suite of database searching algorithms. It also contains tables of useful parameter settings linking various alignment scoring systems and E value calculations based on EVD.*

51. **Arratia R & Waterman MS** (1994) **4**, 200–225.

52. **Smith TF, Waterman MS & Burks C** (1985) *Nucleic Acids Res.* **13**, 645–656.
53. **Waterman MS & Vingron M** (1994) *Proc. Natl. Acad. Sci. U.S.A.* **91**, 4625–4628.
54. **Pearson WR** (1998) *J. Mol. Biol.* **276**, 71–84.
55. **Lawless JF** (1982) In *Statistical Models and Methods for Lifetime Data*, pp. 141–202. John Wiley & Sons, New York.
56. **Mott R** (1992) *Bull. Math. Biol.* **54**, 59–75.
57. **Collins JF & Coulson AF** (1990) *Methods Enzymol.* **183**, 474–487.
58. **Sturrock SS & Collins JF** (1993) *MP_SRCH Version 1.3*. Biocomputing Research Unit, University of Edinburgh, UK.
59. **Barton GJ** (2002) *SCANPS Version 2.3.9 User Guide*. University of Dundee, UK.
60. **Gribskov M, McLachlan AD & Eisenberg D** (1987) *Proc. Natl. Acad. Sci. U.S.A.* **84**, 4355–4358.
61. **Sjolander K, Karplus K, Brown M, et al.** (1996) *Comput. Appl. Biosci.* **12**, 327–345.
62. **Bruno WJ** (1996) *Mol. Biol. Evol.* **13**, 1368–1374.
63. **Henikoff S, Endow SA & Greene EA** (1996) *Trends Biochem. Sci.* **21**, 444–445.
64. **Tatusov RL, Altschul SF & Koonin EV** (1994) *Proc. Natl. Acad. Sci. U.S.A.* **91**, 12091–12095.
65. **Baldi P, Chauvin Y, Hunkapiller T & McClure MA** (1994) *Proc. Natl. Acad. Sci. U.S.A.* **91**, 1059–1063.
66. **Krogh A, Brown M, Mian IS, Sjolander K & Haussler D** (1994) *J. Mol. Biol.* **235**, 1501–1531.
67. **Bucher P, Karplus K, Moeri N & Hofmann K** (1996) *Comput. Chem.* **20**, 3–23.
68. **Hulo N, Bairoch A, Bulliard V, et al.** (2006) *Nucleic Acids Res.* **34**, D227–D230.
69. **Staden R** (1990) *Methods Enzymol.* **183**, 193–211.
70. **Barton GJ & Sternberg MJ** (1990) *J. Mol. Biol.* **212**, 389–402.
71. **Eddy SR** (1996) *Curr. Opin. Struct. Biol.* **6**, 361–365.
★ 72. **Schaffer AA, Aravind L, Madden TL, et al.** (2001) *Nucleic Acids Res.* **29**, 2994–3005. – *In this paper, some important updates of the popular PSI-BLAST database engine are presented.*
73. **Edgar RC** (2004) *Nucleic Acids Res.*, **32**, 1792–1797.
74. **Tomii K & Akiyama Y** (2004) *Bioinformatics*, **20**, 594–595.
75. **Wang G & Dunbrack RL Jr** (2004) *Protein Sci.* **13**, 1612–1626.
76. **von Ohsen N, Sommer I, Zimmer R & Lengauer T** (2004) *Bioinformatics*, **20**, 2228–2235.
77. **Ginalski K, von Grotthuss M, Grishin NV & Rychlewski L** (2004) *Nucleic Acids Res.* **32**, W576–W581.
78. **Edgar RC & Sjolander K** (2004) *Bioinformatics*, **20**, 1309–1318.
79. **Capriotti E, Fariselli P, Rossi I & Casadio R** (2004) *Proteins*, **54**, 351–360.
80. **von Ohsen N, Sommer I & Zimmer R** (2003) *Pac. Symp. Biocomput.* 252–263.
81. **Sadreyev R & Grishin N** (2003) *J. Mol. Biol.* **326**, 317–336.
82. **Mittelman D, Sadreyev R & Grishin N** (2003) *Bioinformatics*, **19**, 1531–1539.
83. **Ginalski K, Pas J, Wyrwicz LS, von Grotthuss M, Bujnicki JM & Rychlewski L** (2003) *Nucleic Acids Res.* **31**, 3804–3807.
84. **Jaroszewski L, Rychlewski L & Godzik A** (2000) *Protein Sci.* **9**, 1487–1496.
85. **Rychlewski L, Jaroszewski L, Li W & Godzik A** (2000) *Protein Sci.* **9**, 232–241.
86. **Soding J** (2004) *Bioinformatics*, **21**, 951–960.
87. **Sadreyev RI, Baker D & Grishin NV** (2003) *Protein Sci.* **12**, 2262–2272.
88. **Chung R & Yona G** (2004) *BMC Bioinformatics*, **5**, 183.
89. **Yona G & Levitt M** (2002) *J. Mol. Biol.* **315**, 1257–1275.
90. **Ohlson T, Wallner B & Elofsson A** (2004) *Proteins*, **57**, 188–197.
91. **Edgar RC & Sjolander K** (2004) *Bioinformatics*, **20**, 1301–1308.
★ 92. **Simossis VA, Kleinjung J & Heringa J** (2005) *Nucleic Acids Res.* **33**, 816–824. – *This is the first method for multiple profile alignment, where each profile represents a given query sequence with putative homologous sequences gathered by the method PSI-BLAST.*
93. **Heringa J** (2000) *Curr. Protein Pept. Sci.* **1**, 273–301.
94. **Heringa J** (1999) *Comput. Chem.*, **23**, 341–364.
95. **Simossis VA & Heringa J** (2003) *Comput. Biol. Chem.*, **27**, 511–519.

96. **Sonnhammer EL & Durbin R** (1994) *Comput. Appl. Biosci.* **10**, 301–307.
97. **Taylor WR** (1998) *J. Mol. Biol.* **280**, 375–406.
98. **Plewniak F, Thompson JD & Poch O** (2000) *Bioinformatics*, **16**, 750–759.
99. **Thompson JD, Plewniak F, Thierry J & Poch O** (2000) *Nucleic Acids Res.* **28**, 2919–2926.
100. **Kuroda Y, Tani K, Matsuo Y & Yokoyama S** (2000) *Protein Sci.* **9**, 2313–2321.
101. **Gouzy J, Corpet F & Kahn D** (1999) *Comput. Chem.* **23**, 333–340.
102. **George RA & Heringa J** (2002) *Proteins*, **48**, 672–681.

CHAPTER 4

Gene prediction

Marie-Adele Rajandream

1. INTRODUCTION

This chapter provides a guide to the identification of protein-coding genes and their structures with examples using tools available over the Internet. Two separate strategies are discussed here for eukaryotic and prokaryotic sequences, respectively. Ways of identifying noncoding RNAs are not included (see Chapter 5 for these), although some sites relevant to this have been listed in section 4.

Prokaryotic and eukaryotic genomes are generally very dissimilar in terms of size and organization. Compared with eukaryotic genomes, prokaryotic genomes tend to be small and compact with a high density of genes. Splicing is extremely rare in prokaryotes, whereas most eukaryotes have some proportion of spliced genes.

As a consequence, there are important differences in the approaches used for identifying genes in prokaryotes and eukaryotes. In both cases, coding bias and sequence motifs are central to the prediction process. The motifs, or signals, can include promoters, stop codons, and start codons. Eukaryote-specific signals include splice donors and acceptors and polyadenylation signals, whilst prokaryote-specific signals include ribosome-binding sites.

Recognition of prokaryotic genes is a less complex problem, as all genes must be encoded within a single open reading frame (ORF) – a tract of sequence in which at least one of the reading frames is free of stop codons. However, not all ORFs are coding, and the number of spurious (noncoding) ORFs is determined by the G+C content, which affects the frequency of stop codons (which are A+T-rich) in noncoding sequence. Hence, gene prediction involves identifying which of the ORFs are coding sequences (CDSs) for proteins, and where the CDS starts within the ORF (it will not always be the first apparent start codon). This can be done by using other signals such as ribosome-binding sites, coding, and nucleotide biases. With eukaryotes, the association between ORFs and potential protein-coding regions is less strong due to the presence of introns, which will disrupt the reading frame. The main challenges here are to locate the gene and then to determine its precise structure (I. M. Meyer, personal communication).

Bioinformatics: *Methods Express* (Paul H. Dear, ed.)
© Scion Publishing Limited, 2007

It cannot be emphasized strongly enough that gene predictions should always be treated as an approximation and should be corroborated against evidence from other types of analysis. Using several gene prediction tools, ideally based on different methods, is also advisable, as these often complement each other. Most programs attach probability scores to their results, and these can be useful indicators of the quality of the predictions. Even where gene predictions have already been made for the organism or region you are interested in, carrying out your own gene predictions can still be useful for evaluating the interpretation supplied (1).

Gene prediction is an evolving discipline, so it is important to follow the literature to keep up to date with the latest developments and tools.

1.1 *Ab initio* methods

For a region to qualify as a candidate gene, the organization of the signals associated with it (including the boundaries of protein-coding regions and any intron–exon structure) has to correspond to that of a valid gene structure. Prediction tools work on this principle, but, in addition, more-specific criteria are needed for accurate detection of genes. Coding bias, splicing motifs, and the sizes of introns and exons can vary considerably among organisms, so apart from conforming to the basic rules defining an apparent stereotypical gene structure, a region has to display properties shared by known genes for that particular organism. This is checked by comparing the region's properties against a set of sequences known to be genes typical for the organism. This set of sequences is known as the 'training set', and the process of creating a statistical model against which to evaluate possible new genes is known as 'training'.

Gene prediction tools trained for a one species can be used on another, but they will always do better when trained for the relevant organism. The training set needs to be sufficiently large and representative, and to consist as far as possible of genes that have been well characterized and carefully reviewed – otherwise there is the risk of errors in the training set being propagated and compounded. In practice, of course, it is not always possible to have such a set available. Some prokaryotic prediction tools are designed to extract and use their own training sets.

As predictions are made by testing the sequence against statistical evidence and a number of rules for a valid structure with no use of independent prior knowledge about the location of the gene, this approach falls in the category known as '*ab initio*'. Different methods exist for integrating gene structure information and compositional features in order to train an algorithm and make it generate predictions. Among them are hidden Markov models (HMMs), generalized HMMs, and neural networks.

Using the approach discussed above will exclude regions from further consideration that do not fit the requirements for a gene sequence in terms of structure, significant signals, and coding bias. This often still leaves numerous possible gene structures for a given region of sequence. The algorithm will try to choose the structure that best reflects the statistical model derived from the common features of the genes in the training set. This inevitably means that alternatives with characteristics under-represented in the training set are less likely to be chosen, although they may represent *bona fide* genes.

1.2 Comparative methods

Precision can be increased by making use of other evidence, especially information on the presence of conserved regions among species. The underlying assumption is that conservation in protein-coding regions is on average higher than in non-protein-coding regions. Rating alternative boundaries for a valid gene in a region according to conservation will reduce the margin of error in determining the exact location of the gene. Comparative gene prediction methods can be further subdivided into three categories: homology-assisted methods, homology-based methods, and the comparative *ab initio* methods (I. M. Meyer, personal communication).

Homology-assisted methods start with the *ab initio* approach, but then also attempt to align a query sequence to a sequence from a related organism known to contain the same gene. Such sequences are often referred to as collinear. Initially, a rapid alignment is performed with programs such as TBLASTX (a program from the BLAST suite – see Chapter 3), which compares the six-frame translation of a nucleotide query sequence against the six-frame translation of a nucleotide sequence database (2). The results are then used to generate a mathematically optimal alignment between the query and reference sequence, and this is taken into account when predictions are ranked to identify the best one.

In homology-based methods, the criteria for what typifies a likely gene have been extended beyond coding bias and signal strength to include the expected pattern of conservation between functional elements in two related genes. Therefore, training includes the extra step of giving the algorithm examples of significant alignments representing corresponding genes both in an annotated reference organism and in the organism on which it needs to predict. Homology-based methods will attempt to create a gene structure as they align the unannotated sequence to a collinear annotated reference gene sequence (amino acid or DNA). This means that the alignments from homology-based methods tend to be better than those from methods that cannot factor in the possible presence of splice sites when making use of homology. Equally, the process of recognizing gene structures is more powerful as it comprises alignment.

Comparative *ab initio* methods again make use of sequence similarity to refine *ab initio* predictions, but they are not guided by any annotation when inferring the nature of functional elements and the distance between signals. This puts them at a slight disadvantage compared with homology-based approaches, although they generally do better than noncomparative *ab initio* methods (I. M. Meyer, personal communication).

A weakness of all comparative algorithms is that they can only find genes in the query sequence for which a counterpart exists in the related organism. Also, many of them require a set of collinear sequences in their training set. A bonus is that they are more accurate and that they can be trained with sets smaller than those required for training noncomparative algorithms. It should also be explained that only a minority of comparative prediction tools for genomes in which splicing occurs can handle cases where the number of exons in two corresponding regions is different, i.e. where exons in one genome appear to have fused or split with respect to the other.

There are comparative methods in existence and under development that approach the problem using a wider context than those described above. The property they share is that they take more than one related sequence into account when constructing a prediction. They rely on different techniques (e.g. the integration of phylogenetic information), although some of their mechanisms are similar to those used by the methods previously outlined. It should be emphasized that nearly all prediction tools (comparative and noncomparative) focus on identifying the protein-coding parts of genes.

2. METHODS AND APPROACHES

In commercial packages such as LASERGENE (3), VECTOR NTI (4), and MACVECTOR (5), gene prediction programs have been incorporated into an environment that includes other analysis tools. There are noncommercial packages suited to a desktop environment to customize and carry out sequence analysis (including gene prediction) such as BASys (6) (for bacterial genomes) and Pegasys (7) and BioMoby (8). At present, installation of the latter two would still require help from a programmer, but BASys is a web server that does not need extension or special configuration of the local software to use it. Some operating systems such Mac OS X will allow you to set up some of the standard analysis tools described here on your local machine.

The examples below were carried out using freely accessible web tools. The drawback with this is that file formats sometimes need to be converted to standard ones, although it does give more flexibility in choice of method. As shown in the examples, servers can be used for converting formats, but often results returned by gene identification software prove to be in nonstandard formats. Depending on the volume of data, programmatic conversion can be preferable in such situations. There are libraries of code that can be used for this (e.g. BIOPERL, BIOJAVA), but it will involve having to write scripts. The on-line material for *Protocol 16* includes a PERL script using BIOPERL. A number of packages listed in *Tables 1–3* can only be run from a local machine, but for some of these, the installation procedure is relatively simple.

Two gene prediction tools feature in the examples that follow: SGP2 (10), a homology-assisted method, and GENEMARK (12), a noncomparative HMM-based method. *Tables 1–3* list some other packages available. Note that many genomes contain repetitive elements. Some gene prediction methods screen these out automatically, whilst others require low-complexity sequence to be masked prior to launching the prediction software. Documentation should always be checked

Table 1. Eukaryotic noncomparative prediction tools

Tool	Site	For organism type:	Server	Software available*	Trainable by user
HMMGENE (9)	http://www.cbs.dtu.dk/services/HMMgene/ [4.63]	Mammalian and *Caenorhabditis elegans*	Yes	Yes (L)	No
GENEID (9, 10)	http://www1.imim.es/geneid.html [4.64]	Various eukaryotes	Yes	Yes	No
NETGENE2 (11)	http://www.cbs.dtu.dk/services/NetGene2/ [4.65]	Various eukaryotes	Yes	Yes (L)	No
GENEMARK.HMM (12)	http://opal.biology.gatech.edu/GeneMark/ [4.66]	Various eukaryotes	Yes	Yes (L)	No
GENSCAN (11)	http://genes.mit.edu/GENSCAN.html [4.67]	Various eukaryotes	Yes	Yes	No
GENLANG (13)	http://arete.ibb.waw.pl/PL/html/gene_lang.html [4.68]	Various eukaryotes	Yes	No	No
SNAP (14)	http://homepage.mac.com/iankorf/ [4.69]	Various eukaryotes	No	Yes	Yes
GLIMMERHMM (15)	http://cbcb.umd.edu/software/glimmerhmm/man.shtml [4.70]	Various eukaryotes	No	Yes	Yes
FGENESH+ (16)	http://sun1.softberry.com/berry.phtml [4.71]	Various eukaryotes	Yes	Yes (C)	No
AUGUSTUS	http://augustus.gobics.de [4.72]	Various eukaryotes	Yes	Yes	Yes

*L, license agreement; C, commercial package.

for how the gene prediction process could be influenced by the presence of repeats and whether low-complexity filtering is built in or not.

It is strongly recommended that the results of the gene prediction methods discussed below are viewed using a suitable sequence viewer and annotation tool with a graphical interface. ARTEMIS (35) is ideal for this purpose and is available for a wide variety of platforms. Chapter 2 discusses the installation and use of ARTEMIS, and further information is available on the ARTEMIS web page (http://www.sanger.ac.uk/Software/Artemis/ [4.1]). The examples that follow were all carried out using a PC; there may be some very minor differences when using other platforms (such as Mac OS X).

2.1 Predicting eukaryotic genes

The protocols that follow are connected and should be worked through in the order given. We will analyze a sequence from *Mus musculus*; where comparative data is needed, we will use a collinear *Homo sapiens* sequence as the informant. Each of *Protocols 1–11* will provide a piece of evidence regarding which part of the sequence is protein coding. Using all results in conjunction should allow you to come up with the best possible approximation of the gene structure(s) in the query sequence.

We begin by downloading both sequence files and viewing the mouse sequence in ARTEMIS.

Table 2. Eukaryotic comparative prediction tools

Tool	Method*	Site	For organism type:	Server	Software available†	Trainable by user
AUGUSTUS (17)	HA	http://augustus.gobics. de/[4.72]	Various eukaryotes	No	Yes	Yes
GRAILEXP (18)	HA	http://compbio.ornl. gov/Grail-1.3/[4.73]	Various eukaryotes	Yes	Yes	No
FGENESH_C (16)	HB	http://sun1.softberry. com/berry.phtml[4.71]	Various eukaryotes	Yes	Yes (C)	No
FGENESH-2 (16)	AI	http://sun1.softberry. com/berry.phtml[4.71]	Various eukaryotes	Yes	Yes (C)	No
SGP2 (19)	HA	http://genome.imim. es/software/sgp2/index. html[4.74]	Various eukaryotes	Yes	Yes	No
WEBGENE (20)	HA	http://www.itb.cnr.it/ sun/webgene/[4.75]	Various eukaryotes	Yes	No	No
TWINSCAN (21)	AI	http://genes.cs.wustl. edu/[4.76]	Various eukaryotes	Yes	Yes	No
TWAIN (22)	HB	http://www.tigr.org/ software/pirate/twain/ twain.html[4.77]	Any eukaryote	No	Yes	Yes
PROJECTOR (23)	HB	http://www.sanger. ac.uk/Software/ analysis/projector/[4.78]	Various eukaryotes	Yes	Yes	No
DOUBLESCAN (24)	AI	http://www.sanger. ac.uk/Software/ analysis/doublescan/[4.79]	Various eukaryotes	Yes	Yes	No
AAT (25)	HA	http://genome.cs.mtu. edu/aat/aat.html[4.80]	Any eukaryote	Yes	No	No
GENEWISE2 (26)	HB	http://www.ebi.ac.uk/ Wise2/[4.81]	Human and *Caenorhabditis elegans*	Yes	Yes	No
GENOMESCAN (27)	HA	http://genes.mit.edu/ genomescan/[4.82]	Various eukaryotes	Yes	No	No
EVOGENE (28)	AI-M	http://www.birc.dk/ Software/evogene/[4.83]	Any eukaryote	No	Yes	No
SLAM (29)	AI	http://baboon.math. berkeley.edu/~syntenic/ slam.html[4.84]	Mammalian	Yes	Yes	No

*AI, *ab initio*; HA, homology-assisted; HB, homology-based; M, can use more than one reference organism for a single prediction.
†C, commercial package.

Table 3. Prokaryotic prediction tools

Tool	Method*	Site	For organism type:	Server	Software available†
ORPHEUS (30)	NC-AI-ST	http://www.cbs.dtu.dk/services/ HMMgene/[4.85]	Bacterial	N	Y (L)
GLIMMER (31)	NC-AI-ST	http://www.cbcb.umd.edu/software/ glimmer/[4.86]	Microbial and some eukaryotes	N	Y
EASYGENE (32)	NC-AI-ST	http://www.cbs.dtu.dk/services/ EasyGene/[4.87]	Prokaryotes	Y	Y (L)
GENEMARK.HMM (12)	NC-AI	http://opal.biology.gatech.edu/ GeneMark/[4.66]	Various prokaryotes	Y	Y (L)
GENEMARKS (33)	NC-AI-ST	http://opal.biology.gatech.edu/ GeneMark/[4.66]	Various prokaryotes and also viruses	Y	Y (L)
FGENESB (34)	HA-ST	http://sun1.softberry.com/berry. phtml[4.71]	Various prokaryotes	Y	Y (C)

*NC, noncomparative; AI, *ab initio*; ST, self-training; HA, homology-assisted.
†L, license agreement; C, commercial package.

Protocol 1

Downloading and viewing eukaryotic sequence files

1. Visit the web site for this book, navigate to the files for this chapter, and, from the folder Protocol_1, download the files human.fasta [4.2] and mouse.fasta [4.3] to your own computer.

2. Use the **File** menu in the ARTEMIS release window to open the mouse.fasta [4.3] file[a]. If you cannot see your file listed (or if it is listed but cannot be selected), change the filter from **Artemis files** to **All Files**. A new window will open, divided into three sections. The top section shows the stop codons (as short vertical lines) of all six reading frames for the complete sequence (5874 bp). The middle section shows a zoomed-in view of the start of the sequence in both orientations, together with the six-frame translation; the stop codons TAA, TGA, and TAG are indicated by '#', '*', and '+', respectively. The bottom section displays annotations and is currently empty. Each section can be closed or opened by clicking the double arrows in its top left corner. It is possible to open another section at the top that will show just the features of the sequence (none, at the moment).

> **Note**
>
> [a]When opening analysis or annotation from a file, ARTEMIS will return warnings if it finds any feature keys or qualifiers that it cannot interpret. The information can still be used, but it is advisable to check the warnings.

In the following protocol, we will use sɢᴘ2 to make gene predictions in the mouse sequence, based on comparison with the human sequence. We will also view and edit the results in ᴀʀᴛᴇᴍɪs.

Protocol 2

Comparative gene prediction using SGP2

1. Visit http://genome.imim.es/software/sgp2/sgp2.html[4.4]. Upload or paste the mouse.fasta[4.3] file (see *Protocol 1*) into **Sequence 1 (query)** and the human.fasta[4.2] file into **Sequence 2 (subject)**.

2. Scroll down the page and, under the section **Prediction options**, select *Homo sapiens* versus *Mus musculus* as organism and **Sequence 1 (Query)** as **Predictions on**. Select **GFF** under the section **Output Format**. Opt for the ᴛʙʟᴀsᴛx output to be included (by checking the adjacent box), as this may contain some useful evidence. Press **Submit** (center of page).

3. The results consist of two sections: a small section at the top with the gene predictions in generic file format (GFF) and a larger one below with the full ᴛʙʟᴀsᴛx output. Copy the GFF section to a separate file (for example, copy and paste the text into a Word document – but be sure to save as 'Text only') – a copy of the GFF file that you should now have is included as mouse_human_sgp2.gff[4.5] in the Protocol_2 folder for this chapter. This GFF file can be read as it is into the most recent version of ᴀʀᴛᴇᴍɪs. However, if you encounter difficulties (particularly using older versions of ᴀʀᴛᴇᴍɪs), modify the GFF file by replacing 'Terminal', 'Internal', and 'First' with 'exon' and deleting 'gene_1' from each line. (A copy of this simplified GFF file is given in the Protocol_2 folder as mouse_human_sgp2_simple.gff[4.6].)

4. The ᴛʙʟᴀsᴛx output will need to be converted with ᴍsᴘᴄʀᴜɴᴄʜ (36) so that it can be read in by ᴀʀᴛᴇᴍɪs. To do this, copy all of the text in the ᴛʙʟᴀsᴛx section of the screen and then go to http://bioweb.pasteur.fr/seqanal/interfaces/mspcrunch.html[4.7] and paste it into the box named **Actual data**. (You will also need to enter your e-mail address in the box at the top of the screen.) Scroll down and, under **output options**, enter a name for the resulting file (or leave as it is: **mspcrunch.txt**) and select **Produce exblx output (for easy parsing) (-x)**. Click **Run mspcrunch** and wait until the next screen appears.

5. Under **Results**, there should be a link to your results file. Save the file (as plain text, not as HTML format). A copy of the file that you should now have is given in the Protocol_2 folder as sgp2_mouse_seq.xtblx[4.8].

6. In ᴀʀᴛᴇᴍɪs, with the mouse.fasta[4.3] file open (see *Protocol 1*), choose **Read an entry into** from the **File** menu, then **mouse.fasta**. Choose the GFF file and then repeat the process to read in the converted ᴛʙʟᴀsᴛx output. You should now be able to see the predicted exons (in yellow) and the nine regions of significant similarity to human (in red).

7. So far, the gene predicted by sɢᴘ2 is shown in the form of three exons; in the following steps, we will create a single CDS (gene) feature in ᴀʀᴛᴇᴍɪs. Hold down the Shift key, and click on each of the three exons to select them (you can click either on the yellow boxes depicting the exons, or on the annotations in the bottom section of the window).

8. From the **Edit** menu, choose **Duplicate selected features**. Each exon will now be duplicated and the new copy of each one will be selected in the annotation list; click on any other annotation line to deselect the duplicated exons for the moment.

9. Select the first of the duplicated exons and then choose **Edit Selected Features** from the **Edit** menu. Using the **Key** menu at the top left of the new window, change the feature type from

its current one (either 'first', 'terminal' or 'internal', or just 'exon') to 'CDS'. Make sure you press **Apply** after you have made the change and then close the editing window. Repeat this for the other two duplicated exons. You should now have (in addition to the original three exons) three corresponding CDS features shown in blue (you may not be able to see the yellow exon features in the upper sections of the window, as they may be masked by the CDS features.)

10. Select all CDS features together (again by clicking on them holding down the Shift key). Create a gene structure feature from these by choosing **Merge selected features** from the **Edit** menu. A new CDS feature (comprising all three exons linked together) will be created; you will be asked if you want to delete old features – click **Yes** to delete the duplicated CDSs. The newly created gene should be selected.

11. Now look at the translation of the gene prediction by choosing **View amino acids of selection** from the **View** menu. It has stop codons in it. Look carefully at the exons in the top section of the ARTEMIS viewer – they appear to be out of frame. Close inspection of the 5′ (right-most) exon shows that its first full codon starts two positions after the start of the gene. When viewing the feature choosing **View Selected Features** from the **View** menu, you will see that the number 2 in the last column of the first exon reflects this (see *Fig. 1a*).

12. Further checking will reveal that the first position of the first codon should be in frame 0. Choose **Edit Selected Features** from the **Edit** menu. This allows you to see the feature in ARTEMIS format (see *Fig. 1b*) and modify it.

(a)

```
11  sgp2_v1.0    CDS    2858    3003    25.43    -    2
11  sgp2_v1.0    CDS    994     1034    25.43    -    .
11  sgp2_v1.0    CDS    490     566     25.43    -    .
```

(b)

```
FT   CDS          complement(join(490..566,994..1034,2858..3003))
                  /gff_seqname=11
                  /gff_seqname=11
                  /gff_seqname=11
                  /gff_source=sgp2_v1.0
                  /gff_source=sgp2_v1.0
                  /gff_source=sgp2_v1.0
                  /score=25.43
                  /score=13.58
                  /score= 7.90
                  /codon_start=3
                  /codon_start=2
```

Figure 1. Gene structure created from the SGP2 results.
Gene structure shown using (a) **View Selected Features** from the ARTEMIS **View** menu; and (b) **Edit Selected Features** from the **Edit** menu.

13. Delete the 'codon_start' lines and other redundant information (such as '/gff_seqname=11', but keep the scores as they were calculated per individual exon. Click on **Apply** and then **OK** to finish editing the CDS feature. You should now have a gene prediction in the correct reading frame (the blue CDS should be superimposed on the original yellow exons and will be free of stop codons).

14. Save the changes by choosing **Save An Entry As** from the **File** menu. Select your GFF format file, but select EMBL as the preferred format and save it under a new name. (A copy of the file you should now have is given in the Protocol_2 folder as mouse_human_sgp2.embl [4.9].)

Performing a global alignment of the query sequence against one or more related genomes can yield supplementary evidence on the location of possible protein-coding regions. The following protocol illustrates this.

Protocol 3

Finding conserved regions through global alignment using MVISTA[a]

1. Visit http://genome.lbl.gov/vista/index.shtml [4.10]. Click on the **Submit** link for MVISTA (37).

2. On the resulting page, enter '2' in the box for specifying **Total number of sequences** and press the **Submit** button[a].

3. On the next page, fill in the mandatory sections (your e-mail address). For 'Sequence #1', use the **Browse** button[b] to select the example file human.fasta [4.2] (see *Protocol 1*); similarly, select mouse.fasta [4.3] as 'Sequence #2'. Click the **Submit** button (leaving **Additional options** unaltered).

4. You will receive an e-mail after a few minutes, containing a link to the results page. Click on this link (or paste it into your browser).

5. On the resulting page, click on the link for **VISTA Browser** for 'Sequence #2'. Alternatively (for example, if your web browser does not have Java enabled), click on the link **PDF** for 'Sequence #2' to view a pdf document showing the same information in a slightly simpler format. (A copy of the pdf document is also given in the Protocol_3 folder as mvista.out.pdf [4.11].)

6. No exons are reported, but there are several regions of strong conservation between the sequences. Compare these regions with the exons proposed by the other methods (including those from *Protocol 2*).

Notes

[a]MVISTA can also use multiple reference sequences for producing an alignment.
[b]On some web browsers, these buttons will appear as **Choose file**.

So far, we have used SGP2 to carry out comparative gene prediction on the mouse sequence and seen how this prediction relates conserved regions using MVISTA. We will now use a second gene prediction package and see how its results compare with those of SGP2.

Protocol 4

Noncomparative gene prediction

1. Visit http://opal.biology.gatech.edu/GeneMark/eukhmm.cgi[4.12]. Paste in or upload the query sequence (mouse.fasta[4.3] – see *Protocol 1*). Select the relevant species (in this case *M. musculus*). Select **Generate PDF graphics (screen)** as the output format option and click on **StartGeneMark.hmm**.

2. You should get a results page that includes a table of coordinates for the predicted genes – in this case, two (one with three exons and one with only one exon). Save this section into a file, which should look like this:

Gene #	Exon #	Strand	Exon Type	Exon Range		Exon Length	Start/End Frame
1	3	-	Internal	506	566	61	2 2
1	2	-	Internal	994	1034	41	1 3
1	1	-	Initial	2858	3003	146	2 1
2	1	+	Initial	5790	5856	67	1 1

(A copy of the file is also included in the Protocol_4 folder as mouse_GeneMarkHMM. out[4.13].)

3. This gene description cannot be read directly by ARTEMIS. You therefore need to create a small text file (a feature table) describing these predicted genes, which *can* be read by ARTEMIS. The feature table is modeled on the standard EMBL format, which is defined at http://www.ebi.ac.uk/embl/Documentation/FT_definitions/feature_table.html[4.14], but with the addition of 'FT' plus three spaces) at the beginning of each line. The feature table should look like this:

```
FT      CDS               complement(join(506..566,994..1034,2858..3079))
FT                        /note="predicted with GeneMarkHMM"
FT                        /codon_start=2
FT                        /colour=12
FT      CDS               5790..5856
FT                        /note="predicted with GeneMarkHMM"
FT                        /colour=12
```

(A copy of this file is also included in the Protocol_2 folder as mouse_GeneMarkHMM. tab[4.15].)

4. It is essential to have the correct number of spaces (not tab characters): there should be three spaces before 'CDS' and thirteen after, or 19 in total in the other lines. The first gene (CDS) is on the negative strand ('–' in the GENEMARK.HMM output), hence the use of 'complement' in the feature table; the 'join' phrase specifies the three predicted exons. For this CDS, '/codon_start=2' indicates the offset at which the first complete codon can be found, relative to the first base of that feature (corresponding to a 'start' value of 2 for the first exon of this gene in the GENEMARK.HMM output). Neither 'complement' nor the 'codon_start' qualifier are needed for the second, single-exon gene that lies on the '+' strand and starts in the first reading frame. The 'colour' qualifier simply determines the color with which the features are displayed in ARTEMIS.

5. Save the feature table file (if saving from Word, remember to save as 'Text only'). In ARTEMIS, open the mouse.fasta[4.3] file and then use **Read an entry into** (under the **File** menu) to read in the feature table from the file you have just created. You will see the two predicted genes displayed on their respective strands, with the exons of the first gene linked.

6. The first (right-most) exon of the first predicted gene includes a stop codon. This may be correct if it lies in the 5′-untranslated part of the gene. However, the results returned by the prediction tool do not make this clear, so there is also a possibility that there could be a problem with the prediction. For the moment, we will note this and see whether further evidence clarifies things.

7. Note that comparison with an expressed sequence tag (EST) database may help identify additional exons or confirm predicted exons.

In the following protocol, we will try using BLASTN to look for possible matches between our mouse sequence and mouse ESTs.

Protocol 5

Using BLASTN to find ESTs matching a genomic region

1. Visit http://www.ncbi.nlm.nih.gov/BLAST/[4.16]. Select **Nucleotide-nucleotide BLAST (blastn)**. On the resulting form, paste the contents of the mouse.fasta[4.3] file (see *Protocol 1*) into the query sequence (the top-most text box). Choose **est_mouse** as the database in the pull-down menu underneath. Scroll down the page and, under **Format**, select **Hit Table** from the **Alignment View** dropdown menu.

2. Click **Blast!** and, on the resulting page, click **Format!** to see whether your results are ready.

3. The results will be returned to you in a tabular format[a]. You need to save this as a text file (for example, by copying all of the text from '# BLASTN' onwards). An example file (blast_mouse_versus_ncbi_mouse_ESTs.res[4.17]) is provided in the Protocol_5 folder.

4. The file of BLASTN hits can be read directly into the annotated mouse sequence in ARTEMIS to see how they relate to the gene predictions[b]. You will see that the hits fall on many parts of the mouse sequence; some ESTs align with several parts of the sequence.

Notes

[a]Although default settings were used here for BLASTN, choosing these rarely gives optimum results so it is always advisable to try and calibrate the search. The BLAST book is a good source of protocols for specific types of searches (38). However in some cases a different alignment algorithm needs to be applied to increase the sensitivity of a particular search.

[b]The BLASTN matches are displayed in ARTEMIS on the frame lines as BLASTCDS, but it would have been more appropriate to display them on the central bars as BLASTN_HIT.

Clearly, the simple BLASTN search gives us many potentially informative ESTs, but does not give a clear indication of how they relate to the structure of the predicted genes. In the next protocol, we will try to refine the alignment of one of these EST hits to the genome, using the splice-site-aware alignment tool EST2GENOME (39) to see whether it can provide better evidence of the gene structure.

Protocol 6

Looking for evidence of a potential gene structure in a match using EST2GENOME

1. In ARTEMIS, viewing the results of *Protocol 5*, select one EST (by clicking on it in any of the sections of the ARTEMIS window) that gives nonoverlapping hits or hits with only marginal overlaps between them to different ESTs.

2. Find the unique EMBL/GenBank identifier for this chosen EST from the BLAST hits in the annotation section of the screen. The identifier is the last part of the second column. For example, if the content of the second column is 'gi|27167631|dbj|BY742253.1|', then 'BY742253' represents the unique identifier (note that the number after the decimal point should be dropped).

3. If you want to use the same example, you can search in ARTEMIS for this particular hit by using **Feature selector** (in the **Select** menu); you will need to search for items whose key is 'BLASTCDS', and with a qualifier of 'note' containing the text 'BY742253'. You will also need to **allow partial match** and to make sure that you are searching both forward- and reverse-strand features. Clicking **Select** will then select and highlight the three hits to the EST BY742253.

4. Visit http://www.ebi.ac.uk/[4.18] and type the identifier into the box next to **Database Search for**. Press the left-most **Go** button. On the resulting page, click on the ID under the column called **'EMBL'** to see the actual entry. Once the entry is displayed, press on the link **Sequence** towards the top-right corner. This will take you to a section showing the sequence in FASTA format. Copy this.

5. Visit http://liv.bmc.uu.se/cgi-bin/emboss/est2genome[4.19]. Paste in the FASTA format sequence of the EST into the first of the two large windows. Upload the query sequence (mouse.fasta[4.3]) into the section for the genomic sequence.

6. Scroll down (leaving the various settings at their default values) and press the button **Run est2genome**. When you get the result page (which will take a minute or more), click on the link **Outfile** (top right) to get the correct format for downloading results. Then use 'Save as' (or the equivalent) in your browser to save the file. The expected result is shown in *Fig. 2(a)*, and a copy of the file is also given in the Protocol_6 folder as EST2genome_BY742253_vs_mouse_pde6g.res[4.20].

7. This output file describes a four-exon gene, which, according to EST2GENOME, best reflects the alignment between the EST and the genomic sequence.

8. You will need to convert this output file into a format that can be read by ARTEMIS, and this must be done manually in a suitable text editor or word processor. (If using Word, remember to save the result as 'Text only'). *Fig. 2(b)* shows the file as it should appear, and a copy is given in the Protocol_6 folder as EST2genome_BY742253_vs_mouse_pde6g.tab[4.21]. If (as in this case) the header indicates that the gene is in reverse orientation, you will need to ensure that the feature is specified to be on the reverse strand. The feature key ('misc_feature') should begin six characters in from the left margin and your coordinates and qualifiers (e.g. '/colour') should start 22 characters in from the left margin. The line width should not exceed 79 characters.

9. Read the file into ARTEMIS[a], so that the newly defined feature can be compared with the other gene prediction results. The alignment to EST BY742253 does not conflict with the CDS identified by SGP2. It does not fully agree with the genes predicted by GENEMARK.HMM.

```
Note Best alignment is between reversed est and forward genome, but
splice sites imply REVERSED GENE
Exon         296  99.3    267    566 11                    1    300 BY742253
Mus musculus adult retina cDNA, RIKEN full-length enriched library,
clone:A930027A04, 5' end partial sequence. ...
-Intron      -20   0.0    567    993 11
Exon          41 100.0    994   1034 11                  301    341 BY742253
Mus musculus adult retina cDNA, RIKEN full-length enriched library,
clone:A930027A04, 5' end partial sequence. ...
-Intron      -20   0.0   1035   2857 11
Exon         222 100.0   2858   3079 11                  342    563 BY742253
Mus musculus adult retina cDNA, RIKEN full-length enriched library,
clone:A930027A04, 5' end partial sequence. ...
-Intron      -20   0.0   3080   5804 11
Exon          66 100.0   5805   5870 11                  564    629 BY742253
Mus musculus adult retina cDNA, RIKEN full-length enriched library,
clone:A930027A04, 5' end partial sequence. ...

Span         565  99.7    267   5870 11                    1    629 BY742253
Mus musculus adult retina cDNA, RIKEN full-length enriched library,
clone:A930027A04, 5' end partial sequence. ...

Segment      296  99.3    267    566 11                    1    300 BY742253
Mus musculus adult retina cDNA, RIKEN full-length enriched library,
clone:A930027A04, 5' end partial sequence. ...
Segment       41 100.0    994   1034 11                  301    341 BY742253
Mus musculus adult retina cDNA, RIKEN full-length enriched library,
clone:A930027A04, 5' end partial sequence. ...
Segment      222 100.0   2858   3079 11                  342    563 BY742253
Mus musculus adult retina cDNA, RIKEN full-length enriched library,
clone:A930027A04, 5' end partial sequence. ...
Segment       66 100.0   5805   5870 11                  564    629 BY742253
Mus musculus adult retina cDNA, RIKEN full-length enriched library,
clone:A930027A04, 5' end partial sequence. ...
```

(a)

```
FT    misc_feature   join(complement(267..566),complement(994..1034),
FT                   complement(2858..3079),complement(5805..5870))
FT                   /colour=8
FT                   /query_id="mouse_pde6g"
FT                   /hit_id="BY742253, Mus musculus adult retina cDNA,
FT                   RIKEN full-length enriched library,
FT                   clone:A930027A04, 5 prime end partial sequence."
FT                   /hit_start=1
FT                   /hit_end=629
FT                   /note="Note Best alignment is between reversed est
FT                   and forward genome, and splice sites imply REVERSED
FT                   gene"
```

(b)

Figure 2. EST alignments returned by the EST2GENOME server.
(a) The key part of the data is the upper half, describing the four exons that the server identifies by alignment with the EST data. Note that each exon and intron is described on a single line: the long lines have been wrapped to fit the window. The second column holds the score of each exon match and the third column holds the percentage identity. The fourth and fifth columns give the start and end positions of the exon (e.g. 267 and 566 for the first exon). Columns 7 and 8 show the start and end of the match in the subject (EST) sequence (e.g. 1 and 300 for the first exon). Columns 6 and 9, respectively, hold the query and subject IDs and the remaining columns hold the subject description. The line starting with Span shows that the match goes across the complete length of the EST. (b) The same information, converted to ARTEMIS format describing the inferred gene. The exon coordinates should be entered in the same order as listed in the original output. Additional information has been added (for example, using the qualifiers 'hit_start' and 'hit_end').

10. Try this process with several ESTs, preferably hitting in different places on the genome sequence, to see whether one or more particular alignments from EST2GENOME confirm any of the gene predictions. When inspecting EST matches, it is important to remember that many ESTs include untranslated 5′ and 3′ regions (including untranslated exons). The mapping of ESTs to genomic sequence is also covered in Chapter 7.

Note

[a]Only the most recent versions of ARTEMIS will recognise the qualifiers 'hit_id', 'hit_start', and 'hit_end'. If this causes problems, these lines can be deleted from the file.

A BLASTX search (in which a nucleotide sequence is first translated in all possible reading frames before searching against protein sequences) against databases such as UniProt is useful for checking for evidence of exons not identified by any of the previous approaches. This is illustrated in the following protocol.

Protocol 7

Comparison of genomic query sequence to a protein database using BLASTX

1. Visit http://www.ebi.ac.uk/blastall/[4.22] to launch such a search. Ensure that, under **Database**, **Protein** is selected as the molecule type to search against and **Uniprot** is selected as the database. Upload or paste the query sequence (mouse.fasta [4.3]) in the lower part of the screen, and use the pull-down menu to indicate that it is a DNA/RNA file. Leave the other settings at their defaults and click **Run Blast.**

2. You will get a summary of the results in tabular format, detailing about 50 hits (correct at the time of writing). Click on the **Blast Result** button to see the verbose output. When you get this, click on the **Printable** button. Select all content, copy it, and paste it into a new text file (if working in Word, remember to save this as 'Text only'). A copy of the file you should now have is given in the Protocol_7 folder as mouse_EBI_uniprot_full_output.blastx [4.23].

3. Next, as in *Protocol 2*, we need to use MSPCRUNCH to convert this blast output into a summary file readable by ARTEMIS. Visit http://bioweb.pasteur.fr/seqanal/interfaces/mspcrunch.html [4.24], fill in your e-mail address as requested, and upload the file containing the BLASTX output (from the previous step), or just paste the text into the box named **Actual data**. Select the output option **Produce exblx output (for easy parsing) (-x)** and click **Run mspcrunch.** On the resulting page, click the link to the text file (mspcrunch.txt, unless you specified a different name). Save the result into a file as plain text (not HTML) format. (A copy of the file you should now have is given in the Protocol_7 folder as mouse_EBI_uniprot_summarised.blx [4.25].)

4. In ARTEMIS, go to **Read An Entry** in the **File** menu and select this summary file. Compare the position of the hits with the location of the predictions and the other evidence. One exon reported by EST2GENOME, GENEMARK, and SGP2 is not confirmed in the any of the best hits. The exon is small and there are no significant hits in the vicinity of it. This could be why it was not reported. A different type of BLAST search may have been more effective. (The last protocol in this section includes a suggestion for this, but only look at this after you have done the other examples).

5. Alternatively, visit http://www.ncbi.nlm.nih.gov/BLAST/[4.16]. Select the link **Translated query vs. protein database (blastx)**. On the resulting form, select BLASTX from the **Choose a translation** drop-down list and nr from the **Choose database** drop-down list. Under **Advanced options** select hit table from the **Output Alignment View** drop-down menu. The results will be returned to you in a tabular format, which can be read directly into ARTEMIS.

In the protocols described so far, we have looked at various pieces of evidence to help identify genes in the segment of mouse genomic sequence. It is now time to try and review and evaluate the evidence and build a plausible gene model. Naturally, no rigid 'protocol' can describe this process of evaluation, but *Protocol 8* can be considered an illustrative case study.

Protocol 8

Making your own interpretation

1. We have seen that the CDS prediction made by SGP2 does not contradict the alignment of EST BY742253 by EST2GENOME. However, where EST2GENOME reports a possible untranslated 5′ exon, the prediction made by GENEMARK.HMM has a small, separate gene on the opposite strand. Also, the end position of the last exon of the larger gene reported by GENEMARK does not agree with EST2GENOME or SGP2: it seems to be upstream of a potential donor site, GTATGG.

2. The conserved regions with the collinear *H. sapiens* sequence reported by MVISTA coincide roughly with the EST2GENOME alignment.

3. The best BLASTX hit confirms the outer exons of the SGP2 prediction. Thus, when all evidence is evaluated jointly, the following gene appears to be the most plausible:

Gene (i.e. the spliced transcript):

```
complement(join(267..566,994..1034,2858..3079,5805..5870))
```

CDS (i.e. the translated part of the gene):

```
complement(join(490..566,994..1034,2858..3003))
```

The most 5′ exon of the gene (nt 5805–5870) is untranslated, as is the first part of its second exon (nt 3004–3079). The last part of the fourth and final exon (nt 267–489) is the 3′ untranslated region.

Now that we are fairly satisfied with our gene prediction, the next three protocols illustrate ways of further checking the results. We will start by using INTERPROSCAN, which analyses a protein sequence for conserved domains and functional sites.

Protocol 9

Conserved domains and functional sites in the predicted protein

1. In ARTEMIS, click on the CDS of your gene and go to the **Write** menu, choosing **Amino Acids of Selected Features**. This will let you create a file in FASTA format of its translated sequence. Alternatively, you can choose **View amino acids of selection as FASTA** from the **View** menu, and cut and paste from the resulting window. A copy of the file that you should have created is also given in the Protocol_9 folder as Mouse_own_annotation.aa [4.26].

2. Next, visit the INTERPROSCAN (40) server (http://www.ebi.ac.uk/InterProScan/ [4.27]). Upload the amino acid sequence file (or paste the amino acid sequence into the window), leave everything else at its default setting, and press **Submit Job**.

3. Once the results are returned, the best hits are shown in a diagram. Pressing the **Raw Output** button shows the exact coordinates of the matches. (A copy of the expected raw output file is included in the Protocol_9 folder as InterProScan.results.txt [4.28].)

4. There are at least three significant hits across almost the complete amino acid sequence, all of them to retinal cGMP phosphodiesterase. This gives us confidence that our gene prediction is likely to be substantially correct.

We can also verify our predicted gene by searching the predicted protein against a protein database such as UniProt (41). This method is more oriented towards finding complete matches than to finding significant hits to protein signatures. This can make it more comprehensive but also less flexible than INTERPROSCAN, as it relies solely on protein sequence similarity. We will do this in the following protocol.

Protocol 10

Checking the predicted protein using protein–protein BLAST

1. Go to http://www.ebi.ac.uk/blastall/ [4.22] and select protein–protein BLAST (BLASTP).

2. Paste or upload the amino acid sequence of the predicted gene into the window and ensure that the menu immediately above it is set to 'PROTEIN'. Click **Run Blast**.

3. The results page will show about 40 (correct at the time of writing) good matches in a summary format. Clicking on **Show alignments** will display the matches at the amino acid level.

4. In this case, our example gene has a perfect match to two mouse entries. Clearly, this is unlikely to be the case if we were making gene predictions on a truly novel sequence!

5. There are also many very good matches to proteins from other vertebrates. One might not always expect to see such strong conservation, but in this case, it suggests that our CDS prediction is almost certainly correct.

The majority of the evidence obtained fits well with the gene created. *Fig. 3* (also available in the color section) is a snapshot from ARTEMIS, showing the key parts of this evidence. However, some pieces of information may have been missed. In this example, the mouse sequence we chose had already been extensively annotated. So, although normally it would not be possible to check the predicted gene against pre-existing annotation, it is done here to demonstrate the benefits of a thorough approach and to illustrate how the gene structure can be improved by re-examining the existing evidence.

Protocol 11

Validation of results

1. Visit Vega at http://vega.sanger.ac.uk/Mus_musculus/index.html [4.29] (42). Type in 'pde6g' in the search box in the top right-hand corner and press **Go**. Click on the gene identifier (**Vega Gene: OTTMUSG00000004174**) under the first reported match in the *M. musculus* gene index. The Curated Locus Report shows a gene structure to the model created (scroll down the page).

2. In the navigation bar on the left, click on **Export EMBL file** to download the annotation. On the form returned, tick the feature types for **Contig** and **Gene information**, opting for the text output format, and press **Continue**. Save the resulting page to a text file; if there are any spurious headers or footers, remove these: the file should begin with 'ID' and end with the DNA sequence followed by '//'. A copy of the file as it should appear is included in the Protocol_11 folder as mouse_pde6g.embl[a] [4.30].

3. Now import it into ARTEMIS[a] using **Read An Entry** from the **File** menu so that it is visible alongside your own analysis and annotation. The entry shows that the annotated mRNA goes beyond the 5' end of the gene feature created. On closer inspection, there appears to be a possible poly(A) tail ending at complement 34.

4. Go back to the BLASTN results versus est_mouse and look at the hits covering the region downstream of our gene. Hits to CK619967 and BB280527 are possible evidence (see files in the Protocol_11 folder - EST2genome_CK619967_mouse_pde6g.tab [4.31] and EST2genome_ BB280527_mouse_pde6g.tab [4.32]).

5. Repeat the EST2GENOME alignment steps to see whether the alignment for these can be improved. The alignment for CK619967 overlaps to a large degree with that for BY742253 in example 7 but extends it further at the 3' end.

6. The alignment with reversed BB280527 would extend it even further but is on the main strand. However, the unreversed BB280527 aligns well with the same region on the complementary strand (see leftmost feature on *Fig. 3*).

7. Both ESTs are from retinal cells. Taken together, this makes a good case for extending the gene to:

   ```
   complement(join(1..566,994..1034,2858..3079,5805..5870))
   ```

8. Note that the annotated exon features representing the coding segment of the gene are not in frame.

Note

[a]The Vega annotation, like all databases, is liable to be updated. If you experience problems, use the archived version – mouse_pde6g.embl [4.30] – in place of the file you have just created.

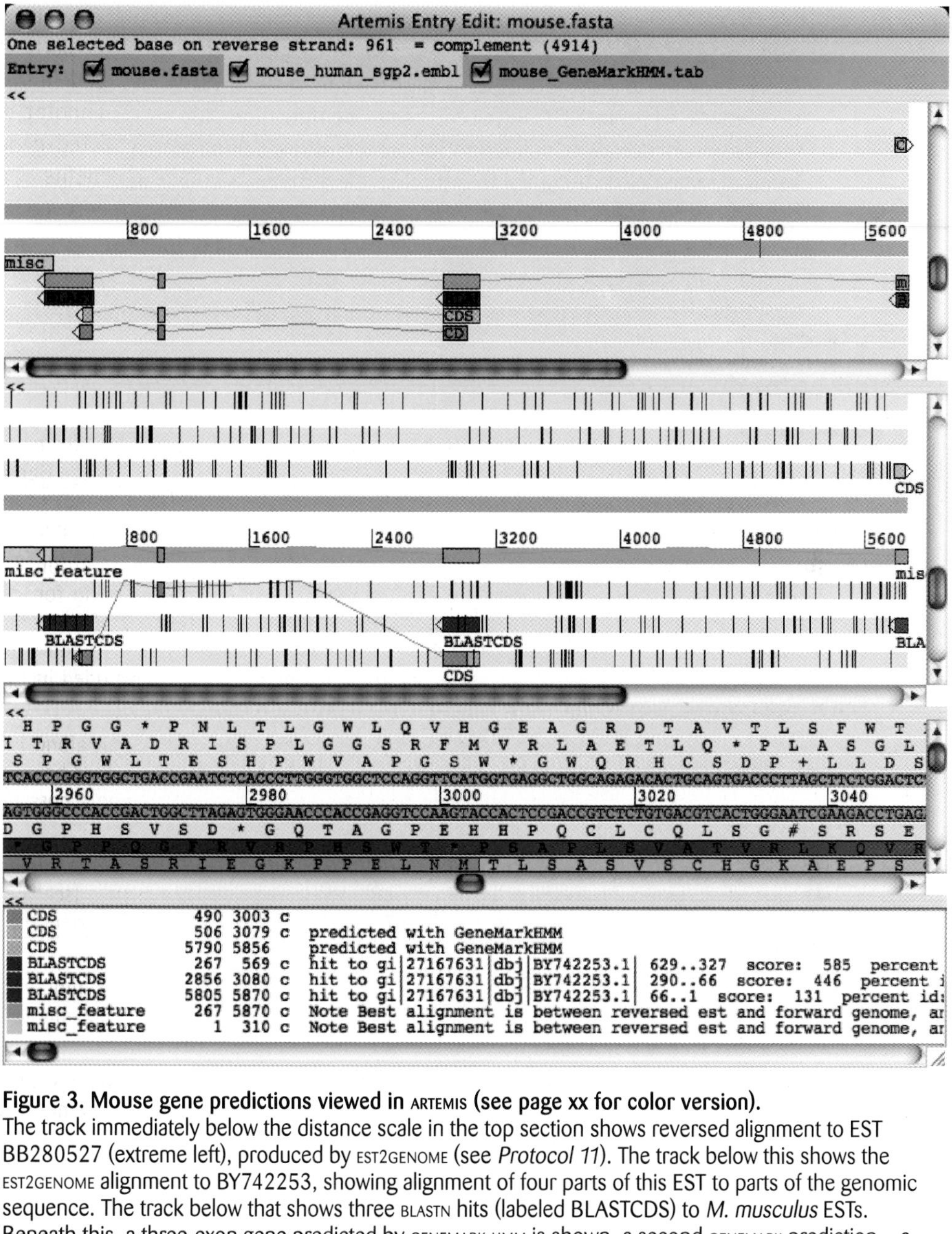

Figure 3. Mouse gene predictions viewed in ARTEMIS (see page xx for color version).
The track immediately below the distance scale in the top section shows reversed alignment to EST
BB280527 (extreme left), produced by EST2GENOME (see *Protocol 11*). The track below this shows the
EST2GENOME alignment to BY742253, showing alignment of four parts of this EST to parts of the genomic
sequence. The track below that shows three BLASTN hits (labeled BLASTCDS) to *M. musculus* ESTs.
Beneath this, a three-exon gene predicted by GENEMARK.HMM is shown; a second GENEMARK prediction – a
small single-exon gene – appears at the extreme right on the forward strand (shown above the distance
scale). The lowest track in this section (just above the top-most horizontal scroll bar) shows the gene
predicted by SGP2. The second section of the screen (between the top two horizontal scroll bars) shows
the same region, but with the three reading frames represented as individual lines above (forward strand)
and below (reverse strand) the distance scale; features are shown in the relevant reading frame; vertical
lines are stop codons. The third section shows a close-up view of part of the sequence (nt 2960–3040)
including nucleotide and amino acid sequences. The bottom section lists the annotated features.

2.2 Predicting prokaryotic genes

Given that prokaryotic genes generally are not spliced, there are some major differences in approach between gene identification for prokaryotes and eukaryotes. Hence, gene prediction for prokaryotes is demonstrated in a separate series of steps below, designed to introduce the different methods used to identify genes in prokaryotic genomic sequences (J. Parkhill, personal communication).

As with the mouse example, the protocols should be followed in order, building up successive pieces of evidence for gene predictions. The analysis will again revolve around an ARTEMIS entry, to which we will add successive layers of information.

Protocol 12

Downloading prokaryotic sequences and configuring ARTEMIS for bacterial genomes

1. Visit ftp://ftp.sanger.ac.uk/pub/pathogens/leprae/ [4.33] and download the unannotated copy of the ±3.3 Mb *Mycobacterium leprae* genome: ML.dna [4.34] (a copy of this file is also provided in the Protocol_12 folder). Open the sequence in ARTEMIS.

2. As the sequence is large, create a smaller sequence from this entry, comprising 40 kb. From the Select menu, choose **Base Range ...** Type '1..40000' in the box for specifying the base range and press **OK**. Then choose **Bases of Selection** from the **Write** menu, specify FASTA as the required format, and save the file as ML_1st_40kb.dna [4.35] (a copy of this file is also present in the Protocol_12 folder). Close the current entry and open the smaller sequence entry.

3. Click on the first ARTEMIS window (this is the small window that appears when ARTEMIS is started, showing release and copyright information. If you cannot see this window, move or minimize the window showing the *M. leprae* entry). Under the **Options** menu (which will appear at the top of the screen), select **Bacterial and Plant Plastid**; this defines the translation table that will be used, which differs among groups of organisms. (ARTEMIS defaults to the standard translation table, which is applicable to the nuclear genome of higher eukaryotes.)

As a first step, we will use a quick, but crude, method to identify all possible genes – finding all of the ORFs in the sequence. This method should only be used in short regions with subsequent manual editing – gene prediction programs should be used for longer sequences.

Protocol 13

Finding the open reading frames

1. In ARTEMIS (having opened the 40 kb of *M. leprae* sequence – see *Protocol 12*), select **Mark open reading frames** from the **Create** menu. You will be asked to set a minimum ORF size (in amino acids): choose 150. Remember that the number of ORFs in a sequence is determined by the G+C content (see above) and that not all (and often not the full length) of the ORFs are CDSs.

2. Bacterial transcripts normally begin with ATG or, less often, GTG or TTG. We will therefore trim all of the ORFs so that they start with one of these codons. Go to the **Select** menu and choose **All** to select all of the ORFs and then choose **Trim Selected Features to Any** from the **Edit** menu.

3. Two ORFs contained no suitable start codons and hence cannot be trimmed. Select these and delete them (using **Delete Selected Features** from the **Edit** menu).

4. This will leave you with all possible long ORFs that begin with a valid bacterial start codon. In order to distinguish them from other features that will be added in the following protocols, we will change their color. To do this, select all of the features and choose **Change Qualifiers of Selected** from the **Edit** menu. Select the qualifier **color** from the drop-down menu and type '/colour=10' in the text box. Press the **Replace** button and then **Close**. The ORFs should now be shown in brown.

Sequence composition – particularly in relation to the reading frame – can be another important clue in identifying CDS. The G+C content in each position of the reading frame of a piece of DNA is often different in coding regions compared with noncoding regions due to the limitations imposed by codon usage. In the next protocol, we will use ARTEMIS to display this positional G+C content.

Protocol 14

Positional G+C content as a clue to CDS

1. In ARTEMIS (starting where the previous protocol left off), go to the **Graph** menu and select **GC frame plot**. A chart will appear with each of the colored graphs representing the G+C content of one frame (i.e. of every third base in the sequence, but starting from the first, second, or third base for the red, green, and blue plots, respectively). The vertical scroll bar on the right of the graph determines the window size – in effect, the smoothing of the plot. This is useful when trying to optimize the graphs.

2. Compare this with the overall G+C content by choosing **GC content (%)** from the same menu.

3. In many bacterial genomes – particularly those with a high overall G+C content – the constraints imposed by codon usage result in a marked difference in G+C content between some coding and noncoding regions. (The frame in which the CDS starts, relative to the start of the sequence, will dictate which of the three plots shows the anomalous G+C content.)

4. In this case, it is difficult to interpret the G+C plots at present, so we will move on to a more-complex analysis.

If G+C content alone is not helpful, we can try a more sophisticated positional analysis, namely codon usage. Most organisms have a strong preference for certain codons over others in a translated sequence. Codon usage plots show the extent to which the genomic sequence matches the expected pattern of preference, when considered in each possible reading frame.

Protocol 15

Codon usage as a clue to CDS

1. To obtain a codon usage table for *M. leprae*, visit the Codon Usage Database (http://www.kazusa.or.jp/codon/ [4.36]). Type 'Mycobacterium leprae' into the query box. Press **Submit**. Select the relevant organism (here, *Mycobacterium leprae* [gbbct]) and you will be shown a codon usage table. Copy the codon usage table (the table itself, not the text above and below it), paste this into a text file, and save this (as 'Text only', if you are using Word). A copy of the file that you should have made is given in the Protocol_15 folder as M_leprae.cu.txt [4.37].

2. From the **Graph** menu, choose **Add Usage Plot** to read in the codon usage file you saved. A chart will appear showing the codon usage in each of the possible reading frames (relative to the start of the genomic sequence) and in both directions. (If the G+C content plots from the previous protocol are overcrowding the screen, you can remove them by selecting the relevant items in the **Graph** menu.)

3. Moving the right-hand scroll bar controls the degree of 'smoothing' of the peaks and troughs by changing the window size (indicated above each of the plots).

4. The codon usage plots can provide suggestive (although not conclusive) evidence about coding regions. If a given ORF is indeed coding, one might often expect to see a corresponding peak in the codon usage plot for one of the three reading frames on that strand. For example, the very first ORF lies on the positive strand, and one of the three reading frames shows a peak in codon usage on the positive strand (but not the negative strand); this would suggest that this ORF is likely to be a genuine CDS.

Clearly, neither positional G+C content nor codon usage is sufficient for robust identification of true CDSs from amongst the many ORFs. We will therefore add the output of gene prediction programs to our repertoire of evidence. Many gene prediction tool internet sites have sets of pre-calculated gene predictions for particular organisms. In this example, we will start by using a set of automated predictions pre-computed by GENEMARK.HMM trained for *M. leprae*; examining these critically will help us to decide which of our ORFs are likely to be real genes.

Protocol 16

Non-comparative gene prediction

1. Visit http://opal.biology.gatech.edu/GeneMark/prokaryotes_database/index.cgi [4.38]. Select the link for *M. leprae* from the list of organisms shown.

2. On the resulting submission form, under the heading **Complete Genome Download Option**, tick **Coordinates of Prediction** and press **Submit**. This should produce a text file showing the approximately 4000 predicted genes on this genome.

3. From the resulting page containing the predictions, select the header lines and all of the feature lines with coordinates within the first 40 kb from the 5′ end. Copy this, paste it into a new file, and save this (as 'Text only' if using Word). A copy of the file you should now have is provided in the Protocol_16 folder as ML_1st_40kb_canned_genemark.res [4.39].

4. This file must be converted into ARTEMIS format; this can be done manually or programmatically. For the purpose of this example, the converted file (named ML_1st_40kb_canned_genemark.embl [4.40]) has been provided in the Protocol_16 folder – looking at this file will reveal how the original file has been edited. The program (a Perl script) that was used to do the conversion is also provided in the Protocol_16 folder (process_canned_genemark.perl [4.41]), for users familiar with Perl.

5. In ARTEMIS, use **Read an Entry** from the **File** menu to read in the converted (EMBL) file (either your own or the example provided) so you can see the GENEMARK predictions[a].

6. In many cases, these are superimposed on the ORFs generated in the earlier protocols, so that one feature obscures another. To overcome this, right click (or, if you are using a one-button mouse in Macintosh, control click) on any one of the features in the upper section of the window. A pop-up menu will appear: click on **One line per entry**, and the features will no longer be superimposed. You can show all of the evidence on separate lines by right clicking on the frame lines and selecting one line per entry.

7. Compare the different predictions, and using the plot information (especially codon usage) remove any features from the set of trimmed ORFs that you think are not real genes. For example, look at the right-most part of the sequence (from about nt 36200 onward). On the top strand, there is one large ORF (in reading frame 3; if you double click on it in the upper section of the display, it will be highlighted and you will see that it is on the third line of amino acids) and a smaller ORF in reading frame 1 nested within it. On the bottom strand, there are three smaller ORFs. The codon usage plot shows a broad peak in one frame on the top strand, and GENEMARK has also predicted a large forward gene in this region (albeit with a slightly different start position). You might therefore infer that the large top-strand ORF is well supported. The smaller ORF within it, and those on the bottom strand, are less well supported by codon usage and do not match GENEMARK predictions, so you might be justified in deleting these as candidate genes.

8. Not all cases are as clear cut. You can be conservative at this stage – it is best to retain a feature as a possible gene at this stage and perhaps reject it later.

Note

[a]You may get a warning that 'CDS can't have method as a qualifier'. If so, remove all of the complete lines containing '/method' from the file and try again.

The methods discussed for prokaryotes so far have only looked at sequence properties to predict whether a region is coding. As we saw in the eukaryotic example, using homology evidence is another important way of establishing this. Here, homology data will be obtained in the first instance through a BLASTX comparison of the *M. leprae* genomic sequence with the protein set of the related organism *Mycobacterium tuberculosis*. In the following example, BLASTX is used to compare all possible translations of the *M. leprae* genomic sequence against the complete protein set of the related organism, *M. tuberculosis*. This is more sensitive than a DNA–DNA comparison, as protein sequences tend to be conserved even when nucleotide sequences have diverged.

Protocol 17

Comparison of genomic query sequence with a protein database using BLASTX

1. Visit http://www.sanger.ac.uk/cgi-bin/blast/submitblast/m_tuberculosis[4.42]. Upload or paste your sequence (the first 40 kb of the *M. leprae* genome – see ML_1st_40kb.dna[4.35]) into the **Query data** section. Under options, select **M. tuberculosis predicted proteins** as the database and **Blastx** as the type of search. Leave the other settings at their defaults and press **Start Blast**.

2. Retrieve the results from the link indicated on the resulting web page. Copy the BLAST results, leaving out the header and footer of the submission form. (The copied text should begin with 'BLASTX 2.0MP-WashU...' and end with 'Start:...'). Paste this into a new text document and save this (as 'Text only', if using Word).

3. As in *Protocol 2*, we will use MSPCRUNCH to process this file into a summary form that ARTEMIS can read. Go to http://bioweb.pasteur.fr/seqanal/interfaces/mspcrunch.html[4.7], enter your e-mail address where requested, and either paste or upload the copied BLASTX output into **Actual data**. Select output option **Produce exblx output (for easy parsing) (-x)** and click **Run mspcrunch**.

4. On the results page, select the text file (mspcrunch.txt, unless you gave it a different name). Save the result into a file as plain text and not HTML format (a copy of the expected file is given in the Protocol_17 folder as ML_1st_40kb_vs_TB_prots.xblx[4.43]).

5. In ARTEMIS, go to **Read An Entry** in the **File** menu and select the summary file you have just saved. You should now be able to see the BLASTX results alongside the GENEMARK predictions and your original set of ORFs. (If features are superimposed on one another, switch to **One line per entry** – see step 6 of *Protocol 16*).

6. You should now review your set of ORFs again, taking into account the BLASTX results. For example, there is an ORF on the reverse strand at positions 11393–12006 (unless you removed it during the previous protocol). Part of it is supported by a GENEMARK prediction. It also clashes with a larger ORF on the top strand, again partially supported by a GENEMARK prediction – it is unlikely that both ORFs are correct (highly overlapping antiparallel genes are rare). Codon usage tends to support the forward-strand prediction more than the bottom-strand one, but is not decisive. However, a BLASTX hit agrees almost perfectly with the forward-strand ORF. (You can view the details of the hit by selecting the BLASTX feature and using **View selected features** from the **View** menu). On this basis, we can be confident in the forward-strand ORF and delete the reverse-strand one.

7. As before, be conservative – if in doubt, keep a poorly supported ORF rather than discarding it at this stage. Homology evidence carries considerable weight, especially where (as above) the choice is between two ORFs that clash and only one is supported by BLAST evidence. However, lack of homology alone is not sufficient to reject an otherwise plausible ORF that does not clash with other predictions – if it were, no novel genes would be predicted!

We now have a good set of candidate genes – ORFs that are supported (or at least not rejected) by a range of evidence, which may include sequence composition and codon usage, agreement with GENEMARK.HMM predictions, and amino acid similarity to proteins from a related species. In the next protocol, we will use BLASTP to search more widely for homology to all known proteins.

Protocol 18

Comparison of predicted proteins to all known proteins using
BLASTP

1. Click on the ORF you want to perform your search on (in this case, we will use the ORF that runs from nt 11195 to 12106 on the top strand).

2. Under the **Write** menu, choose **Amino Acids of Selected Features** to create a file in FASTA format of its translated sequence.

3. Next, visit http://www.ebi.ac.uk/blastall/[4.22] to launch the search. Choose BLASTP as the BLAST program and UniProt as the database. Upload the amino acid FASTA file (or paste its contents) into the window and ensure that 'PROTEIN' is selected from the pull-down menu immediately above this. Click **Run blast**.

4. When the results are ready, you will see a summary table. In this example, the top hit is for a hypothetical *M. leprae* protein (correct at the time of writing) – we will ignore this: it is essentially the same prediction that we are trying to confirm!

5. Many of the other good hits are also identified as 'hypothetical', 'putative', or 'possible'. However, this is still supporting evidence for our prediction: the fact that similar proteins have been predicted in a wide range of other species strongly suggests a conserved amino acid sequence and hence a true CDS. Any functional assignments accompanying significant hits require close scrutiny, as these may not always be correct and may have been derived from incorrect annotation. To avoid the risk of propagating annotation errors, therefore, it is best to go back to the paper where the link between the sequence and function was first made and check its validity.

6. Use a similar approach to examine other ORFs of which you are unsure. Again, be conservative and avoid rejecting a poorly supported but otherwise plausible ORF unless it clashes with another well-supported one.

7. Further evidence about the accuracy of a predicted gene can be obtained by searching its amino acid translation for conserved domains and functional sites (see *Protocol 9*). This approach is useful for finding supporting evidence for less well-conserved proteins.

In order to achieve the best possible interpretation of the sequence, all of the results should be inspected in context, and any problems raised through apparent inconsistencies should be investigated further. Here, to highlight regions for which the evidence so far may not be correct, we will compare our predictions with those made (43) by experienced annotators drawing on all of the available evidence. Such regions will require additional work to improve the accuracy of the analysis. Normally, of course, it would not be possible to check the analysis against the annotation, but here it is done to show how looking at the data with a critical eye is good practice.

Protocol 19

Comparison of results with manual annotation

1. Visit ftp://ftp.sanger.ac.uk/pub/pathogens/leprae/[4.33] and download the annotated copy of the *M. leprae* genome: ML.embl[4.44] (a copy of the same file is provided in the Protocol_19 folder, but that on the Sanger site may be updated). Load the entry into ARTEMIS using the **File** menu in the small ARTEMIS release window (do not read it into the existing entry).

2. We need to create a smaller section from this entry – the first 40 kb of it – to compare against our own predictions. From the **Select** menu, choose **Base Range** ... ; type '1..40000' in the box for specifying the base range and press **OK**.

3. Again from the **Select** menu, choose **Features Overlapping Selection** to select all features lying within the first 40 kb.

4. From the **Create** menu, select **New Entry**, which will produce a blank entry called 'no name'. From the **Edit** menu, now select **Copy Selected Features To** and choose 'no name' as the destination.

5. Untick the ML.embl entry in the top bar of the entry window. Only the features of 'no name' are displayed now.

6. Remove any features that start within the first 40 kb but extend beyond it (by clicking on any such feature, then choosing **Delete Selected Features** from the **Edit** menu).

7. From the **File** menu, choose **Save An Entry As**, choose EMBL format, and choose 'no name' to save this entry. At this point, you can give it a more descriptive name.

8. Switch to the entry window containing your analysis. Go to the **File** menu and choose **Read An Entry** to view the manual annotation alongside your own.

9. The manual annotation also includes features other than protein-coding ones. For the purpose of this exercise, we will filter these out. Choose **All CDS Features** from the **Select** menu. Then copy them into a new entry, choosing **New Entry** from the **Create** menu, followed by **Copy Selected Features To** with destination 'no name'. Save this entry as an EMBL format file, giving it a more descriptive name. (A copy of the file that this should produce is given in the Protocol_19 folder as ML_1st_40kb_annotated_CDSs.embl[4.45], but remember that this may not be as up to date as the file you create from the current Sanger annotation.)

10. Scroll along the sequence so that the top pane shows the region 32000 to 39239. This section contains six genes (see *Fig. 4*, also available in the color section). It may help to navigate amongst genes if you right click (or control click) on any feature in the annotation section of the screen and choose **Show gene names** from the pop-up menu.

11. Where gene ML0027 has been annotated, GENEMARK.HMM has predicted two small genes. The BLAST matches versus the *M. tuberculosis* protein set do not go completely across the annotated gene. There was no ORF of the minimum size of 150 amino acids in the region. Look at the note accompanying the gene. It has been annotated as a pseudogene.

12. The annotated gene ML0028, the created ORF, the gene prediction, and the match to the *M. tuberculosis* protein are consistent with each other, differing only by one amino acid at the start of the protein.

13. Annotated gene ML0029 does not have an overlapping hit to a *M. tuberculosis* protein. GENEMARK.HMM has predicted two genes in this region, only separated by three bases. The note in the annotation reveals that it might be a disrupted ancestral CDS. (Use **View Selected Features** from the **View** menu to see the complete annotation for any selected feature.)

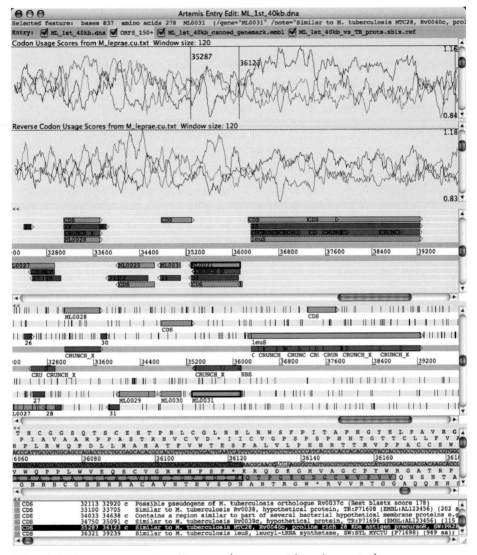

Figure 4. *M. leprae* genome viewed in ARTEMIS (see page xxi for color version).
The top section shows the codon usage plots for the forward (upper) and reverse (lower) strands
in each frame. In the section below this, the tracks immediately above and below the distance
scale show the CDSs of the published annotation (forward and reverse strands). The next tracks
(working outwards from the distance scale) show the BLASTX matches (labeled 'CRUNCH X') to the
M. tuberculosis protein set. The next tracks show the GENEMARK.HMM predictions (numbered) and
then the trimmed ORFs of >150 amino acids (labeled CDS). The lower sections of the screen
show the same information in greater detail, with the annotations listed in the bottom section.

14. ML0030 appears to have been predicted correctly by GENEMARK.HMM. However, no ORF
    for it was marked (its length being only 113 amino acids) nor has a hit been reported to
    a *M. tuberculosis* protein. However, the annotation mentions a significant match to
    M. tuberculosis. Select the gene and choose **View Bases of Selection** from the **View** menu:
    it is clear that the sequence is low complexity (with tracts of predominantly Cs and Gs).

Repeating the BLASTX search (see *Protocol 17*) with the filter for low complexity switched off will result in a match being reported for its region. The composition of the gene reflects its hydrophobic properties.

15. The start of ML0031 does not coincide with that of the trimmed ORF or the GENEMARK.HMM prediction. The match to the *M. tuberculosis* protein is a more accurate indication of its true start. There is a likely ribosome-binding site at complement(36133..36136), which is shown on the six-frame translation in the center pane.

16. ML0032 (LeuS) has been correctly predicted by GENEMARK.HMM.

Although the emphasis in the above is on finding the protein-coding regions of genes, it is important to realize that predictions should be looked at in a broader framework. The ribosome-binding sites of different bacterial mRNAs display two common features: the start codon and within 10 bases upstream of this the Shine–Dalgarno sequence corresponding to part of the hexamer 5'-AGGAGG-3' (44). Identification of likely Shine–Dalgarno sequences linked to a candidate gene can be used as another aid in the gene prediction process. In addition, bacterial mRNAs can be polycistronic containing a cluster of multiple adjacent genes, an operon, controlled as a single genetic unit. Evidence showing that a gene model is part of an operon can help confirm a prediction. Many gene prediction tools make use of these facts.

It is clear in the *M. leprae* example that there are some real genes that are below the minimum ORF size (150 amino acids) specified in the example. The initial cut-off size is used to make the number of potential coding regions to check manageable. If other evidence indicates the presence of a possible gene in a region, the ORF can still be marked at a later stage (using **Mark ORFs in Range** from the **Create** menu).

M. leprae is a particularly difficult genome to perform accurate gene predictions on, because of the high proportion of pseudogenes (degraded gene remnants) that it contains. These contain coding signals but are full of stop codons, frameshifts, and deletions. Certain annotated CDSs have little supporting evidence (e.g. ML0023). In some cases, the annotation has to remain a best guess until new evidence becomes available.

3. TROUBLESHOOTING

Good gene prediction relies both on the judgment of the annotator and on a very wide range of software tools and resources. Below are a number of tips – some general, some specific – that may help to resolve some of the common problems you may encounter.

- As gene prediction is a fast-changing field, tools and the web sites associated with them are not always of a permanent nature. If, after several attempts, a site appears to be missing or broken, it is usually possible to find the new location, or an alternative, with the help of a search engine.

- Output in GFF does not always comply with the endorsed standard. Compare it with the definition (see section 4) to see whether there are any differences. If complete columns need to be taken out, use a spreadsheet program to do this, but save the content as 'Text only'.
- When doing BLAST searches, it is important to use filtering options if your query contains a considerable proportion of low-complexity sequence. Queries searched with the BLASTN program are filtered with DUST. The other BLAST programs use SEG (http://www.ncbi.nlm.nih.gov/blast/blast_FAQs.shtml [4.46]).
- If a predicted gene model seems to be inconsistent with other analyses, inspect it carefully, in particular the boundaries. For eukaryotic gene models, introns should start with GT and end with AG. (Less often, introns can start with GC and very rarely with AT. Introns can also end with AC instead of AG.) In such cases, you can use ARTEMIS to adjust the gene model interactively. Select **Enable direct editing** under **Options** in the ARTEMIS release window if it is not switched on already. Click and drag the last/first base of each exon until they border canonical splice junctions. Furthermore, a prediction may not be on the correct strand. Always look at both orientations when inspecting a prediction.
- If a file does not read into ARTEMIS, try removing any single or double quotes from its content.
- If your BLAST search fails to finish, it could be due to the length of the query sequence. Ensure that it does not exceed 50 kb. To avoid having to adjust coordinates retrospectively, use WUBLAST's nwstart and newlen parameters to select a region of the query. When using this, make sure any complexity filtering (if required) is applied to the full-length sequence first, as the filtering programs do not have the capability to work on sections of the sequence only (38).

4. ADDITIONAL WEB RESOURCES

Tools for finding non-protein-coding genes

- Vienna RNA Package: http://www.tbi.univie.ac.at/~ivo/RNA/ [4.47]. Resources for RNA secondary structure prediction and comparison.
- Sean Eddy's laboratory at Janelia Farm: http://selab.janelia.org/servers.html [4.48]. Several resources for RNA analysis including Rfam.

Comprehensive servers for sequence analysis

- Sourceforge: http://emboss.sourceforge.net/ [4.49]. A suite of computational tools for molecular biology.
- Institute Pasteur biological software: http://bioweb.pasteur.fr/intro-uk.html [4.50]. A suite of web-based computational tools for molecular biology.

Interactive annotation software

- ARTEMIS: http://www.sanger.ac.uk/Software/Artemis/ [4.1]. Java-based sequence viewer and annotation tool with a graphical user interface.

Sequence comparison resources

- EBI BLAST server: http://www.ebi.ac.uk/blastall/[4.22].
- NCBI BLAST server: http://www.ncbi.nlm.nih.gov/BLAST/[4.16].
- Sanger Institute BLAST servers: http://www.sanger.ac.uk/DataSearch/blast.shtml[4.51]. Summary page of BLAST servers at the Wellcome Trust Sanger Institute.
- EST2GENOME: http://liv.bmc.uu.se/cgi-bin/emboss/est2genome[4.19]. Server for aligning an EST to a genomic sequence.
- VISTA: http://genome.lbl.gov/vista/index.shtml[4.52]. Comprehensive suite of programs and databases for comparative analysis of genomic sequences.

Tools for predicting protein signatures

- INTERPPROSCAN: http://www.ebi.ac.uk/InterProScan/[4.27]. Server for comparing a sequence with conserved domains and functional sites in InterPro.

Sequence statistics databases

- Codon Usage Database: http://www.kazusa.or.jp/codon/[4.36]. Codon usage table resource for many organisms.

Programming and scripting resources

- JAVA: http://java.sun.com/j2se/[4.53]. Java software and resources.
- BIOTEAM Mac OS X: http://bioteam.net/MacOSX/[4.54]. Operating system that can accommodate standard bioinformatics tools in a desktop environment.
- BIOPERL: http://bioperl.org/wiki/Main_Page[4.55]. Modular computational biology code base in the Perl language.
- BIOJAVA: http://www.biojava.org[4.56]. Modular computational biology code base in the Java language.

Annotation pipeline software

- BASys: http://wishart.biology.ualberta.ca/basys/cgi/programs_and_dbs.pl[4.57]. Desktop machine-compatible analysis pipeline software.
- PEGASYS: http://bioinformatics.ubc.ca/pegasys/[4.58]. Desktop machine-compatible analysis pipeline software.
- Biomoby: http://www.biomoby.org/[4.59]. Analysis pipeline framework useable in a desktop environment.

Definition standards for formats commonly used in bioinformatics

- GFF specification: http://www.sanger.ac.uk/Software/formats/GFF/GFF_Spec.shtml[4.60]. Definition of GFF specification.
- DDBJ/EMBL/GenBank feature table definition: http://www.ebi.ac.uk/embl/Documentation/FT_definitions/feature_table.html[4.61]. Explanation and definition of these three feature-table formats.

Online catalog of molecular biology databases

- The Molecular Biology Database Collection, 2007 update: http://www.oxfordjournals.org/nar/database/cap/[4.62]. An extensive listing of biological databases with links.

Acknowledgements

I wish to thank the Wellcome Trust for funding, and the following persons for their helpful input and support: Julian Parkhill, Matthew Berriman, Irmtraud M. Meyer, Barclay G. Barrell, and Timothy J. Carver.

5. REFERENCES

★ 1. Bork P (2000) *Genome Res.* **10**, 398–400. – *This is a useful illustration of what to take into consideration when using evidence based on existing functional assignments.*
2. Altschul SF, Gish W, Miller W, Myers EW & Lipman DJ (1990) *J. Mol. Biol.* **215**, 403–410.
3. Burland TG (2000) *Methods Mol. Biol.* **132**, 71–91.
4. Gorelenkov V, Antipov A, Lejnine S, Daraselia N & Yuryev A (2001) *BioTechniques*, **31**, 1326–1330.
5. Rastogi PA (2000) *Methods Mol. Biol.* **132**, 47–69.
6. Van Domselaar GH, Stothard P, Shrivastava S, *et al.* (2005) *Nucleic Acids Res.* **33**, W455–W459.
7. Shah SP, He DY, Sawkins JN, *et al.* (2004) *BMC Bioinformatics*, **5**, 40.
8. Wilkinson M, Schoof H, Ernst R & Haase D (2005) *Plant Physiol.* **138**, 5–17.
9. Krogh A (1997) *Proc. Int. Conf. Intell. Syst. Mol. Biol.* **5**, 179–186.
10. Parra G, Blanco E & Guigo R (2000) *Genome Res.* **10**, 511–515.
11. Hebsgaard SM, Korning PG, Tolstrup N, Engelbrecht J, Rouze P & Brunak S (1996) *Nucleic Acids Res.* **24**, 3439–3452.
12. Borodovsky M & McIninch J (1993) *BioSystems*, **30**, 161–171.
13. Dong S & Searls DB (1994) *Genomics*, **23**, 540–551.
14. Korf I (2004) *BMC Bioinformatics*, **5**, 59.
15. Majoros WH, Pertea M & Salzberg SL (2004) *Bioinformatics*, **20**, 2878–2879.
16. Salamov AA & Solovyev VV (2000) *Genome Res.* **10**, 516–522.
17. Stanke M, Schoffmann O, Morgenstern B & Waack S (2006) *BMC Bioinformatics*, **7**, 62.
18. Xu Y & Uberbacher EC (1997) *J. Comput. Biol.* **4**, 325–338.
19. Parra G, Agarwal P, Abril JF, Wiehe T, Fickett JW & Guigo R (2003) *Genome Res.* **13**, 108–117.
20. Milanesi L, D'Angelo D & Rogozin IB (1999) *Bioinformatics*, **15**, 612–621.
21. Korf I, Flicek P, Duan D & Brent MR (2001) *Bioinformatics*, **17** (Suppl. 1), S140–S148.
22. Majoros WH, Pertea M & Salzberg SL (2005) *Bioinformatics*, **21**, 1782–1788.
23. Meyer IM & Durbin R (2004) *Nucleic Acids Res.* **32**, 776–783.
24. Meyer IM & Durbin R (2002) *Bioinformatics*, **18**, 1309–1318.
25. Huang X, Adams MD, Zhou H & Kerlavage AR (1997) *Genomics*, **46**, 37–45.
26. Birney E, Clamp M & Durbin R (2004) *Genome Res.* **14**, 988–995.
27. Yeh RF, Lim LP & Burge CB (2001) *Genome Res.* **11**, 803–816.
28. Pedersen JS & Hein J (2003) *Bioinformatics*, **19**, 219–227.
29. Alexandersson M, Cawley S & Pachter L (2003) *Genome Res.* **13**, 496–502.
30. Frishman D, Mironov A, Mewes HW & Gelfand M (1998) *Nucleic Acids Res.* **26**, 2941–2947.
31. Delcher AL, Harmon D, Kasif S, White O & Salzberg SL (1999) *Nucleic Acids Res.* **27**, 4636–4641.

32. **Larsen TS & Krogh A** (2003) *BMC Bioinformatics*, **4**, 21.
33. **Besemer J, Lomsadze A & Borodovsky M** (2001) *Nucleic Acids Res.* **29**, 2607–2618.
34. **Tyson GW, Chapman J, Hugenholtz P, et al.** (2004) *Nature*, **428**, 37–43.
35. **Rutherford K, Parkhill J, Crook J, et al.** (2000) *Bioinformatics*, **16**, 944–945.
36. **Sonnhammer EL & Durbin R** (1994) *Comput. Appl. Biosci.* **10**, 301–307.
37. **Frazer KA, Pachter L, Poliakov A, Rubin EM & Dubchak I** (2004) *Nucleic Acids Res.* **32**, W273–W279.
★ 38. **Korf I, Yandell M & Bedell J** (2003) BLAST – *An Essential Guide to the Basic Local Alignment Search Tool.* O'Reilly & Associates Inc., Sebastopol, CA. – *A thorough, step-by-step explanation of the principles that the* BLAST *method is based on. It provides clear guidance on how to use this tool correctly and how to evaluate the results accurately. It also includes protocols for types of comparison for which* BLAST *is commonly used.*
39. **Mott R** (1997) *Comput. Appl. Biosci.* **13**, 477–478.
40. **Quevillon E, Silventoinen V, Pillai S, et al.** (2005) *Nucleic Acids Res.* **33**, W116–W120.
41. **Bairoch A, Apweiler R, Wu CH, et al.** (2005) *Nucleic Acids Res.* **33**, D154–D159.
42. **Loveland J** (2005) *Brief. Bioinform.* **6**, 189–193.
43. **Cole ST, Eiglmeier K, Parkhill J, et al.** (2001) *Nature*, **409**, 1007–1011.
44. **Lewin B** (1997) *Genes VI.* Oxford University Press, Oxford, UK.

CHAPTER 5

Prediction of noncoding transcripts

Alex Bateman and Sam Griffiths-Jones

---

1. INTRODUCTION

The central dogma states that DNA is transcribed into RNA, which is then translated into protein. However, for many genes, the RNA, rather than a translated protein, is the functional product. Well-known examples of these functional RNA genes include tRNAs and rRNAs; further examples of RNA genes are given in *Table 1*. Like DNA, RNA is made up of four nucleotides: adenine (A), cytosine (C), guanine (G), and uracil (U) (which replaces thymine used in DNA). Whilst DNA usually forms the classic double-helical structure, RNA can adopt a much wider range of shapes. Functional RNAs form complex secondary structures composed of single- and double-stranded regions. The double-stranded regions are formed by intramolecular base pairs including the classic Watson–Crick GC and AU pairs and also a number of possible non-Watson–Crick base pairs of which GU is by far the most important. A particular feature of RNA genes is that their primary sequence is often poorly conserved, whilst their secondary structure is maintained by compensatory mutations in the helical regions. For example, a mutation from C to U is often seen with a compensatory mutation of G to A in the base pairing position as shown in *Fig. 1*. This phenomenon is also known as *covariation* and is exploited by many tools for the detection and analysis of RNA genes. Covariation has been used by experts to predict correctly the secondary structure of noncoding transcripts when enough sequences are available (1).

Computational prediction of protein-coding genes is a mature field (see Chapter 4). However, this is not the case for noncoding RNA genes, which for many years have been regarded as curiosities by the gene prediction community. This is largely because they lack many of the signals that make protein gene prediction tractable. For example, RNA genes are not translated, so they do not have a start, stop, or any other type of codon. It was only in 2001 that the first large-scale computational screens for noncoding RNA genes in bacteria were performed (2–5). Many of these approaches are motivated by the idea that there exist many important RNAs that are yet to be discovered. Discoveries of a wealth of small nucleolar RNAs (snoRNAs) and microRNAs (miRNAs) that had lain undetected

Bioinformatics: *Methods Express* (Paul H. Dear, ed.)
© Scion Publishing Limited, 2007

Table 1. Examples of vertebrate RNA gene diversity

Function	RNA class
Protein synthesis	tRNA
	SSU (18S) rRNA
	5.8S rRNA
	LSU (28S) rRNA
	5S rRNA
Splicing	U1
	U2
	U4
	U4atac
	U5
	U6
	U6atac
	U11
	U12
Telomere maintenance	Telomerase RNA
Translational regulation	miRNA
RNA processing	RNAse P
	RNAse MRP
	C/D snoRNA
	H/ACA snoRNA
X chromosome inactivation	Xist
Protein export	SRP (7SL) RNA
Unknown	Y RNA
	vRNA
	7SK
	H19
	mRNA-like ncRNAs

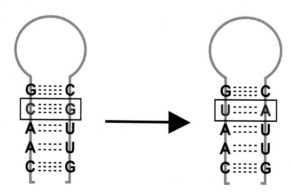

Figure 1. Example of a compensatory mutation in an RNA hairpin.
A mutation of C to U is compensated by a G to A mutation, preserving the base pairing of the hairpin.

for many years (6, 7) suggest that we are far from characterizing the complete complement of RNA genes.

Both computational and experimental approaches have been used to annotate known and novel noncoding RNAs in genomes. Notably, Sean Eddy and Tom Jones enumerated all known human RNA genes found in the literature and used BLAST to identify homologs of these in the draft human genome sequence (8). This study located 706 putative RNA genes, but also several thousand related matches that could be pseudogenes. More recently, the RNAZ program (9) has been used to estimate that there are around 35 000 RNA genes and other structured elements in the human genome (10). Experimental approaches have also been used to survey vertebrate genomes for RNA gene diversity. Genome tiling microarrays suggest that the majority of genomic DNA is transcribed (11) and recent work on the FANTOM3 project suggests that there are between 3600 and 34 000 noncoding RNA transcripts (albeit overlapping in some cases) in mouse (12). Thus, increasing evidence surprisingly suggests that there may be as many noncoding RNAs in the genome as protein-coding genes.

2. METHODS AND APPROACHES

2.1 *Ab initio* versus family-specific searches

In the context of complete genomes, there are two different approaches for identifying structured RNA genes:

- *Ab initio* detection of novel structured RNAs
- Detection of members of known classes of RNA gene

Several tools have been designed for *ab initio* detection of RNA structure including QRNA (3) and RNAZ (9). The use of such tools in recent comparative whole-genome studies has predicted that around 35 000 regions of the human genome (10), and over 3500 regions of the worm genome (13) have patterns of conservation that suggest coding for structured RNAs. However, the relationship between such computational predictions and actual expression and biological function is poorly understood. This approach is therefore currently of limited use for RNA annotation on a genome-wide scale. Tools for *ab initio* RNA gene prediction are not discussed further here.

Detecting homologs of known RNA classes is more straightforward. Computationally, there are two main approaches to this problem:

- General approaches to find homologs of any given RNAs, based largely on sequence similarity
- Specific tools to identify members of specific RNA classes, based largely on specific features characteristic of each class

Both approaches aim to identify regions of genomes that share significant similarity with a known RNA gene or gene class. The simplest 'general' approach is to use sequence similarity methods such as BLAST to search known RNA genes against a sequence. This approach is significantly limited by two factors. First, sequence similarity alone is ineffective at recognizing structured RNA homologs. Secondly, databases cataloging known RNA genes have only recently begun to emerge (NONCODE, 14; RNAdb, 15), and there is as yet no comprehensive RNA equivalent of the Swiss-Prot database for proteins.

Several tools have been developed to detect members of specific classes of RNA. For example, TRNASCAN-SE is a highly sensitive and specific method for computational prediction of tRNA genes (16). Other tools include SRPSCAN to identify signal recognition particle RNA (17), SNOSCAN to predict snoRNAs (6), BRUCE to identify tmRNA in bacterial genomes (18), and a number of methods to predict miRNAs (19–21). The advantages of such approaches are clear: the search for family-specific signatures allows heuristics that can be fast, sensitive, and specific. These methods therefore produce the most accurate predicted RNA gene sets on a whole-genome scale. However, development of a novel tool for every class of RNA is labor-intensive and thus cannot adapt well to rapid discovery of novel RNA families.

2.2 Web servers for the detection of single, specific RNA classes

Protocols 1–4 briefly describe the use of some tools designed to detect specific RNA classes. Fortunately, web servers are available for each of these methods, allowing access without the need to download and install any software. We present protocols for TRNASCAN-SE (16), BRUCE (18), and SRPSCAN (17). All of the methods presented have online documentation, which should be referred to alongside this text. It is important to note that the methods described predict RNA genes from sequences and structure information alone and may not be sufficient to describe a *bona fide* functional RNA. Each tool employs a different scoring scheme and the default thresholds should be appropriate. However, the user may find it interesting to retrieve low-scoring matches, for analysis with caution.

2.2.1 Detection of tRNA genes

Typical tRNAs share well-conserved features, making their identification robust in most cases. TRNASCAN-SE is one of the most widely used tools for their prediction.

Protocol 1

Identification of tRNA genes using TRNASCAN-SE (default mode)

1. From the Protocol_1 folder for this chapter on the book's web site, retrieve the file sampleYeastTrna.fa [5.1]. This FASTA file contains a set of five yeast sequences.

2. Go to the TRNASCAN-SE search server at http://lowelab.ucsc.edu/tRNAscan-SE/ [a 5.2].

3. Under **Format**, select **Other**.

4. Paste the complete contents of the file sampleYeastTrna.fa [5.1] into the query sequence window. Alternatively, you can use the **Choose file** (or **Browse**) button to upload the file.

5. The other settings can be left at their default values. This will give a fast and sensitive search in most cases (but see *Protocol 2*).

6. Click **Run tRNAscan-SE**.

7. The results page shows that five tRNAs have been found (one on each of the five sequences in the FASTA file) and gives details of their locations in the input sequences and of their cognate amino acids, anticodons, and introns. Note that the last of the five tRNAs contains an intron from bases 38 to 60. The table also gives the score of the tRNA: any value above 20 is recognized as a tRNA, and the scores in this example are all well above this lower limit.

8. Scrolling down the screen, you will see statistics for the search, and then, under **Summary**, a summary of the results including an anticodon table detailing the numbers of tRNAs found with each possible anticodon.

9. Further down, the predicted secondary structure of each of the tRNAs is shown. The base-paired stems of the structure are shown by nested sets of arrowheads underneath the sequence, with the unpaired bases of the loops as dots (>>>>>........<<<<<). Any bases that do not conform to the consensus model (including introns) are indicated in lower case.

10. For a graphical representation of the secondary structure, scroll back up to the results table and click the **View tRNA** button next to any one of the predicted tRNAs. A new window opens, displaying the predicted stem–loop structure of the tRNA; Watson–Crick base pairs are indicated by red dots and other base pairs (such as GU) by blue dots.

Note

[a]TRNASCAN-SE is highly sensitive and specific. The web server works well for up to 100 kb of nucleotide sequence. If you wish to search larger sequences, such as a vertebrate chromosome, then you should download and install a local copy of TRNASCAN-SE software from http://selab.janelia.org/software.html [5.3]. This software runs on Unix-based operating systems.

The default options of TRNASCAN-SE give a fast and sensitive search. However, there are a number of unusual tRNAs that will not be found with the default method. These can sometimes be detected by using the Cove-only option, although this is extremely slow and should only be tested on very short sequences. As an example, the following protocol shows the use of the Cove-only option to identify an unusual archaebacterial pyrrolysine-encoding tRNA.

Protocol 2

Identification of tRNA genes using TRNASCAN-SE (Cove-only mode)

1. From the Protocol_2 folder for this chapter, download the file MbarkeriTrnaPyl.fa [5.4], which contains the pyrrolysine-encoding tRNA from *Methanosarcina barkeri*.

2. Following the same steps as in *Protocol 1*, run TRNASCAN-SE on this sequence using the default options; no tRNAs are found.

3. Now repeat the search, but this time with the Search Mode set to **Cove only**, and the Source set to **Mixed (general tRNA model)**.

4. The search will take several moments, but will then find a tRNA with a Cove score of 30.2. In this case, TRNASCAN-SE identifies the tRNA as a likely pseudogene.

2.2.2 Detection of tmRNA genes

tmRNAs are bacterial RNA molecules that share features of both tRNAs and mRNAs. They cause the addition of a short 'tag' peptide to unfinished proteins in stalled ribosomes, freeing the ribosome and marking the truncated protein for degradation. *Protocol 3* illustrates the use of a web server to identify tmRNAs.

Protocol 3

Identification of tmRNA genes using ARAGORN

1. From the Protocol_3 folder for this chapter on the book's web site, download the file EcoliSampleRegion.fa [5.5]. This FASTA file contains approximately 12 kb of genomic sequence from *Escherichia coli* O157:H7.

2. Go to the ARAGORN server[a] at https://pcmbioekol-bioinf2.mbioekol.lu.se/ARAGORN1.1/HTML/ [5.6].

3. Using the **Choose file** button (or **Browse**, as it appears in some browsers), upload the EcoliSampleRegion.fa [5.5] file[b].

4. Under **Select options**, set the **Search for** menu to **tmRNA**. The other settings can be left at their defaults (not allowing introns; searching only for linear topology; searching both strands; and giving a standard output) in this case[c].

5. Click the **Submit query** button.

6. The output page reveals that one tmRNA has been found. The secondary structure of the tRNA domain of the tmRNA is shown (with Watson–Crick base pairs displayed as '-' and '!', and non-Watson–Crick basepairs as '+'). Below this, details of the tmRNA sequence, and of the tag peptide that it will add to stalled proteins, are given.

Notes

[a]There is a second server, BRUCE, at https://pcmbioekol-bioinf2.mbioekol.lu.se/ARAGORN1.1/HTML/bruceindex.html [5.7]. The ARAGORN server provides the most recent version of the software and, unlike BRUCE, can also be used to predict tRNAs.
[b]The maximum sequence length for the web server is 15 Mb of nucleotide sequence. Larger sequences, such as a vertebrate chromosome, can be submitted in overlapping segments of 15 Mb. The software is also available for local installation from the web server page.
[c]This search may fail to find some rare circularly permuted tmRNA genes such as are found in, for example, some alphaproteobacteria. Change the sequence topology from linear to circular to find such tmRNAs. In this case, two results may be reported for each match, one permuted and one nonpermuted.

2.2.3 Detection of signal recognition particle (SRP) RNA genes

SRP RNA is a component of the signal recognition particle, responsible for guiding ribosomes that are synthesizing secreted proteins to membranes. SRP RNAs are large and have a complex secondary structure. *Protocol 4* illustrates the use of SRPSCAN to find SRP RNA genes.

Protocol 4

Identification of SRP RNA genes using SRPSCAN

1. From the Protocol_4 folder for this chapter on the book's web site, download the file MjannaschiiSampleRegion.fa [5.8], which contains about 5 kb of genomic sequence from *Methanocaldococcus jannaschii* DSM 2661.

2. Go to http://bio.lundberg.gu.se/srpscan/ [5.9].

3. The default option is for a fast scan (allowing no mismatches) and should be tried first. Under **Select organism/RNA structure subtype**, select **Archaebacteria** (as this is an archaebacterial sequence).

4. Either paste[a] the contents of MjannaschiiSampleRegion.fa [5.8] into the **Input sequence** window, or use the **Choose file** (or **Browse**) button to upload the file.

5. Click **Submit**. The search will take about a minute.

6. The search finds one SRP RNA gene, details of which are reported on the output page. Buttons to the right display the predicted secondary structure using either of two folding models. The **COVES** option finds two possible secondary structures, differing only slightly (see *Fig. 2*), and the graphic files can be downloaded. (A copy of the first structure produced using the COVES folding model is given in the Protocol_4 folder as MjannaschiiSampleRegion.pdf [5.10]).

7. Further test sequences are available by following the **Test sequences** link underneath the input window on the search page; any one of these can be copied and pasted into the input. Note that it is important to select the appropriate option under **Select organism/RNA structure subtype**; for some organisms (eubacteria), there are several subtypes of SRP RNA that can be chosen – if one does not yield results, try the others.

Note

[a]Depending on your operating system and the text processor you use, you may find that one or two 'empty' lines (carriage return or paragraph marks) are added to the end of the sequence when you paste it into the text window of SRPSCAN. In this case, you will get an error message indicating that analysis was not possible. If this happens, use your browser's 'back' button to return to the input page and delete the carriage-return characters at the end of the pasted text before proceeding.

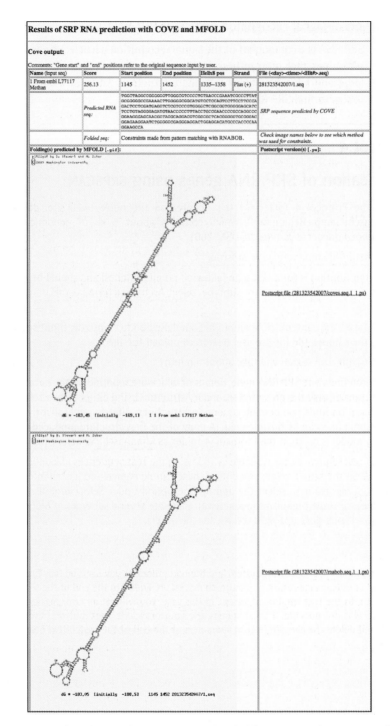

Results of SRP RNA prediction with COVE and MFOLD

Cove output:

Comments: "Gene start" and "end" positions refer to the original sequence input by user.

Name (input seq)	Score	Start position	End position	Helix8 pos	Strand	File (<day>-<time>-<Hit#>.seq)
1 From embi L77117 Methan	256.13	1145	1452	1335--1358	Plus (+)	281323542007/1.seq
	Predicted RNA seq:	TGGCTAGGCCGGGGCGTTGGGCGTCCCCTGTAACCCGAAATCGCCCTTAT GCGGGGGCCGAAAACTTGGGGGCGGCATGTGCCTCCAGTGCGTTCCTTCCCA GACTCCTCGATCAGCTCTCGTCCCGTGGGGCTCGGCGGTGGGGGAGCATC TCCTGTAGGGGAGATGTAACCCCCTTACCTGCCGAACCCGCCAGGCCC GGAAGGGAGCAACGGTAGGCAGGACGTCGGCGCTCACGGGGTTGCGGGAC GGACAAGGAATCTGCGGCCGAGGGAGCACTGAGGCACATGCCCACCCCAA GGAAGCCA	*SRP sequence predicted by COVE*
	Folded seq:	Constraints made from pattern matching with RNABOB.				*Check image names below to see which method was used for constraints.*

Folding(s) predicted by MFOLD [.gif]:	**Postscript version(s) [.ps]:**
plt22gif by D. Stewart and M. Zuker (c)2003 Washington University	Postscript file (281323542007/coves.seq.1_1.ps)
dG = -163.45 [initially -169.11 1 1 From embl L77117 Methan]	
plt22gif by D. Stewart and M. Zuker (c)2003 Washington University	Postscript file (281323542007/rnabob.seq.1_1.ps)
dG = -103.05 [initially -108.53 1145 1452 281323542007/1.seq]	

Figure 2. Structural predictions for an SRP RNA identified by srpscan in the *M. jannaschii* genomic sequence.

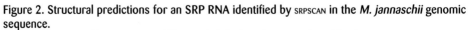

Note that two possible structures, differing slightly, have been predicted for this SNP RNA.

2.3 Web servers for the prediction of multiple RNA classes

More general approaches have been made available through the advent of databases of RNA families such as ERPIN (22) (http://tagc.univ-mrs.fr/erpin/ [5.11]) and Rfam (23, 24) (http://www.sanger.ac.uk/Software/Rfam/ [5.12] and http://rfam.janelia.org/ [5.13]). These resources aim to provide a general approach to identify homologs of all known classes of RNA, thus allowing users to search genomes for a wide range of RNA genes. Both resources are based on multiple sequence alignments of RNA genes, with secondary structure annotation. ERPIN builds a statistical secondary structure profile (SSP) from the alignment, whilst Rfam uses the INFERNAL package (Eddy 2002, http://infernal.janelia.org [5.14]) to build a covariance model or stochastic context free grammar (SCFG). Both SSPs and covariance models can be used to recognize homologs of the RNA family of interest in a given sequence, including a complete genome. The advantage of both general approaches is that models of novel families of RNA can easily be distributed electronically for the purpose of genome annotation, as in the Rfam and ERPIN databases. These general approaches form the first automated methods to annotate RNAs in whole genomes.

Protocols 4 and *5* detail the use of two RNA family databases to find homologs of a wide range of RNA genes in short sequences. These tools can also be downloaded to annotate larger sequences such as complete genomes, but this requires considerable computing resources and should be considered more advanced than the protocols presented here. Please refer to the online database documentation for information on how to do this.

2.3.1 Searching against the Rfam database

Protocol 5

Identification of RNA genes using Rfam

1. From the Protocol_5 folder for this chapter on the book's web site, download the file HsapiensSampleRegion.fa [5.15], which contains 2 kb of human genomic sequence in FASTA format.

2. Go to the Rfam web server at http://www.sanger.ac.uk/Software/Rfam/search.shtml [5.16].

3. Paste the contents of HsapiensSampleRegion.fa [5.15] into the **New search** window.

4. Click **Search Rfam** and follow the prompts to await and retrieve the results (the search will take one or two minutes).

5. On the results page (the top of which is shown in *Fig. 3*), you will see a table showing that five microRNAs were found. Links within this table (under **Family**) take you to more information about the respective RNA families.

6. Below the table is shown the alignment of each of the predicted RNAs with its respective consensus sequence. Arrowheads (< and >) indicate paired regions in the predicted secondary structure, and a more detailed description of the results can be found at http://www.sanger.ac.uk/Software/Rfam/help/gstart.shtml [5.17].

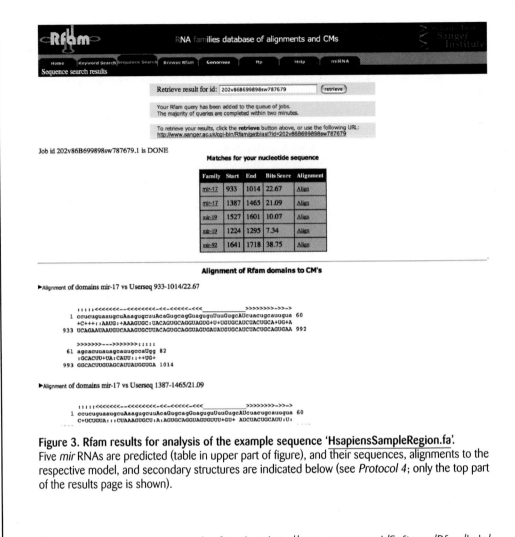

Figure 3. Rfam results for analysis of the example sequence 'HsapiensSampleRegion.fa'.
Five *mir* RNAs are predicted (table in upper part of figure), and their sequences, alignments to the respective model, and secondary structures are indicated below (see *Protocol 4*; only the top part of the results page is shown).

7. Further example sequences can be found at http://www.sanger.ac.uk/Software/Rfam/help/ samples.fa [5.18]; copy and paste any one of them (or, of course, your own sequence[a] – either as FASTA or as pure sequence) into the **New search** window. The input is limited to one sequence of ≤2 kb.

8. You can also try using this protocol to analyze the example sequences used in *Protocols 1–4*. You will see some differences in the results, reflecting differences in the models and search methods used by the different programs.

9. Note that Rfam makes results of many complete genomes available, so check whether it has already been pre-calculated (see *Fig. 4*).

The representation of the chromosome is split into contigs as defined by EMBL annotation. The genomic positions of Rfam hits are shown by black ticks - those on the outside are on the forward strand, and those on the inside are on the reverse strand. Ticks on the scale bar are at 100,000kb intervals. Mouse-over the contigs or entries in the table below to highlight classes of Rfam hits. Click the family names to view details.

Rfam family	Type	Count	View
tRNA	Gene;tRNA;	86	☐
T-box	Cis-reg;	18	☐
S_box	Cis-reg;riboswitch;	11	☐
SSU_rRNA_5	Gene;rRNA;	10	☐
5_8S_rRNA	Gene;rRNA;	10	☐
5S_rRNA	Gene;rRNA;	10	☐
Purine	Cis-reg;riboswitch;	5	☐
THI	Cis-reg;riboswitch;	5	☐
ykkC-yxkD	Cis-reg;riboswitch;	2	☐
ydaO-yuaA	Cis-reg;riboswitch;	2	☐
Lysine	Cis-reg;riboswitch;	2	☐
yybP-ykoY	Cis-reg;riboswitch;	2	☐
RFN	Cis-reg;riboswitch;	2	☐
Intron_gpI	Intron;	2	☐
gcvT	Cis-reg;riboswitch;	1	☐
ykoK	Cis-reg;riboswitch;	1	☐
glmS	Cis-reg;riboswitch;	1	☐

Figure 4. Pre-calculated results available for complete genomes using Rfam.
See http://www.sanger.ac.uk/cgi-bin/Rfam/genome_dist.pl [5.21].

Note

[a]The web server upload is limited by sequence length – currently to 2 kb. If you need to search larger sequences, including whole genomes, for RNA homologs, then you should download the Rfam flatfiles (http://www.sanger.ac.uk/Software/Rfam/ftp.shtml [5.19]) and the INFERNAL software package (http://infernal.wustl.edu/ [5.14]) for local searching. Help and a convenient wrapper script for such searches are available from http://www.sanger.ac.uk/Software/Rfam/help/software.shtml [5.20]. The Rfam/INFERNAL approach to annotate RNA homologs in higher eukaryotes is limited by the computational complexity and the prevalence of RNA-derived pseudogenes and repeats. See section 3 for more details.

2.3.2 ERPIN

ERPIN behaves somewhat differently from the Rfam search site in that it requires you to specify at the outset which type of RNA gene you want to search for. This makes it more difficult to search for 'anything' in a nucleotide sequence, but it also helps to make the searches faster and helps to enable ERPIN to analyze larger sequences than the Rfam search (up to 5 Mb); it can also accept files containing multiple sequences.

Protocol 6

Identification of RNA genes using ERPIN

1. Go to the ERPIN web server at http://tagc.univ-mrs.fr/erpin/[5.11] and click on **Online search**.

2. Into the **Select database** field, paste the contents of the file HsapiensSampleRegion.fa[5.15] (see *Protocol 5*), or use the **Choose file** (or **Browse**) button to upload the file.

3. From the list of motifs on the left, select **miRNA, miR-17-18-20-93-106 family**. Information on this RNA family appears under **RNA Information** on the right; in particular, note the **specificity** line: this indicates how many instances of this RNA family might be predicted per 100 Mb of random sequence. In this instance, only about 12 such 'spurious' hits would be expected per 100 Mb, so any strong hits found in this 2 kb of human genomic sequence are likely to be significant.

4. Click **Run Erpin**. The results page shows that three RNAs in this group have been predicted. For each result, a score and *E* value are given: the *E* value indicates how likely it is that a hit with that score or above would be found by chance. (Lower *E* values, therefore, are more likely to represent significant results.)

5. For each of the predicted RNAs, **draw** buttons will display the predicted secondary structure and the adjacent **save** button will download the corresponding graphics file to your computer.

6. You are encouraged to search this and other sequence files (including those used as inputs for the earlier protocols in this chapter) for a variety of RNA genes using ERPIN. Note that ERPIN's predictions will differ in detail from those of other tools, reflecting differences in the methods and in the underlying models and assumptions on which they depend.

3. TROUBLESHOOTING

3.1 RNA-derived repeats and pseudogenes

A number of eukaryotic repeat elements are evolutionarily derived from specific RNA genes. For example, the B2 repeat in rodents is derived from a single tRNA gene and an SRP RNA is the ancestor of the Alu repeat in human. These relationships significantly complicate the detection of RNA homologs in higher eukaryotes. For example, the best computational tool for identifying tRNAs in complete genomes, TRNASCAN-SE, reports around 25000 predictions in mouse (25) and over 175000 in rat (26). Some classes of RNA genes (often those transcribed by RNA polymerase III) appear to be especially prone to pseudogenization. For example, the Rfam database reports over 800 homologs of the U6 spliceosomal RNA in human, along with over 700 copies of Y RNA, a component of the Ro complex. In these cases, the number of derived pseudogenes overwhelms the expected number of functional RNA copies, but current methods are unable to distinguish between the 'real' genes and their recent pseudogene copies. Automated RNA analysis tools such as those described here should therefore be used cautiously for analyzing complete eukaryotic genomes. For some methods, scores can be indicators of how reliable predictions are. The best possible set of RNA gene predictions involves combining and analyzing the results of several different software packages.

3.2 Computational complexity

The use of current statistical models of RNA sequence and structure to annotate homologs of known RNAs in genomes is a computationally demanding task. In particular, the covariance models employed by the Rfam/INFERNAL approach suffer from $O(n^3)$ time complexity with respect to the length of the RNA. This means that doubling the length of the RNA of interest entails an eightfold increase in the length of time required to search for homologs. This quickly explodes such that it is not computationally feasible to use covariance models to search large eukaryotic genomes for large RNAs such as large-subunit ribosomal RNA. A number of heuristic simplifications allow the problem to be tackled in sensible time, including the ERPIN approach, which allows no inserts or deletions in helical regions. Such heuristics are likely to give rise to significant (but often unknown) decreases in sensitivity.

4. REFERENCES

1. **Gutell RR, Lee JC & Cannone JJ** (2002) *Curr. Opin. Struct. Biol.* **12**, 301–310.
2. **Wassarman KM, Repoila F, Rosenow C, Storz G & Gottesman S** (2001) *Genes Dev.* **15**, 1637–1651.
3. **Rivas E, Klein RJ, Jones TA & Eddy SR** (2001) *Curr. Biol.* **11**, 1369–1373.
4. **Argaman L, Hershberg R, Vogel J, et al.** (2001) *Curr. Biol.* **11**, 941–950.
5. **Carter RJ, Dubchak I & Holbrook SR** (2001) *Nucleic Acids Res.* **29**, 3928–3938.
6. **Lowe TM & Eddy SR** (1999) *Science*, **283**, 1168–1171.
7. **Ambros V** (2001) *Cell*, **107**, 823–826.
★ 8. **Lander ES, Linton LM, Birren B, et al.** (2001) *Nature*, **409**, 860–921. – *Includes a comprehensive analysis of noncoding RNA genes in the human genome.*
9. **Washietl S, Hofacker IL & Stadler PF** (2005) *Proc. Natl. Acad. Sci. U.S.A.* **102**, 2454–2459.
★ 10. **WashietlS, Hofacker IL, Lukasser M, Huttenhofer A & Stadler PF** (2005) *Nat. Biotechnol.* **23**, 1383–1390. – *In our experience, this is the best current method for* ab initio *prediction of structural RNA elements.*
11. **Kampa D, Cheng J, Kapranov P, et al.** (2004) *Genome Res.* **14**, 331–342.
12. **Carninci P, Kasukawa T, Katayama S, et al.** (2005) *Science*, **309**, 1559–1563.
13. **Missal K, Zhu X, Rose D, et al.** (2006) *J. Exp. Zoolog. B Mol. Dev. Evol.* **306**, 379–392.
14. **Liu C, Bai B, Skogerbo G, et al.** (2005) *Nucleic Acids Res.* **33**, D112–D115.
15. **Pang KC, Stephen S, Engstrom PG, et al.** (2005) *Nucleic Acids Res.* **33**, D125–D130.
16. **Lowe TM & Eddy SR** (1997) *Nucleic Acids Res.* **25**, 955–964.
17. **Regalia M, Rosenblad MA & Samuelsson T** (2002) *Nucleic Acids Res.* **30**, 3368–3377.
18. **Laslett D, Canback B & Andersson S** (2002) *Nucleic Acids Res.* **30**, 3449–3453.
19. **Lim LP, Lau NC, Weinstein EG, et al.** (2003) *Genes Dev.* **17**, 991–1008.
20. **Lai EC, Tomancak P, Williams RW & Rubin GM** (2003) *Genome Biol.* 4, R42.
21. **Adai A, Johnson C, Mlotshwa S, et al.** (2005) *Genome Res.* **15**, 78–91.
22. **Lambert A, Fontaine JF, Legendre M, et al.** (2004) *Nucleic Acids Res.* **32**, W160–W165.
23. **Griffiths-Jones S, Bateman A, Marshall M, Khanna A & Eddy SR** (2003) *Nucleic Acids Res.* **31**, 439–441.

★ 24. **Griffiths-Jones S, Moxon S, Marshall M, Khanna A, Eddy SR & Bateman A** (2005) *Nucleic Acids Res.* **33**, D121–D124. – *Currently the most comprehensive collection of RNA genes and families.*

  25. **Waterston RH, Lindblad-Toh K, Birney E,** *et al.* (2002) *Nature,* **420**, 520–562.

  26. **Gibbs RA, Weinstock GM, Metzker ML,** *et al.* (2004) *Nature,* **428**, 493–521.

CHAPTER 6

Finding regulatory elements in DNA sequence

Debraj GuhaThakurta and Gary D. Stormo

1. INTRODUCTION

1.1 Background

Identification of all of the functional elements in genomes, including genes and regulatory elements, is a fundamental challenge now that the complete genomic sequences of a number of prokaryotes and eukaryotes are available. Of the ~5% of the mammalian genome that is estimated to be under evolutionary selection pressure, less than a third is coding (1, 2). The remainder is suggested to belong to untranslated regions, noncoding genes, chromosomal structural elements, and regulatory elements that control a variety of biological processes. However, in contrast to the computational and experimental advances made in identifying protein-coding sequences, identification of genomic regulatory elements still largely remains an unsolved problem. Whereas the number of coding genes in many of the sequenced prokaryotes and eukaryotes can now be reasonably estimated, there is no clear estimate of the number of functional regulatory elements in these genomes, especially in multicellular eukaryotes. The number of coding genes in eukaryotes ranging from *Drosophila* or *Caenorhabditis elegans* to humans is similar (ranging from ~14 000 to ~29 000) (2–5), and it is now thought that organismal complexity may be attributed to phenomena such as alternative splicing, DNA rearrangement, and an increased number of transcription regulatory elements and transcription factors (TFs) (6). The identification of *cis*-regulatory elements controlling gene expression, and the TFs that bind to these elements, thus lie not only at the very heart of elucidating the network of gene interactions at the cellular level, but also of explaining the origins of organismal complexity and development.

It is estimated that there are roughly 2000 TFs in mammalian genomes (2, 4) and about 1000 in *Drosophila* and *C. elegans* (6). However, for only a fraction of these TFs (about 900 in human, 700 in mouse, 200 in *Drosophila*, and 100 in

Bioinformatics: *Methods Express* (Paul H. Dear, ed.)
© Scion Publishing Limited, 2007

C. elegans) is information about DNA binding or involvement in a DNA–protein complex (7) available. Currently, DNA-binding-site models are available for about 500 vertebrate TFs, and fewer than 5000 genomic sites are known in all vertebrates in fewer than 3000 genes. Therefore, identifying novel regulatory elements and deciphering their function is likely to be an important research area in genomics for years to come.

In prokaryotes and simpler eukaryotes such as yeast, these regulatory elements are mostly present in the core promoter region consisting of several hundred base pairs immediately upstream of the transcription start site. However, in multicellular eukaryotes, the regulatory elements can be present in the distant 5′ or 3′ regions as well as within introns (8). In *Drosophila*, such elements can be spread over a region of 10 kb around the genes, whereas the average transcribed DNA is 2–3 kb; in mammals these elements can be scattered over hundreds of kilobases (9). In recent years, comparative sequence analysis (phylogenetic footprinting) has been used successfully to identify sequence elements of conserved regulatory function (10, 11). Despite the utility of cross-species comparisons, the identification and modeling of the individual *cis*-regulatory elements and the binding sites for the regulatory proteins (TFs) remains a challenging problem. This is because TF-binding sites (TFBSs) are often short (6–25 nt) sequences and the TFs can tolerate significant degeneracy in their target sites (12). This chapter is devoted to the methods available for identification and detection of DNA sequence motifs representing transcriptional regulatory elements in the genome.

1.2 An overview of progress in the computational identification of DNA sequence motifs

Computational methods for identifying and modeling DNA sequence motifs and regulatory elements have been developed over the past 25 years (for recent reviews, see 12–18). These methods either use DNA patterns or position weight matrices to model target DNA-binding sites (12). Some of these methods have been used successfully to discover the *cis*-regulatory elements in genes that are thought to be regulated by a common transcription mechanism (19–22). Orthogonal information from comparative genomics or co-regulation at the expression level have been incorporated into these methods to identify *cis*-regulatory sites (11, 23–28). Because many of the regulatory elements are functional *in vivo* only in certain temporal or spatial contexts and require other nearby sites that together function as *cis*-regulatory modules (29–31), methods have also been developed to address composite regulatory elements or regulatory modules that consist of DNA sites bound by multiple regulatory factors (32–40). There are now many programs available for DNA motif discovery and putative regulatory element identification, and several have web interfaces for easy use. In this chapter, we will discuss the methods for modeling and identifying regulatory elements and describe the use of some of the commonly available software. The chapter is not meant to be an exhaustive review of all developments in the field, but rather an overview of useful methods for data analysis for the practicing biologist.

1.3 Modeling and representation of DNA motifs

1.3.1 Representation of DNA motifs with sequence patterns

Motifs in DNA sequences can be modeled by sequence patterns, which are simply 'words' in the four-letter alphabet (A, C, G, T) of DNA. In order to capture the information about degeneracy, where more than one base may occur in a specific position of a binding site, the International Union of Pure and Applied Chemistry (IUPAC) nucleic acid codes (41) are used (see *Fig. 1*). Once a sequence pattern for a binding site is determined, there are very fast methods for finding all occurrences of that pattern in a DNA sequence. The limitation of sequence patterns is that many binding sites do not match the pattern exactly, although they are usually very similar to it. *Fig. 2* shows an alignment of 14 binding sites for the Pho4 TF from *Saccharomyces cerevisiae*, taken from the Promoter Database of *Saccharomyces cerevisiae* (SCPD) (42), and shows that at most positions more than one base is tolerated, even though there is always a predominant, or preferred, base. A *consensus sequence*, which indicates the most common base at each position, is easily determined for this site: ATTCGCACGTGG. However, as the occurrence matrix (see *Fig. 2*) shows, only positions 6–9 always match the consensus sequence, and every other position tolerates variants; in a larger sample size of functional sites, even the apparently conserved positions may show variations. A search for the consensus sequence would miss all of the known sites. If we were to encode every position with the IUPAC degeneracy letter from the observed sites, we would get the pattern HBHHNCACGYKN (invariant positions in bold). A search of the genome with this pattern would find all of the known sites, but would also produce a large excess of nonfunctional sites that match the pattern. For example, the sequence CGATACACGCTA matches the degenerate IUPAC pattern but has almost no similarity to the consensus sequence, matching only the four conserved positions, and is unlikely to be a binding site. Alternatively, we could search using the consensus sequence allowing one mismatch, but we would miss 13 of the 14 known sites that way. The known binding sites have between one and five mismatches to the consensus sequence, so to find all of them in the genome one would have to allow up to five mismatches. However, there are more than 200 000 different sequences that match the consensus allowing for five mismatches, the vast majority of which are unlikely to be true binding sites.

```
W = A or T          ('Weak' base pairing)
S = C or G          ('Strong' base pairing)
R = A or G          (Purine)
Y = C or T          (Pyrimidine)
K = G or T          (Keto group on base)
M = A or C          (Amino group on base)
B = C, G, or T      (Not A)
D = A, G, or T      (Not C)
H = A, C, or T      (Not G)
V = A, C, or G      (Not T)
N = A, C, G, or T   (Any base)
```

Figure 1. IUPAC nucleotide codes for representation of degenerate DNA sequence patterns.
The reason for the choice of each code is given in brackets and serves as a useful *aide memoire*.

(a)

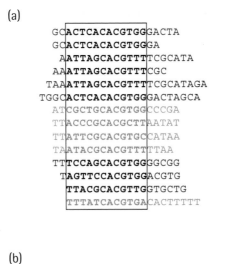

(b)

A	10	0	2	5	3	0	14	0	0	0	0	1
C	1	5	3	7	1	14	0	14	0	1	0	1
G	0	2	0	0	9	0	0	0	14	0	8	7
T	3	7	9	2	1	0	0	0	0	13	6	5

Consensus:	A	T	T	C	G	C	A	C	G	T	G	G
IUPAC:	H	B	H	H	N	C	A	C	G	Y	K	N

Figure 2. Alignment and analysis of yeast Pho4-binding sites.
(a) Alignment of the 14 *S. cerevisiae* Pho4-binding sites generated by the wconsensus program reveals a 12 nt representative DNA motif. The binding site is boxed; sequences shown in gray are on the opposite strand of the DNA and have been reverse-complemented in this figure. (b) The table shows the frequency of each base at each of the 12 positions in the motif. Beneath it are shown the consensus sequence and the IUPAC representation, which shows all degeneracy in the sequence.

1.3.2 Representation with weight matrices

A more general representation of TF-binding sites is a position weight matrix (PWM) (alternative names include weight matrix, position-specific scoring matrix or PSSM, specificity matrix, and others) (12). In this representation (see *Fig. 3*), an *l*-long sequence motif is represented by a $4 \times l$ matrix where each possible base, at each position in the binding sites, is assigned a score. The score of any specific sequence is just the sum of the position scores from the weight matrix corresponding to that sequence. For example, the score of the consensus sequence, 19.41, is the sum of the bold numbers in the matrix (see *Fig. 3d*). Any other sequence can be scored similarly, so that an entire genome can be scanned by a matrix and the score at every position obtained.

Often in a small sample of sites, a few positions of the motif may appear invariant (as is the case for positions 6–9 of the Pho4 motif; see *Fig. 2*), although in reality these positions may vary when a larger collection of sites is available. Motifs built from a small collection of sites, with no variability at certain positions, would severely penalize new sites with variability at those positions. To address this issue, some variations are usually introduced in the invariant positions by adding

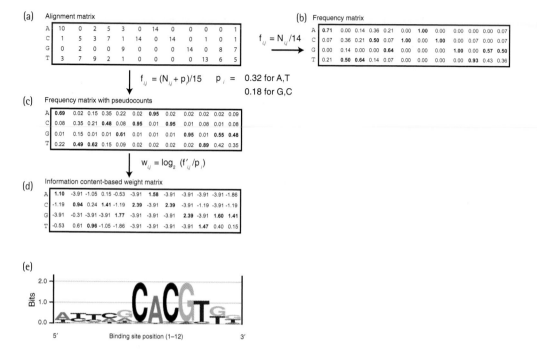

Figure 3. Matrix representations of the 14 Pho4-binding sites in *Fig. 2*.
(*a*) Alignment or occurrence matrix, simply representing the number of times a base, *i*, is observed at each position, *j*, of the alignment of binding sites. (*b*) The frequency matrix, obtained by dividing the number of occurrences of a particular base at each position by the total number of sites in the alignment. The highest values at each of the positions in the matrix are in bold. (*c*) The frequency matrix with pseudocounts (f$_{i,j}$), obtained by adding the frequency of a given base at each position (0.18 for G and C, and 0.32 for A and T), effectively adding a count of 1 to the total number of sites; pseudocounts are added in order to accommodate sites that may be fully conserved in the available dataset but variable in a larger set. (*d*) Information content-based weight matrix. The equations for deriving (*b*), (*c*) and (*d*) from (*a*) are given in each case. (*e*) Sequence logo of the alignment shown in (*a*); the G+C content of the yeast genome was taken as 0.36.

pseudocounts. A common way of introducing pseudocounts is simply to add one site to the collection of known sites and then distribute the base frequency of that artificial site according to the base composition of the genome (see *Fig. 3*). This also avoids taking the logarithm of zeros in the calculation of the weights.

Various methods have been described to set the values in the matrix. In the ideal case, the numbers would represent the contribution of each base to the binding energy between the protein and DNA, so that the total binding energy would be the sum. Note that this works perfectly only in the situation where the protein's binding is completely additive, where the binding energy contributed by each base is independent of the others in the site. Such a simple model is unlikely to be exactly accurate, as demonstrated in a few cases (43, 44), but can be a reasonable approximation for some proteins (45). If additivity is not a valid model, then more complex models can be developed that include nonindependent contributions at two or more positions (45–49). Although such nonadditive models may perform more effectively in some cases, they require more data for training, which is not available in most cases. For the remainder of this chapter, we will focus on additive models, assuming that they are reasonable approximations to the true binding energy for a site.

As every l-long sequence is assigned a score by the PWM, predicting functional sites requires a cut-off score. For the 14 known Pho4 sites, the minimum score is 9.29 (for the site TTTATCACGTGA), so one might use that as a threshold. However, this may result in many false-positive predictions due to the inaccuracy of the PWM in representing the true binding energy. Inaccurate PWMs could be due to the small sample size from which the values are estimated or because of the inadequacy of the additive model. Often one uses a higher threshold, such as one or two standard deviations below the mean site score (50, 51). This usually means that there are some false-negative predictions, true sites that are not above the cut-off, but it also results in fewer false positives. It is also possible that the lowest-scoring sites are only functional due to their specific context in the genome, such as cooperative binding with another factor, and they do not have sufficient binding affinity to function independently. This is one biological reason for false-negative predictions with DNA motif models.

1.3.3 Information content of binding site motifs

It is useful, especially when considering whether a predicted binding site is statistically significant, to have a measure of the 'information content' of a motif. In general terms, information content refers to the probability of observing a site by chance, with high information content being less likely. An exact DNA pattern of 6 nt would be expected to occur about every 4096 bases (assuming the bases are all equally abundant; if the genome is biased towards some bases, this bias can easily be taken into account). Degenerate positions in the pattern can be handled similarly. For a pattern represented by a PWM, its information content (or relative entropy), $I(p)$, is given by the following equation (52–54):

$$I(p) = \sum_{j=1}^{l} \sum_{i=A}^{T} f_{i,j} \log \frac{f_{i,j}}{P_i}$$

where l is the pattern length, i is the index of a base at position j of the pattern, $f_{i,j}$ is the frequency of base i at the position j of the PWM, and P_i is the probability of observing that base in the data. Under some simplifying assumptions, the contribution of a single base at position j to the total binding free energy is proportional to $\log(f_{i,j}/P_i)$ (53, 54), which is the reason for using these values as the weights of the PWM (see *Fig. 3d*). The information content represents the average binding energy of the collection of sites. In the above representation, the information from each position is simply added over the length of the motif to give the total information content, employing the additivity assumption described earlier. It is of note that when $f_{i,j}$ equals P_i for each i, then the information content of the pattern is zero at that position. Statistical methods for computing the *P* value of information content have been defined using large-deviation statistics (55). Several of the popular methods that use PWMs to model DNA motifs use the information content measure to identify the optimal motifs from input sequences (55–58).

Sequence 'logos' (59) are useful ways of representing the information content of a given alignment of sites or a PWM. An example of an alignment matrix and its corresponding sequence logo for the Pho4-binding site in yeast is shown in *Fig. 3.*

1.3.4 Advantages and disadvantages of sequence pattern and PWM representations

Both pattern and PWM representations have been used widely since the early 1980s to identify DNA regulatory elements and serve as complementary approaches. The sequence patterns are simpler and enable fast searches; it is also easy to calculate whether the results of such a search are significant or are expected by chance alone, as described below. PWMs capture more quantitative information and so can be a more accurate representation of the specificity of a TF, although extensive data may be required to get an accurate model. Both methods are also used in motifs discovery algorithms where neither the binding sites nor the motif are known in advance. PWMs require heuristic searches of the possible alignments and are not guaranteed to find the optimal solution. Pattern-based methods can search more thoroughly among the set of allowed patterns, but none of them may be particularly good representations of the true motif. There are several available programs for motif discovery that employ a variety of approaches, but it remains an open problem with substantial room for improvement.

2. METHODS AND APPROACHES

2.1 Searching DNA for known motifs

One common task in DNA motif and regulatory element analysis is to search for potential TFBSs within DNA regions of interest. For instance, one may have a gene, or set of genes, whose expression is of interest and one wants to find potential sites governing their regulation. To accomplish this task, one needs a database of regulatory motifs, some software to search for these motifs in the sequence, and some means of determining whether any matches are likely to be significant. Software for visualization of the location and strengths of the binding sites is also useful.

2.1.1 Databases of regulatory motifs and TF information

There are two comprehensive databases containing information on TFs and their binding sites. The TRANSFAC database (http://www.gene-regulation.com/pub/databases.html#transfac[6.1]) is available in commercial and noncommercial versions (7). It provides extensive data on experimentally characterized TFs in different organisms, known binding sites, the PWM models for TFBSs, genes that are regulated by specific TFs, and references to the literature from which the information has been compiled. The PWMs are integrated with software tools to scan DNA sequences for putative TFBSs (60, 61).

More recently, JASPAR (http://jaspar.genereg.net[6.2]) has been developed as an open-access database for eukaryotic TF-binding profiles (62). The prime differences between TRANSFAC and JASPAR are in redundancy and quality: JASPAR is a smaller set that is nonredundant and highly curated. All profiles are derived from published collections of experimentally defined TFBSs for multicellular eukaryotes. The PWMs from both TRANSFAC and JASPAR have been integrated into several publicly available programs and web servers for analysis of TFBSs. MPromDB (http://bioinformatics.med.ohio-state.edu/MPromDb/[6.3]) provides a useful resource for information regarding TFBS locations, first exons, and promoters (63). In addition to the above generic databases, other organism-specific databases exist that host transcriptional regulation data; several of these are listed in section 3.

2.1.2 Searching DNA sequences for sites using known patterns

Given an input sequence and a DNA sequence pattern, finding the subsequences that match the pattern is straightforward. Even in the case of mismatched alignments, one can score DNA subsequences that match the given sequence pattern with a specified number of mismatches (the Hamming distance).

Several programs are available to scan sequences for given DNA patterns and identify putative sites. One such program, DNA PATTERN, is available at http://rsat.ulb.ac.be/rsat/dna-pattern_form.cgi[6.4]. The software takes an IUPAC DNA pattern, input sequences in FASTA format, and parameters including the number of allowed mismatches and which strand(s) of the sequence to search. Other software for finding DNA site matching patterns are listed in section 3. It is also relatively straightforward to write Perl modules to perform DNA pattern matching.

2.1.3 Searching DNA sequences for sites using PWM models

Given an input sequence and a PWM model representing a DNA motif, one can score any subsequence in the input, reflecting how well it corresponds to the motif, as follows. Suppose the length of the PWM is l and the weight at position j and base i is $w_{i,j}$. Then the score of a subsequence, when aligned with the PWM is given by:

$$\text{Score}_s = \sum_{j=1}^{l} \sum_{i=A}^{T} w_{i,j} \cdot s_{i,j}$$

where $s_{i,j} = 1$ if base i occurs at position j and 0 otherwise. This principle has been implemented in several software packages including PATSER (55), MATCH (60), and MAST (5).

The scoring of sites with PWMs can suffer from high false-positive and false-negative rates as discussed earlier, and it is therefore important to select an appropriate threshold score to filter the hits. One approach to this is based on the information content in the PWM (55). Another approach involves the minimization of false-negative and false-positive rates based on predictions on the sequences that are known to contain the regulatory patterns and on sequences that are not likely to contain these patterns (29). Djordjevic *et al.* have recently described an SVM (Support Vector Machine) method based on thermodynamic principles of TF DNA binding that provides a natural threshold energy (called the chemical potential) for distinguishing real sites from spurious hits (64).

2.1.4 Software for searching for sites with PWM models

There are several programs that match given PWMs with input sequences to identify putative sites. These programs work on the basic principle of aligning PWMs to sequences as described above. However, there are some differences in their scoring schemes and in their determination of thresholds for filtering low-scoring sites.

The PATSER program estimates the *P* value of a predicted site given a PWM (65). Using the sample-size-adjusted information content, PATSER chooses a default cut-off score for which the *In*(probability) (i.e. the probability of observing a score greater than or equal to the cut-off score) matches the information content of the sites. Other choices of cut-off score can be set by the user. A web server for PATSER is available at http://rsat.ulb.ac.be/rsat/patser_form.cgi [6.5]. MAST is another commonly used program for searching input sequences and aligning PWMs to putative sites (5). It can be used through a web server (http://meme.sdsc.edu/meme/mast.html [6.6]), where one can input a PWM and set of sequences. MAST only displays sequences matching a query PWM with *E* values below the given threshold. The output describing the location of the putative sites is available in both hypertext and text format; the hypertext format is useful in visualization of the putative sites in each of the input sequences. Other software suites for prediction of binding sites with PWM models are listed in section 3. In addition, there are useful Perl modules available for analysis of TFBS and DNA elements (66), including binding sites in input sequences given PWMs.

Protocol 1

Using MAST to search for sites with PWM models

1. From the Protocol_1 folder for this chapter on the book's web site, download the files Yeast_Mcm1_Regulated_Genes.fa [6.7] and Mcm1.matrix [6.8]. The first of these is a FASTA file containing a set of 15 promoter sequences from *S. cerevisiae* genes known to be regulated by the TF Mcm1. The second file is a PWM representing the binding site for this TF.

2. Go to the MAST web server at http://meme.sdsc.edu/meme/mast.html [6.6]. Enter your e-mail address as requested.

3. Under **Your motif file** (on the left), upload Mcm1.matrix [6.8].

4. On the right of the screen, you need to specify the sequence(s) that you want to scan for sites. A pull-down menu (A **MAST database**) allows you to choose from a large list of sequence

databases, but, for this example, leave this menu blank and instead upload the file Yeast_ Mcm1_Regulated_Genes.fa [6.7] using the button below the menu.

5. In this example, we will choose to see only hits with an E value of less than 1 (relatively strong hits) – set this using the pull-down menu on the left (**Display sequences with E-value below:**).

6. A number of other options can be set by the user – clicking on the link next to each option will call up a help menu. However, for this example, we will leave the other settings at their default values.

7. Click **Start search**; the page will show a brief summary of the search parameters. After a few moments, you should receive an e-mail with the results attached as an HTML file; download the file and open it in your browser. (A copy of the expected file is given as mast.23405.results. html [6.9] in the Protocol_1 folder.)

8. The first part of the results summarizes the details of the search. There is then a table showing which of the input sequences were found to contain significant hits (in this case, all 15 of the sequences in the input file.). Below this, a diagram indicates the position (and strand) of hits in each of the sequences and, finally, each hit is displayed as an alignment between the target sequence and the motif.

2.2 Discovery of DNA motifs from input DNA sequences

For the majority of TFs, binding site models are not available. One interest of many investigators, therefore, is to identify novel regulatory elements from a set of DNA sequences that are known or assumed to be regulated by the same transcriptional mechanism – for example, a set of promoters belonging to genes found experimentally to be co-regulated at the mRNA level. The challenge here is to identify the most significant DNA motifs, given the sequence data. The following sections discuss computational methods for novel DNA motif discovery.

2.2.1 Discovery of DNA sequence patterns: pattern-driven methods

In these methods, generally one has to define the length of the pattern, l, and then determine, for all possible patterns, the one that is most significant based on its frequency compared with some null model (14, 67–72). This approach is enumerative and guarantees to find optimal solutions or the most significant patterns, but has one significant limitation: the number of patterns to be searched increases exponentially as the fourth power of the length. Therefore, searching for patterns that are long (in practice for $l > 10$ bases; 72, 73) is difficult. Approximate sequence patterns, or patterns that contain degeneracy at one or more positions, may also be identified from the sequence.

When multiple patterns identified from the data are close in terms of their distance, they can be merged into one approximate pattern, thereby forming longer patterns that are otherwise difficult to find from the data given their long search times (70, 74). In addition to the exhaustive enumerative procedures, the use of efficient data structures such as suffix trees have also been employed in DNA pattern-recognition methods (18, 75, 76).

2.2.2 Discovery of DNA motifs with PWMs: alignment-driven methods

In these methods, the challenge is to find simultaneously the site locations and the representative PWMs using only the sequence data (hence these algorithms are also called sequence-driven methods). If the site locations were known, PWMs representing those sites could easily be built (see *Figs 2* and *3*); conversely, if the PWM were known, then the most likely sites could be obtained as the highest-scoring sequences matching the PWM (see above). In the current problem, however, the locations of the sites and their PWMs represent missing information, which may be solved by employing machine learning algorithms (77). Several algorithms have been used on the problem of motif finding, and a few are described below. Unlike the pattern-driven methods where direct enumeration of significance is possible, optimal results cannot be guaranteed by these alignment-driven methods as there are too many possible alignments to consider all of them. However, various optimization techniques can often obtain the correct locations and PWMs, with improved representations over pattern methods.

The first alignment-driven method was the greedy algorithm as implemented in the CONSENSUS program (57, 78). Given a set of sequences and a motif length *l* to be searched, the method first compares pairs of sequences and saves some number (typically 1000–10 000) of the best matches. These 'seed alignments' are then compared with the rest of the sequences and the best matches are accumulated into multiple sequence alignments. This process is iterated until some stopping criterion is reached, for instance when every sequence has contributed a binding site or a maximum of the alignment significance is obtained.

Another method that has been used to optimize PWMs is the expectation–maximization (EM) algorithm (56, 79, 80). The EM algorithm starts with a guess of the PWM, which could be completely random or based on some prior knowledge about the binding sites. Using the PWM, the probability that each subsequence is a binding site is estimated and the PWM is then redetermined based on those site probabilities. This process is iterated until it converges. The EM algorithm has been implemented in the widely used MEME software (56, 79–81).

A stochastic variant of the EM algorithm is the Gibbs sampling method (58, 82), which is a type of Markov chain Monte Carlo (MCMC) algorithm (77). The Gibbs sampling technique tends to give a more robust optimization of the PWMs as it is a stochastic procedure that enables the avoidance of local minima in such search problems. The main difference between the EM and the Gibbs sampling procedures is in the selection of expected sites for updating of the PWM. Based on the weights of the PWM at the $(i-1)$th step, the expected probability of any subsequence or site can be computed in the ith iteration. This expected probability determines the probability with which a site will be used in updating of the PWM at the ith iteration (for details, see 58). Variations of the Gibbs sampling technique have been implemented in several programs that are widely used in the bioinformatics community (19, 47, 50, 83, 84).

2.2.3 The utility of background or negative sequences and finding discriminative motifs

Variation in local base composition can adversely affect sequence alignment. As such variation can be complex, and because binding motifs are often AT- or GC-rich, these adverse effects can be difficult to control using existing masking algorithms. In addition, many sequence patterns are commonly observed in all noncoding or promoter sequences in an organism, especially in eukaryotes, but these ubiquitous patterns are not the TF-binding sites that are sought. It is therefore often useful to control for such effects and detect only those motifs that are specific to the input sequence set, i.e. discriminative motifs. Several programs use a 'negative' or 'background' set as a control, and only those motifs that are more prevalent in the positive ('training') set than in the control set are reported. Background motifs are either modeled using a Markov model (84–86) or by sampling a large number of random sites from the background sequence set (33, 83). In the absence of a control sequence set, background models can be built from all sequences in the training set, for instance by shuffling the training set to remove the patterns of interest whilst maintaining other features of the sequences.

2.2.4 Software for discovery of DNA patterns

There are currently many algorithms and programs that identify patterns from DNA sequences, some of which are now available to run online. The YMF program (http://wingless.cs.washington.edu/YMF/YMFWeb/YMFInput.pl[6.10]) was developed to identify DNA patterns through statistical over-representation (69, 86). A background distribution of patterns is created using a third-order Markov chain from a given set of background sequences (e.g. promoter regions of all genes in a genome). Given this background distribution of patterns, one can calculate the statistical significance of any given pattern and thereby identify the patterns with the highest statistical significance. The input sequence needs to be in FASTA format, and several relevant parameters can be input in the program, such as the length of the pattern sought and, in the event that the pattern may be composed of separate halves of conserved sequence with a nonconserved 'spacer' in the middle, the maximum length of the spacer. The software contains background pattern Markov models for many organisms including *S. cerevisiae*, human, mouse, *Drosophila melanogaster*, *Arabidopsis thaliana*, and *Escherichia coli*. Additional background Markov models can be created with sequences defined by the user. Given the input parameters and the background model, the YMF software can extract the most statistically significant motifs according to their Z-score. The software also contains a plotting program to visualize the DNA sites corresponding to the patterns extracted with the program.

Other pattern-recognition software programs in common use are listed in section 3. Of these, WEEDER and MITRA use efficient data structures (i.e. suffix tree data and mismatch trees) instead of the exact enumeration method.

2.2.5 Software for the discovery of optimal PWMs

In this section, several of the most common methods used to identify PWMs are discussed. These methods use different machine learning techniques to identify an optimal PWM in a given set of sequences. It should be noted that most of the stochastic methods, such as the Gibbs sampling strategy, require multiple runs to be examined. In such cases, it may be easier for the user to download the software for multiple runs, rather than do repeated runs of the programs through the web sites. Amongst the methods that are commonly used, three are described in detail.

The CONSENSUS software was the first to be developed in this field (55, 78). The algorithm builds a multiple alignment of putative binding sites by iteratively adding new sites. In the first step, all pairs of sequences are compared and the best alignments (of a defined length) are saved as the starting point for subsequent iterations. The basic CONSENSUS program requires a set of sequences in the Consensus or FASTA format, the length of the pattern to be identified, the prior probabilities of each DNA base (this can also be calculated from the input sequences), whether or not to consider complementary DNA strands for the search, and whether the motif to be searched for is palindromic. There are other useful options that can be specified such as whether to allow multiple sites in each sequence or to allow some sequence not to contain an instance of the motif. The output gives a number of top-scoring matrices, their information content, P values and expectations, and the locations of the sites in the input sequences. For each matrix, the program estimates a P value, which is the probability of observing the particular information content or higher in an arbitrary alignment of random words having a length equal to the width of the matrix. The ultimate statistical significance of a matrix is determined by multiplying the P value by the approximate number of possible alignments containing the designated number of sequences and having the observed width. This product is referred to as the expected frequency, or E value, of the matrix alignment. The E value allows the comparison of matrices summarizing differing numbers of sequences and having differing widths. WCONSENSUS (55) is an extension that uses the CONSENSUS program, but automatically determines the optimal motif width.

The MEME (81) software identifies DNA motifs using the EM algorithm, which iteratively updates the PWM and the site probabilities until convergence. Several input sequence formats can be used, including FASTA. MEME can identify multiple motifs simultaneously from the input sequence data. Input parameters include the number of distinct motifs to be identified from the input data, the number of expected occurrences in each sequence of the input set of sequences (exactly one occurrence, zero or one per sequence, or any number of repetitions), the minimum and maximum width of the motifs to be searched, the number of total sites expected in the data, whether or not to search for palindromic motifs, and whether or not to search for motifs in both strands of the input DNA sequence. One can also shuffle the input sequences to provide a 'control sequence' for assessing the significance of the results. The output contains the PWMs representing the identified motifs, their consensus sequences, the *E* value of the motif, information content, and the number and location of the sites in each of the input sequences. The MEME software is integrated with MAST (see above), which is used to search DNA sequences with the motifs identified using MEME runs. MAST also provides a useful tool for visualization of the sites.

In the following protocol, we will use MEME to see if we can 'discover' a motif. For this purpose, we will use the same set of yeast promoters that were used in *Protocol 1*, but this time without using prior knowledge of the actual binding site that they share.

Protocol 2

Using MEME to find unknown binding sites

1. Go to the MEME web server at http://meme.sdsc.edu/meme/meme.html [6.11].

2. Enter your e-mail address as requested.

3. On the left of the screen, you are asked for a set of sequences that you believe share a common motif. Upload the file Yeast_Mcm1_Regulated_Genes.fa [6.7] (from the Protocol_1 folder for this chapter on the book's web site).

4. There are a number of options to set. These options are simply best guesses (or informed opinions) as to the characteristics of the motif you are looking for. Under '...occurrences of a single motif are distributed among the sequences?', select 'Any number of repetitions' (top right of window). We will guess that the motif is likely to be between 10 and 20 bases – set these numbers as the 'Minimum width' and 'Maximum width' respectively.

5. Under 'Maximum number of motifs to find', enter '3' (MEME will look only for the best three distinct motifs).

6. Based on the fact that the 15 yeast promoter sequences come from genes known to be co-regulated, we might expect at least half of them, and perhaps all, to contain a common motif. Therefore, under 'Minimum sites' and 'Maximum sites', enter '8' and '30', respectively.

7. Finally, enter an optional and arbitrary title under 'Description of your sequences' and click 'Start search'.

8. After some time, you will receive an e-mail containing your results as an HTML document to be viewed with your browser. (A copy of the expected results is in the Protocol_2 folder for this chapter as meme.5207.results.html [6.12]).

9. The output starts with background information and a summary of the search parameters, followed by the first (most significant) of the three motifs that the program found ('MOTIF 1', with an *E* value of 1.1e−24).

10. The first part of the report for Motif 1 gives its probability matrix and (as a histogram) information content, followed by the consensus sequence. Where the consensus is strong (for example, the 'A' at the 5′ end of the motif), only the consensus base is shown; where the consensus is weaker, the second or third most frequent bases are also shown (for instance 'A', 'C' and 'T' for the second base). The consensus is:

   AAATTTCCTAAAAAGGTAAA

Naturally, the choice of DNA strand is arbitrary; the reverse complement of this consensus is strikingly similar to the known consensus sequence of the Mcm1 matrix:

```
TTTACCTTTTTAGGAAATTT          (from MEME)
TTACCCATTTAGGAAA              (consensus of Mcm1 matrix)
```

11. Scrolling down the screen, additional data on the motif (including its occurrence in the target sequences) is available. A further explanation of the results is at the bottom of the page.

12. The two other motifs found by MEME have much higher E values (4.3e2 and 1.0e5) and are probably not significant.

13. Note that by following links from the results page, one can compare the discovered motifs with known motifs in the JASPER database.

14. The MEME server will also automatically run a MAST analysis with the discovered motifs, mapping the putative sites back to the original sequences. The results of this MAST analysis will be sent to the user in a separate e-mail following the MEME results.

The Gibbs motif sampling algorithm for finding DNA motifs was originally described by Lawrence et al. (58). Over the last decade, variations of this algorithm have been implemented in other programs (19, 50, 83–85, 87). The Gibbs motif sampler developed by Lawrence et al. is available at http://bayesweb.wadsworth. org/gibbs/gibbs.html[6.13]. Motifs with the maximum a priori (MAP) log-likelihood scores are reported. The significance of the PWMs may also be calculated based on the nonparametric Wilcoxon rank test, in which the P value is computed based on the scores of sites in the training set versus a shuffled sequence set. The output consists of the significant PWMs representing the sites in the training set, information content at each position of those sites, the MAP score, and locations of the sites in the input sequences corresponding to the reported PWMs. Other implementations of the Gibbs sampling approach for motif discovery include ANN-SPEC (83), ALIGNACE (19), and MOTIFSAMPLER (85). ANN-SPEC and MOTIFSAMPLER can identify discriminative motifs by utilizing background sequence models.

A recent comparative study of some of the commonly used ab initio motif finders (88) made the important observation that different tools complement each other. For example, the predictions of MOTIFSAMPLER complement the predictions of MEME. Biologists would therefore be well advised to use a few complementary tools in combination, rather than relying on a single one, and to pursue the top few predicted motifs of each, rather than the single predicted motif. If used intelligently, these approaches can yield biologically valuable results. For example, GuhaThakurta et al. (21) described the identification of a novel regulatory element involved in the C. elegans heat-shock response. The ab initio motif finders CONSENSUS and ANN-SPEC identified two strong motifs in a set of promoters from heat-shock-responsive genes. One of these motifs was the previously known heat-shock element (HSE) that is broadly conserved in eukaryotes; the other was a novel motif (heat-shock-associated site or HSAS), which has since been shown to be biologically functional in vivo and to function cooperatively with HSE. Similar studies were used to identify several new regulatory elements in C. elegans that were shown to contribute cooperatively to muscle-selective expression (20) and which, in turn, were highly predictive of additional muscle-expressed genes (89).

2.3 Comparative genomics and phylogenetic footprinting in the search for DNA regulatory elements

In eukaryotes, the genomic region in which regulatory elements can be present is vast, sometimes spanning more than 100 kb around the regulated genes (9). Therefore, orthogonal data that help to limit the search space for these elements are of significant value. Cross-species sequence alignment, or phylogenetic footprinting (90), has been widely used in identifying regulatory elements, especially in higher eukaryotes and mammalian genomes (9, 11, 23, 66, 91–94). Enrichment of regulatory elements has been demonstrated in noncoding DNA that is conserved between human and mouse (23, 66, 95). For example, Wasserman *et al.* (23) demonstrated that 98% of the known muscle regulatory elements occur within the 19% of the sequence that is most conserved between human and mouse. Analysis of motif frequencies and correspondence in conserved regions across multiple species have been used to identify known and novel regulatory elements in yeast (27, 96, 97), as well as in mammals (98). Comparative genomics approaches have also helped in elucidating regulatory elements in prokaryotes (99) and archaea (25).

Despite the utility of phylogenetic footprinting, there are some issues with this approach with respect to identification of the regulatory elements, which are usually variable and short. As many standard methods for phylogenetic footprinting and cross-species genome comparisons work through global sequence alignment algorithms, alignments of these short regulatory elements between distantly related organisms can be missed (21, 100). If organisms are too closely related, however, alignments are extensive and therefore uninformative (100, 101).

There are now numerous programs available for aligning genomic sequences from two or more species and for distinguishing regions that are under selective pressure from neutrally evolving regions (91, 102–112). Results from these programs can be used to limit the search space for finding regulatory elements. Some of these programs have been integrated with motif-finding and visualization programs to provide useful tools for analysis of regulatory elements within cross-species conserved regions (113–115). In addition to the methods for phylogenetic footprinting, more sophisticated programs have been developed that exploit multiple genome sequences in searching for regulatory motifs (92, 94, 116–119).

2.3.1 Software for identifying regulatory elements using cross-species sequence comparisons

The FOOTPRINTER (92) method uses the information on phylogenetic distances between different genomes that are being aligned. This is important information that can help to weight the genomic sequences appropriately for identifying patterns in the conserved regions. The software is available for download or use through a web server (http://wingless.cs.washington.edu/htbin-post/unrestricted/FootPrinterWeb/FootPrinterInput2.pl[6.14]), which also contains example inputs and other instructions (including how to specify a phylogenetic tree; see also Chapter 12). Another program for the discovery of DNA motifs utilizing cross-species sequence conservation is PHYLOCON, which is a modification of the CONSENSUS software and is available for download at http://ural.wustl.edu/~twang/PhyloCon/[6.15].

The RVISTA tool (http://rvista.dcode.org/[6.16]) combines TFBS predictions, cross-species sequence alignments, and binding site cluster analysis to identify putative evolutionarily conserved DNA regulatory regions (113). Two sequences to be examined are used as input in the web server. Pre-calculated pairwise alignments may also be input. Users selecting the TRANSFAC library have the option of specifying the stringency to be used for the PWM identification. Clustering analysis of TFBSs permits the search and subsequent visualization of complex modules consisting of multiple different TFBSs. Some further examples of tools that utilize cross-species (often human–rodent) sequence alignments to identify binding sites are listed in section 3.

Protocol 3 illustrates the use of the CONREAL (114) web server to identify TFBSs by comparison between human and mouse sequences.

Protocol 3

Using CONREAL to find putative regulatory elements by cross-species comparison

1. From the Protocol_3 folder for this chapter on the book's web site, download the file human_ and_mouse_myf6_promoter.fa.txt[6.17]. This FASTA file contains 1000 bp of upstream sequence of the muscle gene *Myf6* from human and the corresponding region from mouse, taken from http://genomebiology.com/2005/6/8/R72[6.18].

2. Go to the CONREAL web server at http://conreal.niob.knaw.nl/[6.19].

3. The top part of the submission form (small text field, pull-down menu and button) allows you to choose genes to analyze from Ensembl. Ignore these and instead upload the file human_ and_mouse_myf6_promoter.fa.txt[6.17] using the button below the large text field.

4. Using the pull-down 'Search parameters' menus and set the threshold for PWMs to 90% (this is the minimum relative score of a TFBS hit to be considered in the analysis). Set the length of flanks to calculate homology to 15 bp and the threshold for homology to 75% (thus, CONREAL will consider only hits that have a minimum of 75% homology, including the 15 bp flanking the hit).

5. Using the check boxes towards the bottom of the screen, choose CONREAL as the only aligner and JASPAR as the only TFBS dataset (i.e. the initial alignment of the two sequences will be done by CONREAL and it will use the alignment to inform its search for TFBSs from amongst those in the JASPAR database).

6. Click the **Search** button. A page will appear summarizing the job and its current status, and this will be refreshed as soon as the results are ready.

7. A copy of the expected results is in the Protocol_3 folder, called Conreal_Result.htm[6.20].

8. At the top of the results page, a histogram gives an overview of the density of putative TFBSs along the aligned sequences (in this case, most of the hits are in the region 600–1000 bp). Below this, the alignment is shown first as an overview and then in detail with the hits shown in upper case.

9. Further down, a results table lists each of the hits, including the motif, positions, and scores on both sequences and the name of the TF that binds to it. For example, motif MA0089 is found at nt 632–637 in sequence 1 (human) and at nt 580–585 in the aligned sequence 2 (mouse); the motif is believed to be bound by TF TCF11-MafG.

10. The final column of the table indicates which of the prediction algorithms found each of the hits; in this case, all of the hits were found only by CONREAL (as we used only this aligner). If multiple algorithms are used, this column reveals the cases where the algorithms agree (increasing confidence in the result) or disagree.

11. The results are linked to the relevant TFBS database (in this case, JASPAR); clicking on the link in the **Matrix** column will display the relevant database entry for that motif.

12. A more detailed description of the search parameters and interpretation of the results is available at http://conreal.niob.knaw.nl/description.html [6.21].

The RVISTA web server also provides example sequences and alignments that are helpful for familiarizing oneself with this software (113).

There are several examples where comparative genome sequence analysis has been used successfully to elucidate regulatory elements in mammalian genomes. In one elegant study, Loots *et al.* (9) demonstrated the utility of phylogenetic footprinting by identifying a regulatory region, conserved between human and mouse, for the *IL-5* gene, which is located about 120 kb away from the gene itself. In a recent study comparing genomic sequences from multiple primates, Boffelli *et al.* (93), identified primate-specific gene regulatory elements for *apo(a)*, a recently evolved primate gene. A region of high conservation adjacent to the transcription start site of the gene was shown to interact with one or more DNA-binding proteins in electrophoretic mobility shift assays using nuclear extracts from liver cells.

2.4 Composite DNA motifs and *cis*-regulatory modules

In eukaryotes, TFs rarely act alone in regulating the expression of a given gene. In most cases, multiple factors bind the DNA, often in close proximity to each other, forming regulatory modules (29–31). By integrating multiple signals, these *cis*-regulatory modules confer an organism-specific spatial and temporal transcription. When scanning sequences with DNA motif models, looking for multiple relevant sites is likely to reduce the large false-positive rate in predictions of individual binding sites and to achieve better predictive specificity by focusing on locally dense clusters of binding sites. Therefore, identification of composite modules and higher-order regulatory structures is currently an important issue in computational analysis of regulatory elements.

2.4.1 Analysis of composite DNA regulatory motifs

The programs that have been developed for analyzing composite motifs fall into two categories:

- Those that identify modules of DNA motifs given a set of PWMs representing sites for certain TFs that are known (or postulated) to act together. These programs rely on hidden Markov models (35, 120, 121) or on observing frequent joint occurrences of sites within defined windows (36, 38, 97, 122, 123).
- *Ab initio* methods that identify pairs of DNA motifs within a certain distance of each other (32, 33, 75, 84, 124, 125).

Most of the programs for analysis of *cis*-regulatory modules are fairly new and under active development and investigation. At this time, the most useful and user-friendly are those that predict modules given a set of motifs representing sites, such as COMET (121), MSCAN (36, 122), and CIS-ANALYST (126).

By using known DNA-binding specificity data for five TFs active in the early *Drosophila* embryo, Berman *et al.* (126) identified genomic regions containing unusually high concentrations of predicted binding sites for these factors using CIS-ANALYST. A significant fraction of these binding-site clusters overlap known *cis*-regulatory modules that are regulated by these factors. In addition, many of the remaining clusters were adjacent to genes that were expressed in a pattern characteristic of genes regulated by these factors. The authors tested one of the newly identified clusters mapping upstream of the gap gene *giant* and showed that it acted as an enhancer that recapitulates the posterior expression pattern of this gene.

2.4.2 Database resources for composite regulatory elements

The TransCompel database (29) provides information about the interaction of TFs and transcription regulatory modules. Each database entry corresponds to individual composite elements within a particular gene and contains information about the two binding sites, their corresponding TFs, and the experiments confirming cooperative action between TFs. This database is a useful resource for information about the composite transcriptional regulatory modules.

3. ADDITIONAL WEB RESOURCES

Databases for TF DNA-binding site models and gene regulation

- JASPAR: http://jaspar.genereg.net/ [6.22]
- TRANSFAC: http://www.gene-regulation.com/pub/databases.html [6.23]
- MPromDb: http://bioinformatics.med.ohio-state.edu/MPromDb/ [6.24]
- TRANSCompel: http://www.gene-regulation.com/pub/databases.html# transcompel [6.25]
- RegulonDB: http://regulondb.ccg.unam.mx/index.html [6.26]
- DPInteract: http://arep.med.harvard.edu/dpinteract/ [6.27]
- TRED: http://rulai.cshl.edu/cgi-bin/TRED/tred.cgi?process=home [6.28]
- SCPD: http://rulai.cshl.edu/SCPD/ [6.29]

Software for scanning DNA sequences with a given DNA motif model

- MAST: http://meme.sdsc.edu/meme/mast.html [6.6]

- PATSER: http://ural.wustl.edu/services.html [6.30]
- TESS: http://www.cbil.upenn.edu/cgi-bin/tess/tess [6.31]
- MATCH: http://www.gene-regulation.com/pub/programs.html#match [6.32]
- OPOSSUM: http://www.cisreg.ca/cgi-bin/oPOSSUM/opossum [6.33] (analysis of over-represented TF-binding sites)

Software for novel DNA pattern discovery: pattern-driven methods

- YMF: http://wingless.cs.washington.edu/YMF/YMFWeb/YMFInput.pl [6.10]
- TEIRESIAS: http://cbcsrv.watson.ibm.com/Tspd.html [6.34]
- WORDSPY: http://cic.cs.wustl.edu/wordspy/index.htm [6.35]
- WEEDER: http://159.149.109.16:8080/weederWeb/ [6.36]
- MITRA: http://www1.cs.columbia.edu/compbio/mitra/ [6.37]
- Oligo/Dyad Analysis: http://rsat.scmbb.ulb.ac.be/rsat/ [6.38]

Software for novel DNA PWM discovery: alignment- or sequence-driven methods

- CONSENSUS (download): ftp://ftp.genetics.wustl.edu/pub/stormo/Consensus/ [6.39]
- CONSENSUS (web server): http://ural.wustl.edu/services.html [6.30]
- Gibbs Sampler (Lawrence): http://bayesweb.wadsworth.org/gibbs/gibbs.html [6.40]
- MEME: http://meme.sdsc.edu/meme/meme.html [6.11]
- MOTIFSAMPLER: http://homes.esat.kuleuven.be/~thijs/Work/MotifSampler.html [6.41]
- ANN-SPEC: http://www.cbs.dtu.dk/~workman/ann-spec/ [6.42]
- ALIGNACE: http://www.psc.edu/general/software/packages/alignace/alignace.html [6.43]

Software for TF-binding site analyses through comparative genomics

- RVISTA: http://rvista.dcode.org/ [6.16]
- CONREAL: http://conreal.niob.knaw.nl/ [6.19]
- FOOTER: http://biodev.hgen.pitt.edu/footer_php/Footerv2_0.php [6.44]
- PAP: http://im-vishnu.wustl.edu:9090/portal/SearchHome.do;jsessionid=61422DB015417493F60B7B5C6AB65F05 [6.45]

Software for sequence logo generation

- ENOLOGOS: http://chianti.ucsd.edu/cgi-bin/enologos/enologos.cgi [6.46]
- WEBLOGO: http://weblogo.berkeley.edu/logo.cgi [6.47]

4. REFERENCES

1. **Chiaromonte F, Weber RJ, Roskin KM, Diekhans M, Kent WJ & Haussler D** (2003) *Cold Spring Harb. Symp. Quant. Biol.*, **68**, 245–254.
2. **Waterston RH, Lindblad-Toh K, Birney E, et al.** (2002) *Nature*, **420**, 520–562.
3. **Adams MD, Celniker SE, Holt RA, et al.** (2000) *Science*, **287**, 2185–2195.
4. **Lander ES, Linton LM, Birren B, et al.** (2001) *Nature*, **409**, 860–921.
5. **Bailey TL & Gribskov M** (1998) *Bioinformatics*, **14**, 48–54.

6. **Levine M & Tjian R** (2003) *Nature*, **424**, 147–151.
7. **Matys V, Fricke E, Geffers R, et al.** (2003) *Nucleic Acids Res.* **31**, 374–378.
8. **Cawley S, Bekiranov S, Ng HH, et al.** (2004) *Cell*, **116**, 499–509.
★ 9. **Loots GG, Locksley RM, Blankespoor CM, et al.** (2000) *Science*, **288**, 136–140. – *This paper presents a nice example of the application of phylogenetic footprinting for identifying distal regulatory sequences in vertebrate genomes.*
10. **Lenhard B, Sandelin A, Mendoza L, Engstrom P, Jareborg N & Wasserman WW** (2003) *J. Biol.* **2**, 13.
11. **Cooper GM & Sidow A** (2003) *Curr. Opin. Genet. Dev.* **13**, 604–610.
★ 12. **Stormo GD** (2000) *Bioinformatics*, **16**, 16–23. – *An important review on the representation and discovery of regulatory sequences.*
13. **Bulyk ML** (2003) *Genome Biol.* **5**, 201.
14. **Brazma A, Jonassen I, Eidhammer I & Gilbert D** (1998) *J. Comput. Biol.* **5**, 279–305.
15. **Brazma A, Jonassen I, Vilo J & Ukkonen E** (1998) *Genome Res.* **8**, 1202–1215.
★ 16. **Wasserman WW & Sandelin A** (2004) *Nat. Rev. Genet.* **5**, 276–287. – *An excellent review of the issues surrounding computational analyses of transcriptional regulatory sequences.*
★ 17. **Pavesi G, Mauri G & Pesole G** (2004) *Brief. Bioinform.* **5**, 217–236. – *Another good review on analysis of regulatory sequences.*
18. **Pavesi G, Mauri G & Pesole G** (2001) *Brief. Bioinform.* **2**, 417–430.
19. **Hughes JD, Estep PW, Tavazoie S & Church GM** (2000) *J. Mol. Biol.* **296**, 1205–1214.
20. **GuhaThakurta D, Schriefer LA, Waterston RH & Stormo GD** (2004) *Genome Res.* **14**, 2457–2468.
21. **GuhaThakurta D, Palomar L, Stormo GD, et al.** (2002) *Genome Res.* **12**, 701–712.
22. **Chen P, Ailion M, Bobik T, Stormo G & Roth J** (1995) *J. Bacteriol.* **177**, 5401–5410.
23. **Wasserman WW, Palumbo M, Thompson W, Fickett JW & Lawrence CE** (2000) *Nat. Genet.* **26**, 225–228.
24. **Gelfand MS, Koonin EV & Mironov AA** (2000) *Nucleic Acids Res.* **28**, 695–705.
25. **Gelfand MS, Novichkov PS, Novichkova ES & Mironov AA** (2000) *Brief. Bioinform.* **1**, 357–371.
26. **Bussemaker HJ, Li H & Siggia ED** (2001) *Nat. Genet.* **27**, 167–171.
27. **Kellis M, Patterson N, Birren B, Berger B & Lander ES** (2004) *J. Comput. Biol.* **11**, 319–355.
★ 28. **Prakash A & Tompa M** (2005) *Nat. Biotechnol.* **23**, 1249–1256. – *A nice overview and comparison of methods for comparative sequence analyses relevant to the identification of regulatory elements.*
29. **Kel-Margoulis OV, Kel AE, Reuter I, Deineko IV & Wingender E** (2002) *Nucleic Acids Res.* **30**, 332–334.
30. **Davidson EH, Rast JP, Oliveri P, et al.** (2002) *Science*, **295**, 1669–1678.
31. **Davidson EH** (2001) *Genomic Regulatory Systems: Development and Evolution.* Academic Press, San Diego.
32. **Eskin E & Pevzner PA** (2002) *Bioinformatics*, **18** (Suppl. 1), S354–S363.
33. **GuhaThakurta D & Stormo GD** (2001) *Bioinformatics*, **17**, 608–621.
34. **Frith MC, Li MC & Weng Z** (2003) *Nucleic Acids Res.* **31**, 3666–3668.
35. **Frith MC, Hansen U & Weng Z** (2001) *Bioinformatics*, **17**, 878–889.
36. **Johansson O, Alkema W, Wasserman WW & Lagergren J** (2003) *Bioinformatics*, **19** (Suppl. 1), i169–i176.
37. **Jegga AG, Gupta A, Gowrisankar S, et al.** (2005) *Nucleic Acids Res.* **33**, W408–W411.
38. **Aerts S, Van Loo P, Thijs G, Moreau Y & De Moor B** (2003) *Bioinformatics*, **19** (Suppl. 2), II5–II14.
39. **Thompson W, Palumbo MJ, Wasserman WW, Liu JS & Lawrence CE** (2004) *Genome Res.* **14**, 1967–1974.
40. **Wasserman WW & Fickett JW** (1998) *J. Mol. Biol.* **278**, 167–181.
41. **Anonymous** (1970) *Eur. J. Biochem.* **15**, 203–208.
42. **Zhu J & Zhang MQ** (1999) *Bioinformatics*, **15**, 607–611.
43. **Bulyk ML, Johnson PL & Church GM** (2002) *Nucleic Acids Res.* **30**, 1255–1261.
44. **Man TK & Stormo GD** (2001) *Nucleic Acids Res.* **29**, 2471–2478

★ 45. **Benos PV, Bulyk ML & Stormo GD** (2002) *Nucleic Acids Res.* **30**, 4442–4451. – *This paper presents discussions on the nonadditivity of intra-site positions within DNA-binding sites for TFs.*

46. **O'Flanagan RA, Paillard G, Lavery R & Sengupta AM** (2005) *Bioinformatics,* **21**, 2254–2263.

47. **Zhou Q & Liu JS** (2004) *Bioinformatics,* **20**, 909–916.

48. **King OD & Roth FP** (2003) *Nucleic Acids Res.* **31**, e116.

49. **Gershenzon NI, Stormo GD & Ioshikhes IP** (2005) *Nucleic Acids Res.* **33**, 2290–2301.

50. **Roth FP, Hughes JD, Estep PW & Church GM** (1998) *Nat. Biotechnol.* **16**, 939–945.

51. **Robison K, McGuire AM & Church GM** (1998) *J. Mol. Biol.* **284**, 241–254.

52. **Schneider TD, Stormo GD, Gold L & Ehrenfeucht A** (1986) *J. Mol. Biol.* **188**, 415–431.

53. **Stormo GD & Fields DS** (1998) *Trends Biochem. Sci.* **23**, 109–113.

54. **Fields DS, He Y, Al-Uzri AY & Stormo GD** (1997) *J. Mol. Biol.* **271**, 178–194.

55. **Hertz GZ & Stormo GD** (1999) *Bioinformatics,* **15**, 563–577.

56. **Lawrence CE & Reilly AA** (1990) *Proteins,* **7**, 41–51.

57. **Hertz GZ, Hartzell GW III & Stormo GD** (1990) *Comput. Appl. Biosci.* **6**, 81–92.

58. **Lawrence CE, Altschul SF, Boguski MS, Liu JS, Neuwald AF & Wootton JC** (1993) *Science,* **262**, 208–214.

59. **Schneider TD & Stephens RM** (1990) *Nucleic Acids Res.* **18**, 6097–6100.

60. **Kel AE, Gossling E, Reuter I, Cheremushkin E, Kel-Margoulis OV & Wingender E** (2003) *Nucleic Acids Res.* **31**, 3576–3579.

61. **Cartharius K, Frech K, Grote K, et al.** (2005) *Bioinformatics,* **21**, 2933–2942.

62. **Sandelin A, Alkema W, Engstrom P, Wasserman WW & Lenhard B** (2004) *Nucleic Acids Res.* **32**, D91–D94.

63. **Sun H, Palaniswamy SK, Pohar TT, Jin VX, Huang TH & Davuluri RV** (2006) *Nucleic Acids Res.* **34**, D98–D103.

64. **Djordjevic M, Sengupta AM & Shraiman BI** (2003) *Genome Res.* **13**, 2381–2390.

65. **Staden R** (1989) *Comput. Appl. Biosci.* **5**, 89–96.

66. **Lenhard B & Wasserman WW** (2002) *Bioinformatics,* **18**, 1135–1136.

67. **Galas DJ, Eggert M & Waterman MS** (1985) *J. Mol. Biol.* **186**, 117–128.

68. **van Helden J, Rios AF & Collado-Vides J** (2000) *Nucleic Acids Res.* **28**, 1808–1818.

69. **Sinha S & Tompa M** (2002) *Nucleic Acids Res.* **30**, 5549–5560.

70. **Rigoutsos I & Floratos A** (1998) *Bioinformatics,* **14**, 55–67.

71. **van Helden J, Andre B & Collado-Vides J** (1998) *J. Mol. Biol.* **281**, 827–842.

72. **Tompa M** (1999) *Proc. Int. Conf. Intell. Syst. Mol. Biol.* **7**, 262–271.

73. **Pevzner PA & Sze SH** (2000) *Proc. Int. Conf. Intell. Syst. Mol. Biol.* **8**, 269–278.

74. **Bussemaker HJ, Li H & Siggia ED** (2000) *Proc. Natl. Acad. Sci. U.S.A.* **97**, 10096–10100.

75. **Marsan L & Sagot MF** (2000) *J. Comput. Biol.* **7**, 345–362.

76. **Apostolico A, Bock ME, Lonardi S & Xu X** (2000) *J. Comput. Biol.* **7**, 71–94.

77. **Baldi P & Brunak S** (1998) *Bioinformatics: the Machine Learning Approach.* The MIT Press, Cambridge, MA.

78. **Stormo GD & Hartzell GW III** (1989) *Proc. Natl. Acad. Sci. U.S.A.* **86**, 1183–1187.

79. **Bailey TL & Elkan C** (1994) *Proc. Int. Conf. Intell. Syst. Mol. Biol.* **2**, 28–36.

80. **Cardon LR & Stormo GD** (1992) *J. Mol. Biol.* **223**, 159–170.

81. **Bailey TL & Elkan CP** (1995) *Mach. Learn.* **21**, 51–80.

82. **Liu JS, Neuwald AF & Lawrence CE** (1995) *J. Amer. Stat. Assoc.* **90**, 1156–1170.

83. **Workman CT & Stormo GD** (2000) *Pac. Symp. Biocomput.* 467–478.

84. **Liu X, Brutlag DL & Liu JS** (2001) *Pac. Symp. Biocomput.* 127–138.

85. **Thijs G, Lescot M, Marchal K, et al.** (2001) *Bioinformatics,* **17**, 1113–1122.

86. **Sinha S & Tompa M** (2003) *Nucleic Acids Res.* **31**, 3586–3588.

87. **Thompson W, Rouchka EC & Lawrence CE** (2003) *Nucleic Acids Res.* **31**, 3580–3585.

★ 88. **Tompa M, Li N, Bailey TL, et al.** (2005) *Nat. Biotechnol.* **23**, 137–144. – *An important article comparing various methods for regulatory element detection. It contains useful advice and guidance for the users of such software.*

89. **Guhathakurta D, Schriefer LA, Hresko MC, Waterston RH & Stormo GD** (2002) *Pac. Symp. Biocomput.* 425–436.

90. Tagle DA, Koop BF, Goodman M, Slightom JL, Hess DL & Jones RT (1988) *J. Mol. Biol.* **203**, 439–455.

91. Siepel A, Bejerano G, Pedersen JS, *et al.* (2005) *Genome Res.* **15**, 1034–1050.

92. Blanchette M & Tompa M (2002) *Genome Res.* **12**, 739–748.

93. Boffelli D, McAuliffe J, Ovcharenko D, *et al.* (2003) *Science,* **299**, 1391–1394.

94. Corcoran DL, Feingold E, Dominick J, *et al.* (2005) *Genome Res.* **15**, 840–847.

95. Levy S, Hennenhalli S & Workman C (2001) *Bioinformatics,* **17**, 871–877.

96. Cliften P, Sudarsanam P, Desikan A, *et al.* (2003) *Science,* **301**, 71–76.

97. Wang T & Stormo GD (2005) *Proc. Natl. Acad. Sci. U.S.A.* **102**, 17400–17405.

98. Xie X, Lu J, Kulbokas EJ, *et al.* (2005) *Nature,* **434**, 338–345.

99. Tan K, Moreno-Hagelsieb G, Collado-Vides J & Stormo GD (2001) *Genome Res.* **11**, 566–584.

100. Cliften PF, Hillier LW, Fulton L, *et al.* (2001) *Genome Res.* **11**, 1175–1186.

101. Tompa M (2001) *Genome Res.* **11**, 1143–1144.

102. Bray N & Pachter L (2003) *Nucleic Acids Res.* **31**, 3525–3526.

103. Schwartz S, Elnitski L, Li M, *et al.* (2003) *Nucleic Acids Res.* **31**, 3518–3524.

104. Schwartz S, Zhang Z, Frazer KA, *et al.* (2000) *Genome Res.* **10**, 577–586.

105. Blanchette M & Tompa M (2003) *Nucleic Acids Res.* **31**, 3840–3842.

106. Brudno M, Do CB, Cooper GM, *et al.* (2003) *Genome Res.* **13**, 721–731.

107. Bray N & Pachter L (2004) *Genome Res.* **14**, 693–699.

108. Frazer KA, Pachter L, Poliakov A, Rubin EM & Dubchak I (2004) *Nucleic Acids Res.* **32**, W273–W279.

109. Shah N, Couronne O, Pennacchio LA, *et al.* (2004) *Bioinformatics,* **20**, 636–643.

110. Zhu J, Liu JS & Lawrence CE (1998) *Bioinformatics,* **14**, 25–39.

111. Schwartz S, Kent WJ, Smit A, *et al.* (2003) *Genome Res.* **13**, 103–107.

112. Kolbe D, Taylor J, Elnitski L, *et al.* (2004) *Genome Res.* **14**, 700–707.

113. Loots GG & Ovcharenko I (2004) *Nucleic Acids Res.* **32**, W217–W221.

114. Berezikov E, Guryev V & Cuppen E (2005) *Nucleic Acids Res.* **33**, W447–W450.

115. Aerts S, Van Loo P, Thijs G, *et al.* (2005) *Nucleic Acids Res.* **33**, W393–W396.

116. Wang T & Stormo GD (2003) *Bioinformatics,* **19**, 2369–2380.

117. Blanchette M, Schwikowski B & Tompa M (2000) *Proc. Int. Conf. Intell. Syst. Mol. Biol.* **8**, 37–45.

118. Sinha S, Blanchette M & Tompa M (2004) *BMC Bioinformatics,* **5**, 170.

119. Siddharthan R, Siggia E & van Nimwegen E (2005) *PLoS Comput Biol.* **1**, e67.

120. Bailey TL & Noble WS (2003) *Bioinformatics,* **19** (Suppl. 2), II16–II25.

121. Frith MC, Spouge JL, Hansen U & Weng Z (2002) *Nucleic Acids Res.* **30**, 3214–3224.

122. Alkema WB, Johansson O, Lagergren J & Wasserman WW (2004) *Nucleic Acids Res.* **32**, W195–W198.

123. Wagner A (1999) *Bioinformatics,* **15**, 776–784.

124. Gupta M & Liu JS (2005) *Proc. Natl. Acad. Sci. U.S.A.* **102**, 7079–7084.

125. Zhou Q & Wong WH (2004) *Proc. Natl. Acad. Sci. U.S.A.* **101**, 12114–12119.

126. Berman BP, Nibu Y, Pfeiffer BD, *et al.* (2002) *Proc. Natl. Acad. Sci. U.S.A.* **99**, 757–762.

CHAPTER 7

Expressed sequence tags

Arthur Gruber

1. INTRODUCTION

Expressed sequence tags (ESTs) are short sequence reads, typically within the range of 100–700 bp (see *Fig. 1*), obtained from randomly selected cDNA clones. The concept was first introduced as a cost-effective approach for the rapid discovery and characterization of expressed genes (1). ESTs are often generated by single-pass sequencing of cDNA clones from one or both ends, usually covering only a part of the transcript sequence, and are relatively prone to error. Despite this latter feature, EST sequencing represents a mainstream methodology for gene surveying. Even nowadays, when whole genome sequences are available for many organisms, ESTs still play an important role in gene identification, transcript mapping, and description of the transcriptional activity of a tissue/cell type. Furthermore, ESTs may represent a very important body of evidence for gene prediction, and an abundant resource of molecular markers for physical mapping (2). Another envisaged application of ESTs is the quantification of gene expression, as the abundance of sequence reads representing each transcript may reflect the steady-state levels of these transcripts in a tissue or cell type (3). Finally, ESTs may

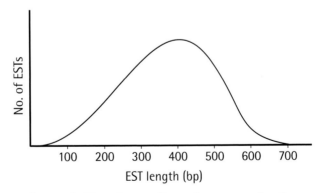

Figure 1. Distribution of EST reads according to the sequence length.
A typical EST sequencing project presents read lengths ranging from 100 to 700 bp, with most reads falling around 400 bp. Nevertheless, values may vary depending on the sequencing equipment, reagents, and protocols utilized.

Bioinformatics: *Methods Express* (Paul H. Dear, ed.)
© Scion Publishing Limited, 2007

provide reagents for downstream applications such as microarray analysis and immunoscreening of potential protective antigens. This chapter will cover some important aspects of EST analysis, including EST clustering, redundancy estimation, automated processing pipelines, and EST database searching. In addition, some important but often neglected methodological details will be also discussed. For additional literature on EST data production and analysis, the reader is advised to consult some reviews (4-9). The role of ESTs in gene prediction and annotation is touched on in this chapter, but the reader is referred to Chapter 4 for a more extensive discussion of gene prediction.

1.1 EST library construction and sequencing

In order to apply effectively the bioinformatics tools that will be covered in this chapter, the reader needs to understand how EST sequences are generated, as this will determine their strengths and weaknesses. We will, therefore, briefly outline the wet-lab side of EST production.

Different methods are currently available for cDNA library construction and will be discussed here in light of some implications on coverage within transcripts. However, it is beyond the scope of this chapter to present the different methods for EST library construction in depth, and the reader is advised to consult some specific reviews (9, 10). *Fig. 2(a)* depicts the most common methods for cDNA synthesis. Purified mRNA is used as a template for reverse transcriptase using either oligo(dT) as a primer for the first-strand synthesis or, alternatively, random hexamer primers (9, 10). After nicking the RNA–DNA hybrid with RNAse H, second-strand synthesis proceeds using the RNA fragments as primers and DNA polymerase I for the extension. Another approach for cDNA library construction is the ORESTES (ORF ESTs) method. The protocol is based on the construction of multiple mini-libraries (11, 12) using low-stringency reverse transcriptase polymerase chain reaction (RT-PCR) amplifications with arbitrary primers. Each mini-library, constructed with a particular primer, results in a heterogeneous population of amplified cDNA products, which correspond to a subset of the expressed gene profile. As each oligonucleotide acts as both forward and reverse primer, the distribution of the amplification products is biased towards the central part of the transcript (11). Whatever the synthesis method chosen, it is good practice to size-fractionate the cDNA fragments to avoid cloning very small inserts.

Conventional libraries can be constructed with specific adapters that permit unidirectional cloning (13). In this case, the researcher can choose which end of the transcript cDNA will be sequenced. On the other hand, because ORESTES libraries employ arbitrary primers, no directionality can be obtained. The synthesis method and the choice of the fragment end to be sequenced have a direct implication on the sequence coverage within transcripts (see *Fig. 2b*). A 5′-end sequencing results in a higher proportion of clones covering the coding regions of the transcripts, which is quite convenient for gene discovery projects (13). Conversely, 3′-end sequencing results in an extensive coverage of the 3′ end of the transcripts. As

(a)

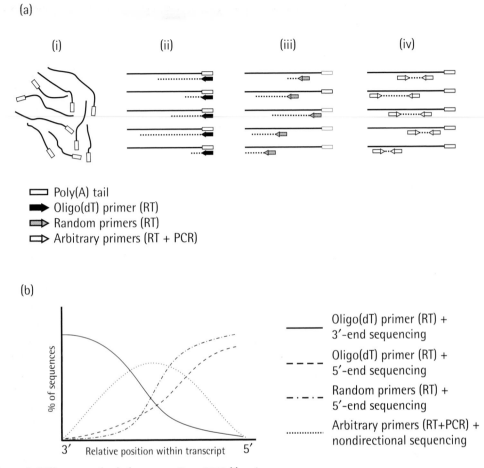

Poly(A) tail
Oligo(dT) primer (RT)
Random primers (RT)
Arbitrary primers (RT + PCR)

(b)

Oligo(dT) primer (RT) +
3′-end sequencing

Oligo(dT) primer (RT) +
5′-end sequencing

Random primers (RT) +
5′-end sequencing

Arbitrary primers (RT+PCR) +
nondirectional sequencing

Figure 2. Different methods for generating cDNA libraries.
The scheme (*a*) shows the first-strand synthesis approaches most commonly used to construct cDNA libraries. The mRNA (*i*) is purified from the total RNA by affinity chromatography using oligo(dT)-linked resins or paramagnetic beads. The most conventional method (*ii*) uses an oligo(dT) primer that anneals to the poly(A) tail and a reverse transcriptase (RT) that catalyzes the cDNA synthesis (9, 10). Daughter strands are of varying lengths and the 5′ ends of the transcripts are usually poorly represented if 3′-end sequencing is carried out (*b*, solid line). Better transcript coverage may be attained by sequencing the 5′ ends (*b*, dashed line). A variation of this method uses random primers instead of the oligo(dT) primer (*iii*). They consist of fully degenerate hexamers (dN$_6$), sometimes linked to an anchor sequence at the 5′ end. This method yields a better coverage of the 5′ end of the transcripts, especially if directional cloning followed by 5′-end sequencing is used (*b*, dash–dot line). In both cases (*a, ii* and *iii*), second-strand synthesis is performed by conventional methods, employing RNAse H for nicking the RNA strand of the DNA–RNA hybrid and DNA polymerase I to replace the RNA segments by DNA (not shown here). An alternative approach, termed ORESTES (11, 12), utilizes an arbitrary primer for both RT reaction and subsequent PCR amplification (*d*). As the same oligonucleotide acts as both the forward and reverse primer, distribution of the amplified products is biased to the central part of the transcripts (*b*, dotted line).

these regions are much more variable than the coding regions, they can be used for the unambiguous identification of the transcripts, thus allowing quantitative gene expression profiling (14–16). In fact, this characteristic has been used in the construction of the UniGene database (14), in which the clusters are anchored by unique 3′-untranslated regions (3′UTRs). In addition, because intronic sequences are rare in 3′UTRs, such ESTs can also be used as sequence-tagged sites (STSs) for genome mapping (17). ORESTES reads, on the other hand, are biased towards the central part of the transcripts (11, 12), which prioritizes the protein-coding information. This aspect makes this method a good choice for gene discovery projects and, furthermore, generates complementary data to 5′- and 3′-end sequencing efforts.

1.2 Representation: normalized and subtracted libraries

The *transcriptome* of an organism can be defined as its complete repertoire of transcripts, including splice variants. Gene expression in all organisms varies over time (for example, during development or in response to changing conditions) and, in complex eukaryotes, from one cell type to another. Therefore, the transcriptome of a particular tissue, sampled at a particular time, will only be a subset of the complete transcriptome of the organism. Of course, much of the value of EST analysis lies in comparing the transcriptomes of different cell types, at different developmental stages or under different conditions.

In principle, a nonbiased cDNA library should faithfully represent the transcriptome of the cell or tissue that it was derived from, provided that a sufficient number of reads has been obtained. However, because some mRNAs are highly expressed whilst many others are only found in tiny amounts, this latter class is under-represented in any cDNA library. If the goal of the EST sequencing project is to obtain a comprehensive survey of the transcriptome, then a very large number of reads would have to be sequenced, making this approach too expensive. In order to avoid missing rare transcripts whilst maintaining a cost-effective approach, normalization techniques have been devised to decrease the relative representation of abundant transcripts whilst increasing that of rare transcripts. The commonest normalization method relies upon reassociation kinetics (9, 18–20): the cDNA is denatured and, because of second-order kinetics, abundant cDNAs tend to renature more quickly than scarce ones. Hydroxyapatite chromatography can then be used for selective purification of the remaining single-stranded molecules, which will be relatively enriched for the scarce transcripts. ORESTES also normalizes the cDNA representation (11, 12) by employing a high number of amplification cycles, so that abundant transcripts attain saturation in the early steps of cycling, whereas rare transcripts continue to be amplified during the later cycles.

Another methodology to change the representation of cDNA libraries is subtractive hybridization (21, 22). This method is used to reduce the representation of transcripts already surveyed in previous libraries, as well as to enrich for sequences that are differentially expressed among specific tissues, cell types, etc.

The cDNA library that will be the target of subtraction (the 'tester' population) is denatured and hybridized to another library that is present in excess (the 'driver' population). Fragments that are common to both populations anneal to each other, whereas the tester-specific products will remain single stranded. The purification of these products can be achieved by either hydroxyapatite chromatography or avidin–biotin binding.

2. METHODS AND APPROACHES

2.1 Overview

EST analysis is a complex multiple-step process that makes use of several distinct programs. In this section, we will give a brief introduction to the theory behind each method and present some specific protocols covering worked examples of widely used methods. As no generic step-by-step recipe can fulfill all specific requirements and characteristics of different EST sequencing projects, we have chosen to offer some typical protocols in a tutorial-like presentation where the reader will be able to adapt the techniques to fit his or her needs with only minor changes. We recommend using a UNIX/Linux machine connected to the Internet through a broadband connection. In our hands, a PC-based server running Linux is the most cost-effective platform and enables one to run analyses of even relatively large numbers of ESTs. Apple Mac OS X 10.x is another recommended choice. For very large datasets (hundreds of thousands to millions of ESTs), a more powerful workstation or a cluster of PCs is recommended. We are assuming that a Perl interpreter (http://www.perl.org[7.1]) has previously been installed on your server, as well as Java Platform 2 (http://java.sun.com/j2se/[7.2]). Other specific programs are listed in each protocol and must be installed on your server before running the tutorials. All software chosen for the protocols is open source or free for nonprofit academic use (other policies may apply for commercial use), and is available on the Internet or on request to the authors. Each protocol lists the required software, corresponding publication (where available), sources for download, and/or author's contact details. Software installation is relatively easy, even for novice users, but may require administrator privileges. Please contact your local server administrator in case you need any help in installing the necessary programs.

Example datasets for this chapter are provided on the book's web site. Because of the large number of subdirectories and files involved, the complete dataset is provided as a single compressed file; *Protocol 1* describes how to install the example datasets.

Protocol 1

Installing the example datasets

1. Download the dataset file protocols.tar.gz [7.3] from the All _Protocols folder for this chapter on the book's web site.

2. Decompress the file with the following UNIX/Linux command: 'tar zxvf protocols.tar. gz'. If this command does not work (depending on your system's configuration), then try the following command:

   ```
   gzip -dc protocols.tar.gz | tar xvf -
   ```

   Either of these commands will extract all files and subdirectories within a directory named protocols.

3. Move your protocols directory to the selected location on your server disk. You may need to have administrator privileges to do this:

   ```
   mv protocols /selected_directory
   ```

   There are four subdirectories within the protocols root directory: Protocol_2, Protocol_4, Protocol_5, and Protocol_6. The required subdirectories and files will be listed in each of the protocol sections below.

2.2 EST databases

In this section, we will review some of the EST resources available and describe how to retrieve data from two of them – dbEST and UniGene.

2.2.1 dbEST

Genbank's EST database (dbEST) is a publicly available repository useful for gene discovery and comparative gene expression studies (17, 23, 24). It constitutes the EST division of Genbank (25) and corresponds to the largest fraction (48%) of entries (source: NCBI web page http://www.nlm.nih.gov/ [7.4]). As of October 2006, dbEST (release 100606) comprised more than 38 million entries from over 1200 organisms, although about half of the entries come from only eight organisms (see *Table 1*). In *Protocol 2*, we will see how to download an organism-specific set of ESTs.

Table 1. Top 20 organisms represented on dbEST (release 100606)

Organism	Number of ESTs
Homo sapiens (human)	7 893 983
Mus musculus + domesticus (mouse)	4 720 064
Oryza sativa (rice)	1 188 565
Zea mays (maize)	1 143 728
Bos taurus (cattle)	1 137 353
Danio rerio (zebrafish)	1 134 553
Xenopus tropicalis	1 044 182
Rattus norvegicus + sp. (rat)	871 144
Triticum aestivum (wheat)	855 066
Ciona intestinalis	686 396
Sus scrofa (pig)	623 929
Arabidopsis thaliana (thale cress)	622 973
Gallus gallus (chicken)	599 141
Xenopus laevis (African clawed frog)	537 424
Drosophila melanogaster (fruit fly)	514 545
Hordeum vulgare + subsp. vulgare (barley)	437 321
Canis familiaris (dog)	365 909
Glycine max (soybean)	359 151
Caenorhabditis elegans (nematode)	346 064
Pinus taeda (loblolly pine)	329 469

Source: NCBI (http://www.ncbi.nlm.nih.gov/ [7.5]).

Protocol 2

Downloading an organism-specific set of ESTs from dbEST

1. Point your web browser to the NCBI's site at http://www.ncbi.nlm.nih.gov/[7.5].

2. Select the **Nucleotide** option (default is **All Databases**) on the **Search** selection box and type 'Plasmodium falciparum [organism]' in the blank form. An organism name followed by the '[organism]' tag will limit the sequence retrieval to this particular organism (see Chapter 1). Click on the **Go** button to retrieve the records.

3. The page will now display the total number of records retrieved (see **All**: approximately 43 000 records were retrieved at the time of writing, although this is of course likely to increase over time) and a summary of the first 20 nucleotide records of *P. falciparum*. Click on the **Search** selection box again and select **EST** (a subset of the nucleotide database) from the pull-down menu. Click on the **Go** button to retrieve the records[a]. The total number of entries (about 21 000 at the time of writing) will be displayed on the top of the record list.

4. Next to **Display** towards the top of the screen, change the setting from **Summary** (the default) to **FASTA** format option. Then select **File** from the **Send to** pull-down menu. The program will ask for confirmation and then the download process will start, creating a large file called sequences.fasta.

5. If you want to confirm that the file is complete and contains all records retrieved from the database (see step 3), you can type the following UNIX command:

```
grep ">" sequences.fasta | wc -l
```

This command extracts all lines displaying a '>' character (present in each FASTA header) and counts the total number of lines obtained: this is a simple way of counting how many sequences are present in the file. The number should be equal to the number of entries displayed in step 3.

6. Rename the downloaded file as P_falciparum.fasta. On UNIX systems, this is done using the command:

```
mv sequences.fasta P_falciparum.fasta
```

7. In the Protocol_2 directory, we provide a P_falciparum.fasta[7.6] file that was downloaded using the above method in October 2006 and contains 21349 sequences in FASTA format. By the time you download yours, the number of sequences will probably be larger, as new entries are continuously being incorporated into dbEST.

Note

[a]If you prefer, you can combine steps 2 and 3 into one: you can specify both the organism and EST subset by typing 'Plasmodium falciparum [organism] AND gbdiv_est [PROP]' in the blank query window.

2.2.2 TIGR Gene Indices (TGI)

TGI, now hosted at the Dana-Farber Cancer Institute (http://compbio.dfci.harvard.edu/tgi/[7.7]), is a database of clustered ESTs from many eukaryotic organisms (26). GenBank coding sequences from genomic and mRNA sequences are clustered by pairwise alignments using a modified version of MEGABLAST (27) and the clusters are then assembled using CAP3 (28). The final result is a set of tentative consensus sequences representing unique transcripts. The tentative consensus sequences are integrated with annotation data into a relational database and can be queried using text and BLAST searches.

2.2.3 STACKdb

STACKdb (29, 30) (http://www.sanbi.ac.za/Dbases.html[7.8]) is a database of clustered ESTs hosted at the South Africa National Bionformatics Institute (SANBI). Clustering is performed with D2_CLUSTER (31) and STACKPACK (30) software. EST sequences are clustered, assembled, and then submitted to a post-assembly analysis protocol in which clusters spanning different regions of a transcript are identified and grouped. Single-nucleotide polymorphisms (SNPs) and splicing variants are also identified.

2.2.4 UniGene

UniGene (http://www.ncbi.nlm.nih.gov/UniGene[7.9]) is an automatic system for partitioning GenBank sequences into a nonredundant set of gene-oriented clusters (14). Each cluster corresponds to a unique transcript, but, unlike TGI and STACKdb, UniGene clusters are not assembled and, therefore, no consensus sequences are available. UniGene clusters also provide some important cross-information such as the tissue types where gene expression was observed and corresponding map locations. UniGene is integrated (32) with other databases such as the IMAGE Consortium (33), a public domain resource of arrayed cDNA libraries, and sequence, map, and expression data.

Protocol 3 illustrates how to search for UniGene clusters and retrieve information of EST clusters, protein similarity and annotation, gene expression, mapping position, and sequence data. For this purpose we will use, as the query, glucagon (a polypeptide hormone secreted by the alpha cells of the pancreas islets of Langerhans in response to hypoglycemia).

Protocol 3

Using UniGene

1. Point your web browser to the NCBI's UniGene site at http://www.ncbi.nlm.nih.gov/UniGene[7,9].

2. Into the text field, next to **Search Unigene for**, enter 'glucagon[All Fields] AND ("Homo sapiens" [Organism])', taking care over the brackets. This query will retrieve information on the glucagon gene, restricted to human sequences. (For more information on querying NCBI databases, refer to Chapter 1, or follow the **Query Tips** link from the web page). Click **Go**.

3. Select the glucagon gene by clicking on the **Hs.516494** link.

4. Now a list of information related to glucagon is displayed. We will start by exploring the link for the Swiss-Prot entry. For this option, click on the **sp:P01275** link in the 'SELECTED PROTEIN SIMILARITIES' section.

5. The new page presents a very comprehensive annotation on the protein, including pertinent bibliographic data, a typical Swiss-Prot description of the protein function, site of production, and pharmaceutical information, plus the corresponding amino acid sequence.

6. Go back to the previous page and click on the **Expression profile** link in the 'GENE EXPRESSION' section. The newly opened page displays a table presenting the gene expression deduced from the analysis of EST counts of different tissues. As expected, pancreas is by far the predominant expression site.

7. Return to the former page and, still in the 'GENE EXPRESSION' section, click on the **GEO profiles** link. This will take you to the Gene Expression Omnibus (GEO) database, a repository of gene expression profiles that includes experimental data of microarray, serial analysis of gene expression (SAGE), and mass spectrometry proteomic experiments. A list of GEO records will be presented, with the respective expression profile data graph on the right.

8. Choose the record corresponding to the experiment with 'Normal tissues of diverse types (SHCN)'. When writing this text, the experiment corresponded to the **GDS1086 record**, but it may have changed by the time you access the site. Click on the expression graph on the right. The graph displays the analysis of glucagon expression in dozens of physiologically normal tissues obtained from various sources. As expected, pancreas tissue presents the highest expression of this gene.

9. Go back to the previous page (listing all of the experiments) and click on the **Record** link for this experiment. This will take you to the corresponding microarray data and the clustering analysis.

10. Go back to the previous page and back once more to the glucagon entry page (this is the page you reached in step 3). Look at the information available at the 'MAPPING POSITION' section (scroll down). It reports that the glucagon gene is located at chromosome 2, with the cytogenetic locus 2q36–q37.

11. Below this information (under 'SEQUENCES'), the page displays a list of all mRNA and EST sequences related to this UniGene cluster. Sequences can be retrieved on an individual basis by clicking on the respective links or, alternatively, can easily be downloaded in a batch by clicking on the **Download sequences** button at the bottom of the page.

12. Repeat the whole analysis, now using the human hemoglobin beta-chain gene. Click on the **Expression profile** link. The expression profile page will show that blood cells, bone marrow, and muscle are the preferential expression sites. Compare the results with those obtained for the glucagon gene.

2.3 Automated EST pre-processing pipelines

Proper multi-step pre-processing of EST raw data is necessary before one can proceed with clustering and assembly (6) or with downstream annotation. *Fig. 3* displays a typical EST processing pipeline scheme. First (*Fig. 3a, b*), the trace files, if available, are submitted to a base caller and quality evaluation program such as PHRED (34, 35). In this step, the nucleotide sequence is extracted and confidence values (also known as 'PHRED values') are then ascribed to all bases. Low-quality sequence can then be filtered out. If only sequence data is available on a public database, then no confidence values can be ascribed and quality filtering cannot be done. Secondly (*Fig. 3c, d*), undesirable parts of the sequence are 'masked' by replacing them with N or X characters, which are disregarded by most clustering and assembly programs. For example, simple sequence repeats (SSRs) can be identified and masked by programs such as DUST (36) and TANDEM REPEATS FINDER (37), and the sequences of vectors and/or primers used in making the EST library can be identified by comparison with specific databases. Poly(A) tracts should also be identified and masked, especially when dealing with 3'-end directional libraries. All of these masking steps are essential to avoid unrelated ESTs being grouped into the same cluster/contig. Thirdly (*Fig. 3e–h*), contaminant sequences must be identified and filtered out. Contaminants may comprise endogenous sources such as ribosomal sequences and transcripts derived from organelles such as mitochondria or plastids, and heterologous sources such as bacteria. Contaminant filtering is done by pairwise alignment of the ESTs against databases containing likely contaminant sequences; a rule of thumb is to perform the filtering steps using the smaller databases at the earlier steps of the pipeline and the larger databases at the later steps. As positive matches are identified and discarded, the overall number of reads that have to be processed by the later slower steps is reduced, thus decreasing the overall pipeline processing time. Finally, the resulting 'clean' EST reads can be assembled into clusters believed to represent distinct transcripts.

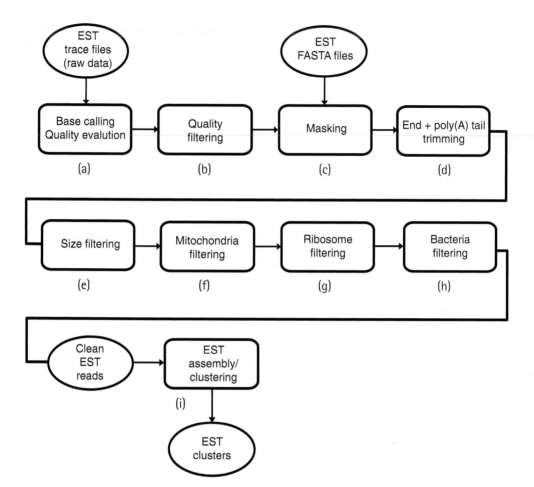

Figure 3. A typical EST processing pipeline.
An EST processing pipeline is composed of several sequential processing steps in which the output data from one step is used as the input for the subsequent one. The scheme shows a typical pipeline. Input data can be loaded into the pipeline using either trace files or FASTA format sequence files. The chromatogram files are processed by PHRED (34, 35), a program that performs base calling and quality evaluation for each base (a). The reads are submitted to a second component (b) that filters out the reads that have an overall quality below a user-defined threshold. (If FASTA sequence files are used instead of trace files, no quality evaluation can be performed and steps a and b are skipped.) Sequences then are processed by a component (c) that runs CROSS_MATCH (38) to mask sequences arising from the cloning vector or from the primers used to generate the ESTs. Sequences are then submitted to a trimming step where low-quality/masked regions and poly(A) tails are trimmed off (d), and the reads whose remaining sequences after trimming fall below a user-defined minimum length are discarded by a size filter (e). Next, sequences are processed by multiple pipeline components that run either CROSS_MATCH (f, g) or BLAST (h, i) to perform similarity searches against databases of undesired sequences and potential contaminants, which include mitochondrial (f), ribosomal (g), and bacterial (h) sequences; all reads presenting alignment blocks above a set of user-defined values are filtered out. Finally, the ESTs are assembled (i) using CAP3 (28). The clean EST assembled clusters are then ready for annotation.

The details of the EST pipeline will, of course, be tailored to suit the dataset being analyzed. *Protocol 4* describes the construction of a typical EST processing pipeline using EGENE, a generic pipeline generation system (39). The pipeline will be used for processing an example dataset consisting of trace files of a cDNA library synthesized through the ORESTES method (12). The suggested parameters of each step are known to work well for this dataset, but should also work reasonably well for most EST sequencing projects. The reader can use this example pipeline as a template and then change the number and order of the processing steps, as well as the corresponding databases, in order to fit specific project requirements. Replacing databases and inserting or removing pipeline components can be performed easily with COED, EGENE's graphical configuration editor (see *Fig. 4*). For more details on how to use EGENE and COED, the reader is referred to the original description (39) and the official web site (see below). All programs or symbolic links should be put on a directory that is specified on the path of the operating system. Should you get any installation problems, please refer to the respective official pages and publications of the software.

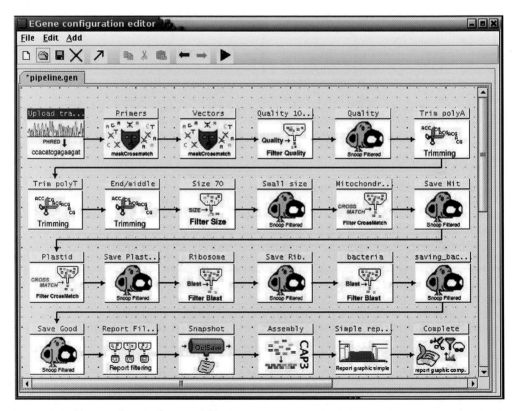

Figure 4. Building pipelines with EGENE and COED.
EGENE is a generic, flexible, and modular pipeline generation system that makes pipeline construction a modular job. Pipelines can be constructed using COED, a Java visual configuration editor. Icons representing each component of the pipeline are selected and landed on the canvas. The different steps are interconnected using an arrow tool and the pipeline can be executed within COED (clicking on the arrowhead button) or directly from the UNIX command line.

The example pipeline to be used in this tutorial will be run on a set of chromatogram files of ORESTES reads of *Eimeria tenella*, an apicomplexan protozoan parasite. All reads will be submitted to contaminant filtering steps against mitochondrial, apicoplast (an ancient plastid-derived organelle), ribosomal, and bacterial sequences. In detail, the pipeline consists of the following steps:

1. Uploading trace files and performing base calling and quality evaluation.
2. Masking primer sequences.
3. Masking vector sequences.
4. Filtering low-quality sequences.
5. Saving sequences invalidated by the quality filter.
6. Trimming off poly(A) and...
7. ...the complementary poly(T)) sequences.
8. Trimming off bases that present a low PHRED quality value and those that are masked.
9. Filtering short sequences.
10. Saving sequences invalidated by the size filter.
11. Filtering mitochondrial sequences.
12. Saving sequences invalidated by the mitochondrial contaminant filter.
13. Filtering plastid sequences.
14. Saving sequences invalidated by the plastid contaminant filter.
15. Filtering ribosomal sequences.
16. Saving sequences invalidated by the ribosomal contaminant filter.
17. Filtering bacterial sequences.
18. Saving sequences invalidated by the bacterial contaminant filter.
19. Saving sequences not previously invalidated by any filter.
20. Generating a report of all filtering steps.
21. Creating an XML snapshot recording all of the processing steps that were performed.
22. Assembling the valid sequences using CAP3.
23. Generating an HTML page with graphical reports.
24. Generating a complete graphical report.

Protocol 4

Building an automated pipeline for EST processing

Software
You will need the following software:

- EGENE (38) (http://www.coccidia.icb.usp.br/egene/ [7.10])
- PHRED, PHRAP, CONSED, and CROSS_MATCH (34, 35, 39, 40). These programs should be requested from the authors. Contact information and instructions are available at http://www.phrap.org [7.11]
- BLAST (41) (http://www.ncbi.nlm.nih.gov/BLAST/ [7.12])

- CAP3 (28) (http://seq.cs.iastate.edu/[7.13])
- Perl interpreter, version 5.6.0 or higher (http://www.perl.org[7.1]), with the GD graphics library (http://www.boutell.com/gd/[7.14]), DBD::Pg
- Java 2 Platform, Standard Edition (J2SE), version 1.4.1 or higher (http://java.sun.com/j2se/[7.2])

Method

1. We have previously constructed a pipeline for this tutorial using COED and saved the configuration file as pipeline.gen [7.15] in the /Protocol_4/config_files directory. In order to run the pipeline, go to the /Protocol_4/dataset directory. This directory contains the chromat_dir subdirectory, which presents a set of 96 trace files.

2. Invoke COED on the UNIX/Linux command line by typing 'coed.pl' and pressing 'enter'.

3. Using the menu bar, choose the **File Open** command and select the pipeline.gen file, located in the /Protocol_4/config_files directory.

4. You should see a graphical display of the pipeline (see *Fig. 4*) with all components represented as interconnected icons.

5. Any component can be deleted simply by selecting it with the mouse and using the **Edit Cut** command. Please refer to the EGENE web site for a complete list of commands.

6. To run the pipeline within COED, click on the green arrowhead button of the toolbar.

7. COED will open up a dialog box asking for the working directory. Click on the blank form with the right button of the mouse and select the directory /Protocol_4/dataset (a full directory path must be specified, according to the directory structure of your local server). Press the **Run** button.

8. COED will now display a small window informing you that your pipeline is executing. Because the multiple processing steps may take a long time to run, the pipeline will be executed in the background. Thus, COED will not report the end of the entire job and can now be closed.

9. To check whether the pipeline processing has finished, you have to use the operating system's command line. Type 'ls -l' to list the files. No temporary directories (identified by a '_temp_' suffix) should be present when the pipeline processing is finished. The pipeline used in this tutorial processes 96 reads and takes around 2 min to run in a PC/Linux with a Pentium 4 2.8 GHz processor.

10. Alternatively, instead of running the pipeline from within COED, you can also save your pipeline as a *.cnf text file using the **Save As EGene file** command.

11. Assuming that the pipeline.cnf configuration file has been stored in the /Protocol_4/config_files directory, and that you are in the /Protocol_4/dataset directory, the UNIX/Linux command to execute the pipeline is:

```
bigou.pl -c ../config_files/pipeline.cnf >/dev/null&
```

This command invokes bigou.pl, an EGENE program that reads the configuration files and starts each processing step of the pipeline. Although it is not necessary, we recommend that you redirect the standard output to a null device (specified by a '>/dev/null' statement at the end of the command) in order not to get your screen jammed with messages. Also, using the '&' character at the end of the command will put the whole process in the background, thus liberating the terminal. At the end of the process, you should find the following additional files in this directory:

```
filtered_by_quality.fasta
filtered_by_size.fasta
filtered_by_mitochondrion.fasta
```

filtered_by_plastid.fasta
filtered_by_ribosome.fasta
filetered_by_bacteria.fasta
filtering_report.html
redundancy_report.html
report_graphic_simple.html
final_snapshot.xml
good_sequences.fasta

and also the additional subdirectories:

assembly_dir
complete_report
images_dir

12. Check the content of the results files containing the prefix 'filtered_by_' by typing 'more name_of_the_file' or loading them onto any text editor. These files contain multiple sequences in a FASTA format and correspond to the reads filtered out in each of the filtering steps. If no read is identified by a filtering step, then the corresponding file will be empty.

13. The file named good_sequences.fasta contains all sequences that passed through the filtering steps and were considered 'good'. These sequences have also fulfilled the minimum quality criteria established on the pipeline. Please note that the sequences are trimmed, so they do not contain any vector, primer, or low-quality bases.

14. The assembly_dir directory contains a typical directory structure required by CONSED to visualize DNA assemblies. Thus, chromat_dir contains the trace files, phd_dir stores the PHD (PHRED-processed) files of the accepted sequence reads, and edit_dir contains the assembly files created by the CAP3 assembler. To inspect the DNA assembly, invoke CONSED within the edit_dir directory. The command varies according to the CONSED version you are using but, in general, it should be 'consed' or 'consed_linux':

```
consed clean.fasta.cap.ace
```

Once CONSED is loaded, select the clean.fasta.cap.ace file. All contigs will be listed. Choose one of them by clicking twice with your mouse. A new window will open up displaying the multiple sequence assembly view. For more information on how to use CONSED, please consult the original documentation of the software.

15. A complete set of the expected results is provided in the /Protocol_4/results directory.

2.4 Transcript reconstruction

One can trace a parallel between an unclustered set of EST reads and a bulk of construction bricks. In the same way that bricks turn out to be much more useful when arranged with each other in a shape of a house, ESTs become much more informative when adequately clustered and assembled in a process known as transcript reconstruction. In fact, EST data has a fragmentary character and presents the following limitations:

- **Short length.** ESTs vary from 100 to 700 bp, with a typical average of 400 bp (see *Fig. 1*). In most cases, this size is not enough to cover whole transcripts.
- **Low-quality data.** ESTs are single-pass reads and only a relatively small portion of the sequence presents a high quality and good confidence level.

- **Nonoverlapping reads may cover the same transcript.** ESTs may cover the same transcript on different nonoverlapping regions, leaving sequence gaps.
- **EST libraries represent subsets of the transcriptome.** EST libraries only reflect the gene expression profile of the cell/tissue used as the source of mRNA, thus representing only a fraction of the transcriptome of the whole organism.
- **Representational bias.** Due to the heterogeneous frequency of transcripts, even a large sampling may still miss some rare transcripts. Biases of cDNA synthesis and cloning contribute towards making this representation worse.

To reduce data fragmentation and extract all potential information, ESTs must be submitted to a process known as transcript reconstruction. It is beyond the scope of this chapter to cover in much detail the different approaches proposed for such a task. For in-depth descriptions, the reader is referred to reviews (5, 42, 43) and the original articles describing the protocols utilized for STACKdb (29-31), TGI (26, 44, 45), and UniGene (14, 46). Here, we will instead discuss general concepts involved with this issue and then introduce the most important EST resources.

EST clustering and assembly are often and erroneously used synonymously. Clustering can be defined as the process of grouping subsets of EST reads that share some sequence among themselves. Thus, clustering involves an all-versus-all comparison using a loose stringency and, preferably, fast algorithms. Once the clusters are established, then a DNA assembler can be used to align the overlapping reads of each cluster and generate consensus sequences. For small sets of ESTs, a single assembly step can be used without previous clustering (6). Bypassing the clustering step, however, may present potential disadvantages. First, because specific clustering programs do not perform a real pairwise sequence alignment, but rather look for sequence words shared by distinct reads (31, 47), they are often much faster than conventional DNA assembly software. This is particularly more pronounced in large EST sets, where a direct assembly process is not always feasible. Secondly, clustering word-sharing sequences makes more sense than going straight to the assembly phase, as sequences representing different paralogous genes or splicing variants can be clustered together and then, after a proper assembly, can be separated into different contigs. This approach preserves the information about which consensus sequences belong to the same cluster and, as such, are closely related. This method (see *Fig. 5*) is employed by stackpack, one of the most complete transcript reconstruction solutions (30, 31, 43). TGI (26, 44, 45, 48), on the other hand, uses a protocol that consists in an all-versus-all sequence comparison using a modified version of megablast (27). Transitive closure is used for building the clusters using a criterion where sequences presenting overlaps of at least 40 bp and 95% identity, with a maximum mismatched overhang of 30 bp, are included in the same cluster. Tentative consensus sequences are then generated for each cluster using the cap3 assembler (28).

Clustering can be performed using either loose or stringent conditions. A loose clustering may result in paralogs being clustered and assembled together, but consensus sequences tend to be longer. Conversely, by using a stringent clustering, one can differentiate paralogs more accurately, but the consensus sequence length of each cluster will be rather shorter. There is no universal recipe

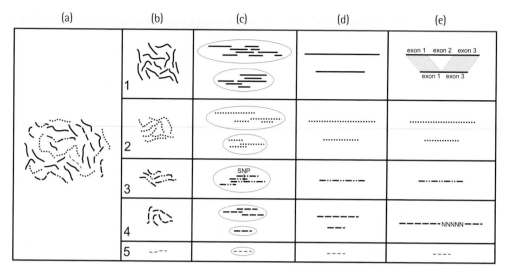

Figure 5. EST clustering and analysis.
Transcript reconstruction is a complex multi-step task. The scheme is based on the STACKPACK (31, 42) pipeline and shows the different steps involved in a process aimed at extracting all potential EST information. The columns of the chart represent the different phases of an EST clustering and analysis process. A set of ESTs (*a*) is submitted to a clustering step (*b*), resulting in clusters (numbered 1–5). Each cluster may be composed of multiple reads, or may consist of a single read (a 'singleton', such as cluster 5). Each cluster is assembled separately (*c*), to generate either a single contig (clusters 3 and 4) or multiple contigs (clusters 1 and 2). Some reads may fail to assemble into contigs, remaining as singletons (as is the case for one read of cluster 4). At this stage, candidate SNPs can be identified in the multiple sequence alignment (*c*, cluster 3). Consensus sequences are derived from each assembly (*d*) and if clone information is available, 5′ and 3′ reads of the same clones can be linked (cluster 4, column *e*). Also, contigs can be aligned with each other for the identification of alternate splicing forms (cluster 1, column *e*).

for establishing a high-confidence clustering without some manual curation, a step needed for checking whether the cluster separation follows a biological sense. The appropriate stringency is very difficult to define *a priori*, as different gene families may present distinct divergences across their respective paralogs due to the different paralogy times. Furthermore, different substitution rates can be observed among gene families. As a result, a gene family may contain paralogs that are much more closely related to each other than those of another family are among themselves.

Following the clustering and assembly steps, a set of clusters is obtained, each comprising one or more contigs composed of multiple reads. The process may also result in some clusters or contigs that consist of only a single read – 'singletons' (see *Fig. 5*). Singletons may correspond to low-expression transcripts that were collected only once in the EST sampling. Alternatively, they may have been originated from an unidentified contaminant source, so a manual inspection must be performed before considering singletons as rare transcript representatives. If unidirectional libraries have been used, and information on sequence direction is available, then it is possible to perform clone linking (see *Fig. 5*). This is analogous to the use of 'read pairs' in shotgun sequence assembly: two clusters can be linked if one contains a sequence read from the 5′ end of an insert and the other contains a sequence read from the 3′ end of the same clone (43). Finally, a deeper inspection may reveal splicing variants and SNPs. We will not cover here methods for SNP surveying, and the reader is advised to consult specific references (49–52).

The following protocol demonstrates the use of the TGI Clustering Tools package (TGICL) and CLVIEW, a graphical interactive tool for visualization of ACE format assembly files generated by CAP3 or PHRAP. TGICL and CLVIEW were developed at TIGR and are freely available for download at the web address below. We will use the *P. falciparum* EST dataset downloaded in *Protocol 2*.

Protocol 5

Clustering ESTs using TGICL

Software
You will need the following software:

■ TGICL and CLVIEW (53) (http://compbio.dfci.harvard.edu/tgi/software/[7.16])

Method

1. Go to the directory where you stored the *P. falciparum* EST dataset that you downloaded from dbEST in *Protocol 2*. Alternatively, go to the /Protocol_5/dataset directory, where you will find P_falciparum.fasta[7.17] (this is identical to the file provided in the Protocol_2 folder, and is a set of *P. falciparum* ESTs downloaded in October 2006). The analysis that follows uses this stored dataset, and the results will differ slightly if you use a more recently downloaded set of ESTs.

2. TGICL is very simple to run in a default mode. Type the following command:

```
tgicl P_falciparum.fasta
```

This process can take several minutes. When finished, you should receive the following message on the screen:

```
tgicl (P_falciparum.fasta) finished on machine in /your_ ↵
    directory/P_falciparum, without a detectable error
```

3. Using UNIX's 'ls' and 'less' commands, identify all files and directories that were created. (A complete set of the results from the dataset provided is given in the /Protocol_5/results directory.)

4. We can now answer the following questions:
   (i) How many clusters were generated?
   TGICL creates a file called *_clusters (where * is the name of the input file), which presents a pseudo-FASTA format where each record is actually a cluster definition and consists of a header line containing a greater than ('>') character. Thus, to count the number of clusters, we just have to use the command:

```
grep ">" P_falciparum.fasta_clusters | wc -l
```

   If you used the dataset provided (P_falciparum.fasta), you should find that there were 2782 clusters.

   (ii) How many singletons were generated?
   Singletons are sequences that are not clustered with others – they represent 'single-

sequence clusters'. TGICL stores a list of all singleton reads in a file called file_name.singletons. If you used the dataset EST.fasta, the file should be P_falciparum.fasta.singletons. As the sequence headers from NCBI use a 'gi' (gene identifier) tag, we can count the number of reads simply by counting how many times that the string 'gi' appears in the file:

```
grep "gi" P_falciparum.fasta.singletons | wc -l
```

Using the dataset provided, you should find that there were 5297 singletons.

(iii) How can one retrieve the singleton sequences?

TGICL creates a *.cidx index file for fast retrieval of cluster sequences. By using the CDBYANK program (provided in the TGICL package), we can extract the singleton sequences from the index file and store them in a file named singleton.seqs using the command:

```
cdbyank P_falciparum.cidx < P_falciparum.fasta.singletons ↻
   > singleton.seqs
```

(iv) How many contigs were generated?

TGICL creates an asm_X subdirectory for each central processing unit (CPU) in a parallel processing setting (where X is a number ≥1) or just one if a single CPU is used. In our case, all assembly files will be stored within the asm_1 subdirectory. All contig sequences are stored in a multiple-sequence FASTA file called contigs. To count the number of sequences, type the following command within the /asm_1 directory:

```
grep ">" contigs | wc -l
```

If you used the dataset provided, you should find that there were 3166 contigs.

(v) How many singlets were generated?

Singlets (as distinct from singletons) are sequences that are initially clustered with others by sharing some level of similarity. However, this similarity is not high enough to allow them to be assembled with any other read into a contig. Singlet sequences are stored in a multiple sequence FASTA file called singlets. To count the number of singlets, just type:

```
grep ">" singlets | wc -l
```

If you used the dataset provided, there should be 133 singlets.

5. We can visualize some contigs using the CLVIEW tool, a graphical program that allows viewing of the ace assembly files. Invoke CLVIEW by typing 'clview' on the command line.

6. A directory tree will be displayed on the left section of the window. Select **All Files** on the **Filter** selection box located at the left bottom corner. Now, using the directory tree, select the ace file stored within the /asm_1 directory and load it.

7. The window (see *Fig. 6a*, also available in the color section) will now present, on the right, a section displaying a consensus sequence (with a yellow background) followed by a stack of aligned sequences (with a blue background). The blue background is in different shades, which indicate the sequence coverage: darker shades represent regions matched by a high number of reads, whilst paler shades are regions matched by fewer reads.

8. The upper part of the right window presents a selection box (labeled **Cluster:**) that allows you to select any cluster for visualization. In the upper-right corner, there are two bars for zooming the assembly (horizontal bar) and to scroll the reads (vertical bar). *Fig. 6(b)* shows the same cluster with a wider (zoomed-out) view.

9. Base discrepancies are represented by red bases (see *Fig. 6a*) or vertical bars (see *Fig. 6b*), depending on the zoom used for visualization. These discrepancies may be due to sequencing errors, but can also represent potential SNPs (see section 3).

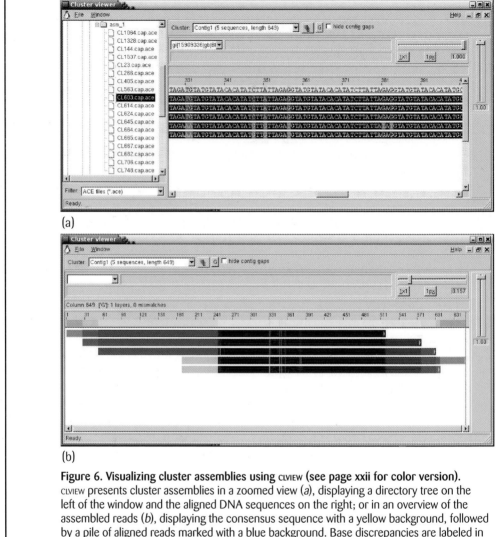

Figure 6. Visualizing cluster assemblies using CLVIEW **(see page xxii for color version).**
CLVIEW presents cluster assemblies in a zoomed view (a), displaying a directory tree on the
left of the window and the aligned DNA sequences on the right; or in an overview of the
assembled reads (b), displaying the consensus sequence with a yellow background, followed
by a pile of aligned reads marked with a blue background. Base discrepancies are labeled in
red and may represent potential SNPs.

2.5 Redundancy estimation

Redundancy assessment of EST libraries generally starts by using assembly programs
such as CAP3 (28) or PHRAP (38), or a clustering step followed by assembly. An initial
analysis may involve an estimation of the proportion of clusters in regard to the
total number of reads of an EST set. By subtracting this value from unity, one can
estimate the overall read redundancy. This value, however, can be misleading, as
it does not tell us precisely how much information is in fact being generated.
Hence, a hypothetical situation where all reads span only a single small region of
a transcript would yield the same redundancy as if these reads were covering the

entire sequence of this transcript. In the latter case, the overall gain of information would certainly be higher that in the former. A possible solution to overcome this limitation is to calculate the base redundancy rather than the read redundancy. In this case, the total number of bases of the consensus sequences would be divided by the sum of bases of the separate reads, thus allowing a measurement of how much novel information has been gained following the incorporation of new reads. If the addition of new reads keeps revealing more transcripts and/or covering novel transcript regions (and hence more bases in total), one should expect base redundancy to be lower than read redundancy. On the other hand, once maximum coverage has been attained, one should expect base and read redundancy to tend towards equity.

Another aspect that has to be taken into account is the complexity of the transcriptome (the total set of transcribed genes). For instance, 1 million reads from a highly representative and unbiased EST library would probably not result in a high redundancy for the human transcriptome (estimated at 30 000 to 100 000 transcripts). Conversely, a similar set of reads would yield a much higher redundancy in a small transcriptome (e.g. 5000 transcripts). *Fig. 7* shows the typical progress of the cluster number as new EST reads are added. At the very early stages of EST sequencing, every single read corresponds to a new cluster. The number of clusters at this stage keeps growing linearly until a point is reached where a new read is as likely to connect two clusters as to form a new cluster and the curve levels off. From this turning point, cluster joining outweighs cluster formation and the number of clusters declines until a final situation is reached in which every cluster corresponds roughly to a reconstructed transcript. However, in real life, there will always remain some gaps in the reconstructed transcripts (inflating the number of clusters), whilst some rare transcripts may not be represented at all (reducing the number of clusters) – the balance between these effects will lead to an eventual over- or underestimate of the total transcriptome size. Finally, it is also important

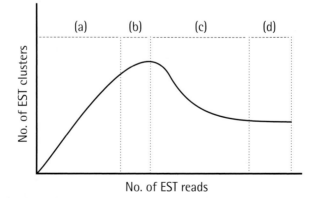

Figure 7. Progress of the cluster number in a typical EST sequencing project.
In the early stages of an EST sequencing project (*a*), most of the reads do not overlap one another and so each new EST initiates a new cluster, resulting in a linear increase in the number of clusters. As more reads are accumulated (*b*), overlaps are found and cluster joining begins to offset the initiation of new clusters. Later (*c*), cluster joining outweighs initiation and the total number of clusters decreases towards an eventual plateau (*d*), where all new ESTs fall into existing clusters.

to take into account the fact that splicing variants can make this scenario even more complex.

2.6 Electronic gene expression profiles

The relative abundance of EST reads may roughly reflect the gene expression profile of the respective tissue/cell type used as the source of mRNA. This assumption must be considered with caution, as several factors can interfere with the quantification. First, the relative abundance of EST reads only reflects the gene expression profile if the EST library has not been normalized. Any significant normalization, as well as construction and cloning biases, may hamper quantitative analyses, as they change the original distribution of the different classes of transcript. Secondly, correct clustering is important for the accuracy of electronic gene expression analysis. For instance, if a read is erroneously grouped to a certain cluster, it will contribute to alter the quantification of the corresponding transcript. Finally, it is worth mentioning that transcript sampling may introduce errors, and comparative expression analysis using different libraries has to take into account systematic biases introduced during cDNA generation (54).

2.7 Mapping ESTs to the genome

Gene structure is much more complex in eukaryotes than in prokaryotes. The discontinuous character of genes in the former, characterized by the presence of coding regions intercalated by intronic noncoding intervening sequences, makes gene prediction a difficult task, especially for identifying small exons. This can be still more complicated by the occurrence of alternative splicing events such as exon skipping, in which an exon may be present in one transcript but be skipped in a splicing variant. The fast-growing nature of EST databases makes them an invaluable tool for gene structure determination. Alignment of cDNAs with genomic sequences may provide important evidence to support gene predictions, but is also not a trivial task. The intervening nature of the introns breaks up the sequence alignments and small exons may remain undetected. Furthermore, most pairwise alignment programs such as FASTA (55) and BLAST (41) do not implement any modeling of the exon–intron structure and the presence of canonical acceptor and donor splicing sites. For this reason, such programs may generate alignment blocks whose ends do not correspond to the exact intron–exon boundaries (56). Alternative alignment programs that can handle introns and are used for cDNA mapping include EST2GENOME (57), SPIDEY (58), BLAT (59), SIM4 (53), and EXONERATE (60). Nevertheless, whatever the program utilized for transcript mapping, the appropriate alignment stringency has to be determined case by case, especially if cross-species cDNA mapping is performed.

In the next protocol, we will see how to map cDNA sequences onto genomic sequence using EXONERATE (60), a flexible and configurable program that allows rapid searches thanks to the use of heuristics based on alignment models. For this purpose, we will use three human cDNA sequences: hemoglobin β-chain

(GenBank accession no. BC007075.1), glyceraldehyde 3-phosphate dehydrogenase (NM_002046.3) and calmodulin 1 (NM_006888.3). These sequences will be mapped onto stretches of the human chromosomes 11, 12, and 14, respectively. To map cDNAs onto genomic sequences, we will use the EST2GENOME model of EXONERATE, which includes intron modeling and permits alignment of spliced transcript sequences to the unspliced genomic sequence. (Chapter 4 also covers the mapping of ESTs to genomic sequence, using a web server.)

Protocol 6

Mapping cDNAs onto genomic sequences

Software
You will need the following software:

■ EXONERATE (60) (http://www.ebi.ac.uk/~guy/exonerate/[7.18])

Method
1. Go to the /Protocol_6/dataset directory where you will find two files: cDNAs.fasta[7.19] and genomic.fasta[7.20]. These files contain three cDNA sequences and three genomic sequences, respectively, in FASTA format.

2. EXONERATE can align two single- or multiple-sequence files against one another. The sequences are aligned all-versus-all and the relevant alignments stored in a single output file. To align the three cDNA sequences with the respective chromosomal sequences using EXONERATE, type the following command (as a single line):

    ```
    exonerate cDNAs.fasta genomic.fasta --model est2genome ↵
        --score 300 --showtargetgff > cDNA_X_genomic_map
    ```

    This command invokes the program EXONERATE, specifies the names of the query and subject sequence files, respectively, and redirects the output ('>' sign) to a specified file name, cDNA_X_genomic.txt. The parameter '--model est2genome' specifies intron modeling. A value of 300 (default value is 100), specified in the '--score' parameter, corresponds to a moderate stringency. The parameter '--showtargetgff' generates a generic file format (GFF) output following the alignment and is convenient for downstream annotation.

3. Inspect the content of the newly created file (cDNA_X_genomic_map) using any text editor or the UNIX 'less' command. It contains the corresponding cDNA-to-genome alignments. EXONERATE displays the sequence alignment blocks with a clear indication of the introns found. (A copy of the expected results file is given in the /Protocol_6/results directory.)

3. TROUBLESHOOTING

3.1 Clone chimerism

The typical short length of EST clones encourages insert-to-insert ligation and cloning during library construction. Chimeric clones, characterized by the concomitant cloning of two fragments derived from different genes, are a common finding in EST libraries (61). Detection of such artifacts is a complex task and may require a manual inspection of the sequences. Furthermore, chimerism must be interpreted with caution and discriminated from chimeric spliced mRNAs (62, 63).

3.2 SNPs

SNPs observed in different EST reads must be interpreted with caution, as they may be a consequence of sequencing errors rather than a natural occurrence of polymorphic sites. Read redundancy and base quality must be checked on each candidate polymorphic site to discard potential artifacts.

3.3 Repeat masking

Masking low-complexity sequences such as tandem repeats before clustering and assembling ESTs may improve the accuracy of the process. However, aggressive masking may prevent the correct grouping of legitimate clusters. On the other hand, lack of masking may result in the formation of chimeric clusters, composed of unrelated sequences that share some repetitive sequences (15).

3.4 Contamination

EST processing pipelines can incorporate filters against heterologous and organellar sequences. Conversely, cross-contamination of mRNA sources with surrounding tissues, other cell types, or different developmental stages can barely be discriminated by conventional software. Hence, any assertion regarding expression specificity must be corroborated by specific experimental approaches.

4. ADDITIONAL WEB RESOURCES

- A Science Primer: http://www.ncbi.nlm.nih.gov/About/primer/index.html [7.21]. A series of introductory texts produced by the National Center for Biotechnology Information (NCBI) that cover topics such as bioinformatics, ESTs, SNPs, and genome mapping, among others.
- Submitting sequences to dbEST and GenBank: http://www.ncbi.nlm.nih.gov/dbEST/how_to_submit.html [7.22]. EST submission must follow a standardized format. This page provides a guideline for EST submission to dbEST.
- The GenBank Expression Sequence Tags Database (dbEST) (23, 24): http://www.ncbi.nlm.nih.gov/dbEST/ [7.23]. A public repository with tens of millions of EST entries.
- UniGene (14, 46): http://www.ncbi.nlm.nih.gov/UniGene/ [7.9]. An integrated database of clustered ESTs and cDNA sequences. Each UniGene cluster presents accession numbers to ESTs, IMAGE clones, etc., plus some informative annotations. No consensus sequences are provided.

- TIGR Gene Indices (26, 44): http://compbio.dfci.harvard.edu/tgi/[7.7]. A resource of species-specific EST data of eukaryotic organisms, including animal, plant, protist, and fungal sources. ESTs are clustered, assembled, and annotated. The data is stored in a relational database with a user-friendly web interface. Tentative consensus sequences are provided.
- The Sequence Tag Alignment and Consensus Knowledgebase (STACKdb) (26, 29, 44): http://www.sanbi.ac.za/Dbases.html[7.8]. A database of clustered human EST and mRNA sequences, including some disease- and tissue-based categories, and variation analysis.
- Genome Sequencing Center at the Washington University in Seattle: http://genome.wustl.edu/data/est.cgi[7.24]. A large amount of sequencing data is available on this site, including both genome and EST projects of many organisms. EST trace files are available for download at ftp://genome.wustl.edu/pub/est/[7.25].
- NCBI Reference Sequence (RefSeq) (64): http://www.ncbi.nlm.nih.gov/RefSeq/[7.26]. A comprehensive, integrated, nonredundant set of sequences for major research organisms, including genomic DNA, transcripts (RNA), and protein products.
- NCBI Trace Archive: http://www.ncbi.nlm.nih.gov/Traces/[7.27]. A public repository with more than one billion trace files. Data is exchanged regularly with the Ensembl Trace Server (http://trace.ensembl.org/[7.28]) at the EBI/Sanger Institute in the UK.
- NCBI dbSNP (65): http://www.ncbi.nlm.nih.gov/SNP/[7.29]. A database of nucleotide sequence variation that includes SNPs, deletion/insertion polymorphisms, microsatellite or short tandem repeats, and multi-nucleotide polymorphisms.

5. REFERENCES

★ 1. **AdamsMD, Kelley JM, Gocayne JD, et al.** (1991) *Science*, **252**, 1651–1656. – *The original publication describing the concept and use of ESTs.*

2. **Adams MD, Dubnick M, Kerlavage AR, et al.** (1992) *Nature*, **355**, 632–634.

3. **Adams MD, Kerlavage AR, Fields C & Venter JC** (1993) *Nat. Genet.* **4**, 256–267.

4. **Gill RW & Sanseau P** (2000) *Biotechnol. Annu. Rev.* **5**, 25–44.

★★ 5. **JongeneelCV** (2000) *Brief. Bioinform.* **1**, 76–92. – *A good overview on EST processing and annotation.*

6. **Lindlof A** (2003) *Appl. Bioinform.* **2**, 123–129.

7. **Marra MA, Hillier L & Waterston RH** (1998) *Trends Genet.* **14**, 4–7.

★ 8. **Parkinson J & Blaxter M** (2004) *Methods Mol. Biol.* **270**, 93–126. – *A theoretical and practical text depicting the most important steps in EST processing using* CLOBB *software.*

9. **Ying SY** (2004) *Mol. Biotechnol.* **27**, 245–252.

10. **Ying SY** (2003) *Generation of cDNA Libraries – Methods and Protocols.* Humana Press Inc., Totowa, NJ.

11. **Dias-Neto E, Correa RG, Verjovski-Almeida S, et al.** (2000) *Proc. Natl. Acad. Sci. U.S.A.* **97**, 3491–3496.

12. **Dias-Neto E, Harrop R, Correa-Oliveira R, Wilson RA, Pena SD & Simpson AJ** (1997) *Gene*, **186**, 135–142.

13. **Adams MD, Soares MB, Kerlavage AR, Fields C & Venter JC** (1993) *Nat. Genet.* **4**, 373–380.

★ 14. **Boguski MS & Schuler GD** (1995) *Nat. Genet.* **10**, 369–371. – *The original publication describing UniGene.*

15. **Jongeneel CV** (2000) *Bioinformatics*, **16**, 1059–1061.

16. **Wilcox AS, Khan AS, Hopkins JA & Sikela JM** (1991) *Nucleic Acids Res.* **19**, 1837–1843.

17. **Sikela JM & Auffray C** (1993) *Nat. Genet.* **3**, 189–191.

18. **Soares MB, Bonaldo MF, Jelene P, Su L, Lawton L & Efstratiadis A** (1994) *Proc. Natl. Acad. Sci. U.S.A.* **91**, 9228–9232.

19. **Bonaldo MF, Lennon G & Soares MB** (1996) *Genome Res*, **6**, 791–806.

20. **Patanjali SR, Parimoo S & Weissman SM** (1991) *Proc. Natl. Acad. Sci. U.S.A.* **88**, 1943–1947.

21. **Diatchenko L, Lau YF, Campbell AP, et al.** (1996) *Proc. Natl. Acad. Sci. U.S.A.* **93**, 6025–6030.

22. **Hedrick SM, Cohen DI, Nielsen EA & Davis MM** (1984) *Nature*, **308**, 149–153.

★ 23. **Boguski MS, Lowe TM & Tolstoshev CM** (1993) *Nat. Genet.* **4**, 332–333. – *The original publication describing dbEST.*

24. **Boguski MS, Tolstoshev CM & Bassett DE Jr** (1994) *Science*, **265**, 1993–1994.

25. **Benson DA, Karsch-Mizrachi I, Lipman DJ, Ostell J & Wheeler DL** (2005) *Nucleic Acids Res* **33**, D34–D38.

26. **Quackenbush J, Liang F, Holt I, Pertea G & Upton J** (2000) *Nucleic Acids Res.* **28**, 141–145.

27. **Zhang Z, Schwartz S, Wagner L & Miller W** (2000) *J. Comput. Biol.* **7**, 203–214.

28. **Huang X & Madan A** (1999) *Genome Res.* **9**, 868–877.

29. **Christoffels A, van Gelder A, Greyling G, Miller R, Hide T & Hide W** (2001) *Nucleic Acids Res.* **29**, 234–238.

30. **Miller RT, Christoffels AG, Gopalakrishnan C, et al.** (1999) *Genome Res.* **9**, 1143–1155.

31. **Burke J, Davison D & Hide W** (1999) *Genome Res.* **9**, 1135–1142.

32. **Miller G, Fuchs R & Lai E** (1997) *Genome Res.* **7**, 1027–1032.

33. **Lennon G, Auffray C, Polymeropoulos M & Soares MB** (1996) *Genomics*, **33**, 151–152.

34. **Ewing B & Green P** (1998) *Genome Res.* **8**, 186–194.

35. **Ewing B, Hillier L, Wendl MC & Green P** (1998) *Genome Res.* **8**, 175–185.

36. **Tatusov RL & Lipman DJ** (1998) *The DUST Program:* ftp://ftp.ncbi.nih.gov/pub/tatusov/dust/ (unpublished).

37. **Benson G** (1999) *Nucleic Acids Res.* **27**, 573–580.

38. **Green P** (1997) cross_match *and* phrap. http://www.phrap.org/phredphrapconsed.html (unpublished).

★ 39. **Durham AM, Kashiwabara AY, Matsunaga FT, et al.** (2005) *Bioinformatics*, **21**, 2812–2813. – *The original publication describing* egene, *an automated pipeline-generation system. The reference also describes the preferred protocol used in this chapter for EST pre-processing and assembly.*

40. **Gordon D, Abajian C & Green P** (1998) *Genome Res.* **8**, 195–202.

41. **Altschul SF, Madden TL, Schaffer AA, et al.** (1997) *Nucleic Acids Res.* **25**, 3389–3402.

★★★ 42. **Wolfsberg T & Landsman D** (2001) In *Bioinformatics – A Practical Approach in the Analysis of Genes and Proteins.* Edited by A Baxevanis & B Ouellette. John Wiley & Sons, New York. pp. 283–301. – *A concise but very clear text describing the most important bioinformatic aspects of EST processing. Recommended for beginners.*

★★★ 43. **Hide W, Miller R, Ptitsyn A, Kelso J, Gopalakrishnan C & Christoffels A** (1999) In *Proceedings of the Seventh International Conference on Intelligent Systems for Molecular Biology (ISMB99).* AAI Press, Menlo Park, CA. Electronic version vailable at http://bioinf.mpi-sb.mpg.de/conferences/ismb99/WWW/TUTORIALS/tutorial_6.html. – *An excellent text discussing the most important aspects of EST clustering.*

44. **Lee Y, Tsai J, Sunkara S, et al.** (2005) *Nucleic Acids Res.* **33**, D71–D74.

45. **Pertea G, Huang X, Liang F, et al.** (2003) *Bioinformatics*, **19**, 651–652.

46. **Pontius JU, Wagner L & Schuler GD** (2003) In *The NCBI Handbook.* Edited by J McEntyre & J Ostell. National Center for Biotechnology Information, Bethesda, MD, pp. 1–12 (Section 21).

47. **Ptitsyn A & Hide W** (2005) *BMC Bioinformatics*, **6** (Suppl. 2), S3.

48. **Liang F, Holt I, Pertea G, Karamycheva S, Salzberg SL & Quackenbush J** (2000) *Nucleic Acids Res.* **28**, 3657–3665.
49. **Marth GT, Korf I, Yandell MD, et al.** (1999) *Nat. Genet.* **23**, 452–456.
50. **Buetow KH, Edmonson MN & Cassidy AB** (1999) *Nat. Genet.* **21**, 323–325.
51. **Irizarry K, Kustanovich V, Li C, et al.** (2000) *Nat. Genet.* **26**, 233–236.
52. **Picoult-Newberg L, Ideker TE, Pohl MG, et al.** (1999) *Genome Res.* **9**, 167–174.
53. **Florea L, Hartzell G, Zhang Z, Rubin GM & Miller W** (1998) *Genome Res.* **8**, 967–974.
54. **Liu D & Graber JH** (2006) *BMC Bioinformatics*, **7**, 77.
55. **Pearson WR & Lipman DJ** (1988) *Proc. Natl. Acad. Sci. U.S.A.* **85**, 2444–2448.
★★ 56. **Korf I, Yandell M & Bedell J** (2003) *BLAST.* O'Reilly and Associates, Inc., Sebastopol, CA. – *A reference book on* BLAST *including numerous protocols for EST mapping to a genome, clustering, and annotation.*
57. **Mott R** (1997) *Comput. Appl. Biosci.* **13**, 477–478.
58. **Wheelan SJ, Church DM & Ostell JM** (2001) *Genome Res.* **11**, 1952–1957.
59. **Kent WJ** (2002) *Genome Res.* **12**, 656–664.
60. **Slater GS & Birney E** (2005) *BMC Bioinformatics*, **6**, 31.
61. **Hillier LD, Lennon G, Becker M, et al.** (1996) *Genome Res.* **6**, 807–828.
62. **Romani A, Guerra E, Trerotola M & Alberti S** (2003) *Nucleic Acids Res.* **31**, e17.
63. **Zhang C, Xie Y, Martignetti JA, Yeo TT, Massa SM & Longo FM** (2003) *DNA Cell Biol.* **22**, 303–315.
64. **Pruitt KD, Tatusova T & Maglott DR** (2005) *Nucleic Acids Res.* **33**, D501–D504.
65. **Sherry ST, Ward MH, Kholodov M, et al.** (2001) *Nucleic Acids Res.* **29**, 308–311.

CHAPTER 8

Protein structure, classification, and prediction

Arthur M. Lesk

1. INTRODUCTION

[handwritten margin notes: functions: / structural / regulatory / defense / catalytic]

Encoded in the genomes of living things is a panoply of protein structures that carry out structural, catalytic, regulatory, defense, and other essential functions of life processes. These functions depend on the proteins adopting precise three-dimensional conformations, which are determined by their amino acid sequences. Some proteins adopt fixed conformations and remain rigid during their activity. Others are tiny machines that undergo specific conformational changes for chemical and/or mechanical purposes.

These observations imply two things. First, that making sense of the variety of observed structures is an initial challenge: how can we recognize the important features of the similarities and differences among proteins, as a basis for a rational classification and comparison of the structures, and for illuminating the mechanisms of protein functions? Secondly, that understanding how amino acid sequence determines protein structure is a fundamental problem of biology: ultimately, from a genome sequence we should be able to predict the organism's proteome – structures as well as sequences. If we understood nature's 'algorithm' for protein folding, we should be able to reproduce it in computer programs that allow prediction of protein structures. This would be very useful because we know the sequences of a large number of proteins for which experimental structures have not been determined. Accurate and reliable structure prediction would unlock the immense stores of information contained in the many genome sequences that are now known, to reveal the evolution and development of protein function.

This chapter will present the basic principles of protein structure, describe some of the projects aimed at analyzing and classifying the known protein structures, and discuss currently available methods for predicting protein structure from amino acid sequence, noting their strengths and limitations.

In general, there are two stages to developing a method. The first stage is getting a procedure to work in the hands of its author, who is intimately aware of which aspects can be relied on, which require manual supervision or editing of

Bioinformatics: *Methods Express* (Paul H. Dear, ed.)
© Scion Publishing Limited, 2007

results, and which are in the realms of hope for the future rather than currently implemented. In the second, mature stage, a method is reliable and automatic, not requiring intervention by the author or even the user. When a computational method has reached this stage, it is common to create a server on the web, which will accept sequences and report the results, either interactively or, if the calculation is a lengthy one, by e-mail. Given the ease of interconnectivity of the web, many sites act as agents for other sites. These allow users to submit a sequence once and have it processed by many other sites to which the query is distributed. These are sometimes called 'meta-servers'.

1.1 The chemical structure of proteins

Chemically, protein molecules are long polymers typically containing several thousand atoms, composed of a uniform repetitive *backbone* (or *main chain*) with a particular side chain attached to each residue (see *Fig. 1*). The amino acid sequence of a protein records the order of the side chains. For general references, see (1–3).

The side chains in proteins show a variety of physicochemical features: some are charged, some are uncharged but polar, and others are hydrophobic. The side chains of hydrophobic residues are primarily hydrocarbon in nature and for that reason have thermodynamically unfavorable interactions with water. Different possible conformations of the backbone of a protein bring different types of residue into spatial proximity and expose some but not all residues to the solvent. Every conformation therefore has a different associated energy that depends on the distribution of favorable and unfavorable interactions. The native state of a soluble globular protein is the conformation that optimizes the set of interactions among the residues and between the residues and the solvent.

Focusing on the backbone, in the native state the polypeptide chain follows a curve in space. The general spatial layout of this curve defines a folding pattern. The backbones of related proteins, for instance spinach plastocyanin and cucumber stellacyanin, show recognizably similar but not identical folding patterns (see *Fig. 2*). A letter of the alphabet in different type fonts – for instance b and *b* – illustrates the kinds of topological similarities and differences in detail that we see among even closely related proteins that have undergone evolutionary divergence, which share a common folding pattern. For distantly related divergent proteins, a better

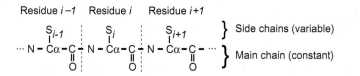

Residue *i*−1 Residue *i* Residue *i*+1

Figure 1. Chemical structure of proteins.
Proteins contain a main chain of constant structure. Attached at regular intervals are side chains of variable structure, each chosen (with few exceptions) from the canonical set of 20 amino acids. Here S_{i-1}, S_i, and S_{i+1} represent successive side chains. Different sequences of side chains characterize different proteins. It is the sequence that gives each protein its individual structural and functional characteristics.

(a)

Spinach plastocyanin [1ag6]   Spinach plastocyanin [1ag6]

(b)

Cucumber stellacyanin [1jer]   Cucumber stellacyanin [1jer]

(c)

Superposition   Superposition

(d)

Well-fitting core   Well-fitting core

Figure 2. Two related proteins that share the same general folding pattern, but differ in detail.
Open circles represent copper ions. (*a*) Spinach plastocyanin [Protein Data Bank entry 1ag6]; (*b*) cucumber stellacyanin [1jer]. Superposition showing (*c*) the entire structures and (*d*) only the well-fitting core. The main secondary structural elements of these proteins are two β-sheets packed face to face. It can be seen in the superposition that several strands of β-sheet are conserved but displaced, and that the helix at the right of the cucumber stellacyanin structure has no counterpart in the spinach plastocyanin structure. Even the (relatively) well-fitting core shows the conservation of folding topology but nevertheless reveals considerable distortion. Images are stereo pairs.

analogy might be to the letters B and R, which share the letter P as a common core but in addition have either a loop (B) or a stroke (R) that differs.

1.2 The hierarchical form of protein architecture

The Danish protein chemist K.U. Linderstrøm-Lang described three levels of protein structure (see *Fig. 3*):

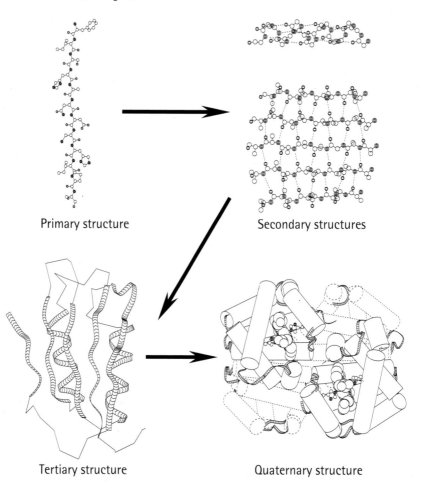

Primary structure	Secondary structures
Tertiary structure	Quaternary structure

Figure 3. Common structural features underlying the great variety of protein folding patterns.
α-Helices and β-sheets are standard elements of the 'parts list' of many protein structures. α-Helices and β-sheets were modeled by Linus Pauling before their experimental observation, on the basis that they provide convenient ways for the residues to achieve comfortable steric relationships and satisfy the requirements for backbone hydrogen bonding, in an (almost) sequence-independent manner. This figure shows the primary structure in terms of a simple extended chain (upper left). The standard secondary structures, the α-helix and β-sheet, are shown at the upper right, with hydrogen bonds indicated by broken lines. Tertiary structure is represented by acyl phosphatase, which contains two α-helices packed against a five-stranded β-sheet. Human hemoglobin, a tetramer containing two copies of two types of chain, illustrates quaternary structure, with the subunits shown in solid or broken lines. (Note that acyl phosphatase is not a subunit of hemoglobin.)

- The amino acid sequence – the set of primary chemical bonds – is the primary structure.
- The assignment of helices and sheets is the secondary structure. The secondary structure of the copper-binding proteins in *Fig. 2* consists primarily of two β-sheets, with a helix in the cucumber stellacyanin structure.
- The assembly and interactions of the helices and sheets is the tertiary structure.

For proteins composed of more than one subunit, J.D. Bernal added the level of:

- Quaternary structure, which is the composition and assembly of the monomers.

1.3 Domains

One way that proteins have evolved increasing complexity is by assembling a large protein from a set of smaller quasi-independent subunits, either by forming stable oligomers, as in hemoglobin (see *Fig. 3*) or ATP synthase, or by concatenating units within a single polypeptide chain. Domains are compact units within the folding pattern of a single chain. Justifications for regarding them as quasi-independent include the observation that domains can be 'mixed and matched' in different proteins and the fact that, in many cases, their folding patterns are similar to those of the homologous monomeric proteins.

Domains form the basis of the higher-level protein structural organization typical of eukaryotic proteins. Modular proteins are multi-domain proteins and often contain many copies of closely related domains. For example, fibronectin, a large extracellular protein involved in cell adhesion and migration, contains 29 domains including multiple tandem repeats of three types of domain, F1, F2, and F3. It is a linear array of the form: $(F1)_6(F2)_2(F1)_3(F3)_{15}(F1)_3$. Fibronectin domains also appear in other modular proteins (see http://www.bork.embl-heidelberg.de/Modules/[8.1] for pictures and nomenclature).

The invention of new domains is an unusual event in the creation of novel proteins. It is far more common to create different combinations of existing domains in increasingly complex ways, as illustrated in *Fig. 4*. This process can occur independently, and take different courses, in different phyla.

2. METHODS AND APPROACHES

2.1 Accessing macromolecular structures on the web

2.1.1 General protein structure databases

Approximately 30 000 protein structures are now known, most of which have been determined by X-ray crystallography or nuclear magnetic resonance (NMR). The Worldwide Protein Data Bank (wwPDB) comprises three primary archival projects collaborating to integrate the archiving and distribution of experimentally determined biological macromolecular structures:

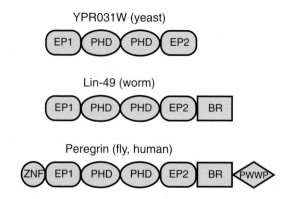

Figure 4. Evolution, by accretion of domains, of molecules related to peregrin – a human protein that probably functions in transcription regulation.
The *Caenorhabditis elegans* homolog, Lin-49, is essential for normal development of the worm. The function of the yeast homolog is unknown. The proteins contain the following types of domain: ZNF = C_2H_2-type zinc finger (not to be confused with acetylene; C and H stand for cysteine and histidine); EP1 and EP2 = enhancer of polycomb 1 and 2; PHD = plant homeodomain, a repressor domain containing the $C_4H_3C_3$-type of zinc finger; BR = bromo domain; and PWWP = domain containing the sequence motif Pro-Trp-Trp-Pro. The author joins the reader in regretting the opacity of the nomenclature.

- The Research Collaboratory for Structural Bioinformatics (RCSB) (USA)
- The Macromolecular Structure Database (MSD) (at the European Bioinformatics Institute (EBI), Hinxton, UK)
- The Protein Data Bank/Japan (Osaka, Japan)

The wwPDB sites accept depositions, process new entries, and maintain the archives. These and many other web sites organize and provide access to these data. Each has its own strengths, based in many cases on the research interests of the contributing scientists. For instance, a feature of the MSD is the Protein Quaternary Structure resource, which gives the probable state of assembly of multi-chain proteins in their biologically active forms.

Many sites offer search facilities to identify structures of interest, based on the presence of keywords (or a logical combination of keywords), or numerical values such as the year of deposition. Different sites also differ in their 'look and feel', and users will discover their own preferences. *Protocol 1* gives a brief introduction to one of these sites, using as an example the enzyme scytalone dehydratase, which catalyzes the conversion of scytalone to 1,3,8-trihydroxynaphthalene in the pathway of melanin biosynthesis.

Protocol 1

Accessing protein structures through RCSB

1. Go to the home page of RCSB at http://www.rcsb.org[8.2].

2. In the small text box towards the top of the page, type 'scytalone dehydratase' and click the adjacent **Search** button to start a keyword search.

3. The resulting page contains a list of known structures, with thumbnail pictures and some basic information for each structure, and links to more detailed descriptions.

4. Select the entry 1idp (F162A mutant of scytalone dehydratase from the fungus *Magnaporthe grisea*) by clicking either on its name or on the thumbnail image.

5. This opens a page from the Structure Explorer, containing a synoptic view of what is known about this protein (see *Fig. 5*, also available in the color section) Data provided include bibliographic information about the publication of the structure, details of the structure determination, and the source of the protein. A picture of the structure appears at the upper right. Tabs towards the top of the page give access to further details. (The page displayed in *Fig. 5* corresponds to the information shown under the **Structure Summary** tab.)

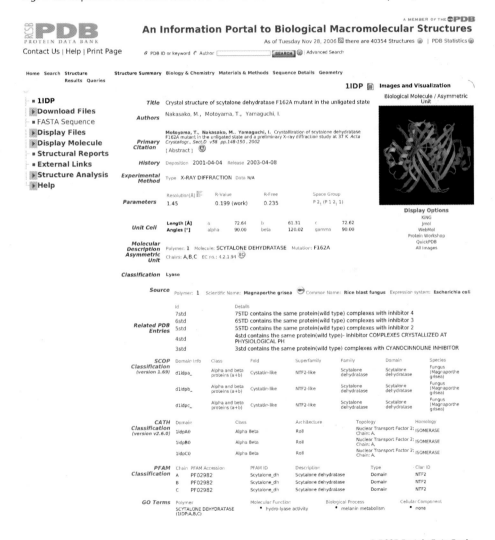

© *RCSB Protein Data Bank*

Figure 5. A page from the RCSB site, showing the Structure Explorer summary page for entry 1idp (*M. grisea* scytalone dehydratase) (see page xxiii for color version).
Bibliographical information is shown, as well as some data about the structure and its determination, links to other databases, and a picture.

6. The page provides links (under **Related PDB Entries**) to related structures, including other determinations of the structure of the same protein, perhaps in different states of ligation.

7. Further down the same page, information is provided about the classification of the structure under SCOP, CATH, and Pfam (see later in this chapter), and classification under gene ontology (GO; see Chapter 9).

2.1.2 Specialized or 'boutique' databases

Many individuals or groups select, annotate, and recombine data focused on particular topics, and include links affording streamlined access to information about subjects of interest. Some examples are as follows:

- The protein kinase resource is a specialized compilation that includes sequences, structures, functional information, laboratory procedures, lists of interested scientists, tools for analysis, a bulletin board, and links relating to the protein kinase family of enzymes (http://www.kinasenet.org/pkr/[8.3]).

- The HIV Protease Database archives structures of the proteinases and their complexes of human immunodeficiency virus 1 (HIV-1), HIV-2, and simian immunodeficiency virus, and provides tools for their analysis and links to other sites with AIDS-related information (http://mcl1.ncifcrf.gov/hivdb/index.html[8.4]).

- In the field of immunology:
  - IMGT, the international ImMunoGeneTics database, is a high-quality integrated database specializing in immunoglobulins, T-cell receptors, and major histocompatibility complex molecules of all vertebrate species (http://imgt.cines.fr[8.5]).
  - IEDB, The Immune Epitope Database and Analysis Resource, containing data on antibody and T-cell epitopes. (http://beta.immuneepitope.org/home.do)[8.6].

2.2 Classification of protein structures

The most general classification of families of protein structures is based on overall features of their secondary and tertiary structure (4), as outlined in *Table 1*.

Within these broad categories, protein structures show a variety of folding patterns. Many proteins with similar folding patterns share enough features of structure, sequence, and function to suggest evolutionary relationships. However, unrelated proteins often show similar structural themes. Several web sites offer hierarchical structural classifications of the entire PDB, and a number of these are listed in *Table 2*. Others are listed at http://www.bioscience.org/urllists/protdb.htm[8.7].

These sites describe projects derived from the primary archival databases of macromolecular structures. They are useful general entry points to protein structural data. For instance, SCOP offers facilities for searching on keywords to identify structures, navigation up and down the hierarchy, generation of pictures, access to the annotation records in the PDB entries, and links to related databases. The following section explores SCOP in more detail.

Table 1. General classification of protein structure families

Class	Characteristic
α-Helix	Secondary structure exclusively or almost exclusively α-helical
β-Sheet	Secondary structure exclusively or almost exclusively β-sheet
α+β	α-Helices and β-sheets separated in different parts of the molecule; absence of β-α-β supersecondary structure
α/β • α/β-linear • α/β-barrels	Helices and sheets assembled from β-α-β units • Line through centers of strands of sheet roughly linear • Line through centers of strands of sheets roughly circular
Little or no secondary structure	–

Table 2. Some sites providing structural classifications of proteins

Resource	URL
SCOP: Structural Classification of Proteins	http://scop.mrc-lmb.cam.ac.uk/scop/[8.23]
CATH: Class/Architecture/Topology/Homology	http://cathwww.biochem.ucl.ac.uk/latest/[8.24]
DALI: based on extraction of similar structures from distance matrices	http://ekhidna.biocenter.helsinki.fi/dali/start[8.25]
CE: a database of structural alignments	http://cl.sdsc.edu/[8.26]

2.2.1 SCOP – Structural Classification of Proteins

SCOP, by A.G. Murzin, L. Lo Conte, B.G. Ailey, S.E. Brenner, T.J.P. Hubbard, and C. Chothia, organizes protein structures in a hierarchy according to evolutionary origin and structural similarity (5). At the lowest level of the hierarchy are individual domains. SCOP groups sets of domains into families of homologs for which the similarities in structure and sequence (and sometimes function) imply a common evolutionary origin. Families containing proteins of similar structure and function, but for which the evidence for an evolutionary relationship is suggestive but not compelling, form superfamilies. Superfamilies that share a common folding topology, for at least a large central portion of the structure, are grouped as folds. Finally, each fold group falls into one of the general classes. The major classes in SCOP are α, β, α+β, α/β, and miscellaneous 'small proteins', many of which have little secondary structure and have structures stabilized by disulfide bridges or ligands.

The box on p. 178 shows the SCOP classification of flavodoxin from *Clostridium beijerinckii* (see *Fig. 6*). For illustrations of the degrees of similarity of proteins grouped together at different levels of the hierarchy and a discussion of other classification schemes, see Chapter 4 of (2).

The latest SCOP release contains 25 973 PDB entries split into 70 859 domains. The distribution of entries at different levels of the hierarchy is given in *Table 3*.

To locate a protein of interest in SCOP, the user can traverse the structural hierarchy or search via keywords such as the protein name, PDB code, function (including Enzyme Commission number), or name of fold (for instance, barrel).

SCOP classification of flavodoxin from *C. beijerinckii*

1. **Root**: scop
2. **Class**: Alpha and beta proteins (α/β)
   Mainly parallel β-sheets (β-α-β units)
3. **Fold**: Flavodoxin-like
   Three layers, α/β/α; parallel β-sheet of five strands, order 21345
4. **Superfamily**: Flavoproteins
5. **Family**: Flavodoxin-related
   Binds FMN
6. **Protein**: Flavodoxin
7. **Species**: *Clostridium beijerinckii*

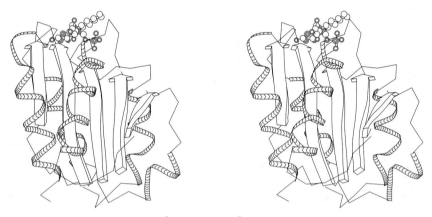

Figure 6. Flavodoxin from *C. beijerinckii* [PDB entry 5nll].
This protein binds a flavin mononucleotide (FMN) group, shown in ball-and-stick representation. It is an α/β protein, containing the characteristic β-α-β secondary structure. (Stereo pair.)

Table 3. Distribution of entries in SCOP

Class	Number of families	Number of superfamilies	Number of folds
All-α proteins	608	376	218
All-β proteins	560	290	144
α/β Proteins	629	222	136
α+β Proteins	717	409	279
Multi-domain proteins	61	46	46
Membrane and cell-surface proteins	99	88	47
Small proteins	171	108	75
Totals	**2845**	**1539**	**945**

For each structure, SCOP provides textual information, pictures, and links to other databases. The web site of SCOP (http://scop.mrc-lmb.cam.ac.uk/scop/index.html[8.8]) provides many useful facilities for searching and visualizing protein structures. *Protocol 2* gives a brief tour of some aspects of SCOP, using the protein scytalone dehydratase as an example

Protocol 2

SCOP

1. Go to the SCOP search page (http://scop.mrc-lmb.cam.ac.uk/scop/search.cgi[8.9]) and type 'scytalone +dehydratase' into the text field (in order to search for both terms, the second must be preceded by a '+'; note also the space after the first term; see *Fig. 7*). Click the **Retrieve information** button to search SCOP.

Structural Classification of Proteins

Search the scop database [scop 1.69]

You can use this search engine to search the SCOP database using several access methods (including *sunid, sid, sccs,* PDB identifiers, and any word that appears in any of the SCOP pages) as well as more sophisticated options. Please read the release notes for a detailed explanation and examples. This kind of search is internal to a SCOP release and therefore will always provide complete results.

By checking the PDB box, you can also search SCOP using the external MSDlite search engine for words that appear in several *text fields* in the corresponding PDB file (including header, author names, abstract, and MeSH terms from the primary citation). Please refer to MSDlite for more details.

scytalone +dehydratase   ✳

⦿ Search the **SCOP** database.
⦾ Search the **PDB** database using MSDlite.
| Retrieve information |
| Clear | the search form.

Copyright © 1994-2005 The scop authors / scop@mrc-lmb.cam.ac.uk
July 2005

Figure 7. Screenshot of the SCOP web page.

2. The search returns one family (scytalone dehydratase) and one protein (from *M. grisea*).

3. Click on the link to the protein. This gives, towards the top of the page, the complete classification (lineage) of the folding pattern. In this instance, it shows that scytalone dehydratase is part of a superfamily of proteins resembling cystatin, with a four-stranded β-sheet against which an α-helix is packed. The nuclear transport factor 2 (NTF2)-like proteins have a β-α-α-β insertion after the common helix.

4. Further down the screen is a detailed list of the relevant PDB domains and their states of ligation. Several structures of this protein, all from *M. grisea*, are known.

5. At the top of the page are icons linking to more detailed information, some but not all of which are pictorially suggestive of what they do. From left to right, the cube returns the user to the SCOP home page, the stamped and addressed envelope provides facilities to send e-mail to SCOP authors, '?' links to a help page, the double triangle takes the user to the top of the hierarchy, and the single triangle goes up one level in the hierarchy (for instance, from

the scytalone dehydratase family to the NTF2-like superfamily). The last two icons expand and condense, respectively, the information presented.

6. Click on the first of the PDB ID codes (in this example, **1idb**) to give access to its PDB entry. Use either the browser's 'back' button, or the **Scop** link at the top of the PDB entry page, to return to SCOP.

7. Small icons to the right of the PDB ID code (and to the right of the links to each of the chains) provide links to graphic representations of the structure, provided that the computer has the necessary plug-ins to display them.

8. Using the facilities to navigate, the user can explore the proteins resembling scytalone dehydratase, either retrieving the individual coordinate sets for further processing and analysis, or going up the hierarchy to identify the proteins that resemble the query structure at different levels of similarity.

2.3 Structural genomics

By analogy with full-genome sequencing projects, structural genomics has the commitment to deliver the structures of the complete protein repertoire (6). X-ray crystallographic and NMR experiments will solve a 'dense set' of proteins, such that all proteins are close enough to one or more experimentally determined structures to model them confidently. More so than genomic sequencing projects, structural genomics projects combine results from different organisms. The human proteome is of course of special interest, as are proteins unique to infectious microorganisms. The goals of structural genomics have become feasible partly by advances in experimental techniques, which make high-throughput structure determination possible, and partly by advances in our understanding of protein structures, which define reasonable general goals for the experimental work and suggest specific targets.

How many structures are needed? The theory and practice of homology modeling suggest that at least 30% sequence identity between the target protein and some experimental structure is necessary. This means that experimental structure determinations will be required for an exemplar of every sequence family, including many that share the same basic folding pattern. Experiments will have to deliver the structures of something like 10 000 domains. In the year 2005, 5436 structures were deposited in the PDB, so the throughput rate is not far from what is required.

2.4 Approaches to protein structure prediction

The amino acid sequence of a protein dictates its three-dimensional structure. In a medium of suitable solvent and temperature conditions, such as provided by a cell interior, proteins fold spontaneously to their active states. Chaperones help proteins to fold properly, but they catalyze the process rather than directing it.

The observation that each protein folds spontaneously into a unique three-dimensional native conformation implies that nature has an algorithm for predicting protein structure from amino acid sequence. Some attempts to understand this algorithm are based solely on general physical principles; others are based on observations of known amino acid sequences and protein structures. A proof of our understanding would be the ability to reproduce the algorithm in a computer program that could predict protein structure from amino acid sequence (7).

Most attempts to predict protein structure from basic physical principles alone try to reproduce the interatomic interactions in proteins, to define a numerical energy associated with any conformation. Computationally, the problem of protein structure prediction then becomes the task of finding the global minimum of the conformational energy function over all possible backbone and side-chain conformations. So far, this approach has not succeeded, partly because of the imprecision of the energy function and partly because the minimization algorithms tend to get trapped in local minima.

The alternative to *a priori* methods is an approach based on assembling clues to the structure of a target sequence by finding similarities to known structures. These empirical or 'knowledge-based' techniques have become very powerful and are currently the most successful methods known. They include:

- Homology modeling. Suppose a target protein, of known amino acid sequence but unknown structure, is related to one or more proteins of known structure. Then we expect that much of the structure of the target protein will resemble that of the known protein. The related protein of known structure can therefore serve as a basis for a model of the target protein. The challenge is to predict how the differences between the sequences are reflected in differences between the structures. This can be thought of as the 'differential' rather than the 'integral' form of the folding problem.
- Attempts to predict secondary structure without attempting to assemble these regions in three dimensions. The results are lists of regions of the sequence predicted to form α-helices and regions predicted to form strands of β-sheet.
- Fold recognition. Given a library of known structures, determine which of them shares a folding pattern with a query protein of known sequence but unknown structure. If the folding pattern of the target protein does not occur in the library, such a method should recognize this. The results are the nomination of a known structure that has the same fold as the query protein, or a statement that no protein in the library has the same fold as the query protein.
- Prediction of novel folds, by either *a priori* or knowledge-based methods. The results are a complete coordinate set for at least the main chain and sometimes the side chain also. The model is intended to have the correct folding pattern, but would not be expected to be comparable in quality to an experimental structure.

D. Jones has likened the distinction between fold recognition and *a priori* modeling to the difference between a multiple-choice question in an exam and an essay question.

We will examine some of these methods in more detail, but first we need to consider how to evaluate techniques critically for structural prediction, so that one can measure progress in the field and know how much confidence to place in each method.

2.4.1 Critical Assessment of Structure Prediction (CASP)

Judging of techniques for predicting protein structures requires blind tests. To this end, J. Moult initiated biennial CASP programs. Crystallographers and NMR spectroscopists in the process of determining a protein structure are invited to (i) publish the amino acid sequence several months before the expected date of completion of their experiment, and (ii) commit themselves to keeping the results secret until an agreed date. Predictors submit models, which are held until the deadline for release of the experimental structure. Then the predictions and experiments are compared.

The results of CASP evaluations record progress in the effectiveness of predictions, which has occurred partly because of the growth of the databanks but also because of improvements in the methods.

2.4.2 Homology modeling

Model building by homology is a useful technique when one wants to predict the structure of a target protein of known sequence, when the target protein is related to at least one other protein of known sequence and structure. If the proteins are closely related, the known protein structures – called the parents – can serve as the basis for a model of the target. It is on homology modeling that we depend to extend the results of structural genomics to the entire protein world.

The completeness and quality of the result depend crucially on how similar the sequences are. As a rule of thumb, if the sequences of two homologous proteins have 50% or more identical residues in an optimal alignment, the structures are likely to have similar conformations over more than 90% of the model. This is a conservative estimate, as the following illustration shows (see *Fig. 8*).

Although the quality of the model will depend on the degree of similarity of the sequences, it is possible to specify this quality before experimental testing. Therefore, knowing how good a model is necessary for the intended application permits intelligent prediction of the probable success of the exercise. Steps in homology modeling are:

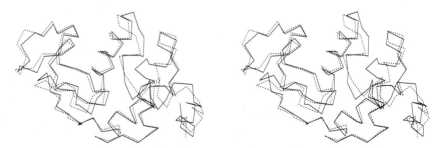

Figure 8. The aligned sequences and superposed structures of two related proteins, hen egg white lysozyme (solid lines) and baboon α-lactalbumin (broken lines).
The sequences are related (37% identical residues in the aligned sequences) and the structures are very similar. This figure shows that each protein could serve as a good model for the other, at least as far as the main chain is concerned. (Stereo pair.)

1. Align the amino acid sequences of the target and the protein or proteins of known structure on which the model is to be based. Usually, insertions and deletions will lie in the loop regions between helices and sheets.
2. Determine main-chain segments to represent the regions containing insertions or deletions. Stitching these regions into the main chain of the known protein creates a model for the complete main chain of the target protein.
3. Replace the side chains of residues that have been mutated. For residues that have not mutated, retain the side-chain conformation. Residues that have mutated tend to keep the same side-chain conformational angles and could be modeled on this basis. However, computational methods are now available to search over possible combinations of side-chain conformations.
4. Examine the model – both by eye and using programs – to detect any serious collisions between atoms. Relieve these collisions, as far as possible, by manual manipulations.
5. Refine the model by limited energy minimization. The role of this step is to fix up the exact geometrical relationships at places where regions of the main chain have been joined together and to allow the side chains to wriggle around a bit to place themselves in comfortable positions. The effect is really only cosmetic – energy refinement will not correct serious errors in such a model.

In most families of proteins, the structures contain relatively constant regions and more variable ones. The core of the structure of the family retains the folding topology, although it may be distorted, but the periphery can entirely refold (see *Fig. 2*; in contrast the structures of hen egg white lysozyme and baboon α-lactalbumin, shown in *Fig. 8*, are very closely related and quite similar in structure). A single parental structure will permit reasonable modeling of the conserved portion of the target protein, but will fail to produce a satisfactory model of the variable portion. Moreover, it will not be easy to predict which are the variable and constant regions. A more favorable situation occurs when several related proteins of known structure can serve as parents for modeling a target protein. These reveal the regions of 16 constant and variable structures in the family. The observed distribution of structural variability among the parents dictates an appropriate distribution of constraints to be applied to the model.

Mature software for homology modeling is available, as are libraries of pre-computed models using such software (8, 9). SWISS-MODEL, developed by T. Schwede, M.C. Peitsch and N. Guex (now at the Geneva Biomedical Research Institute), is a web site that will accept the amino acid sequence of a target protein, determine whether a suitable parent or parents for homology modeling exist, and, if so, deliver a set of coordinates for the target. The web server is available at http://www.expasy.ch/swissmod/SWISS-MODEL.html [8.10], and the pre-computed results of the application of SWISS-MODEL to proteins of known sequence are available through 3DCRUNCH: (http://swissmodel.expasy.org/repository/ [8.11]).

Another program in widespread use, MODELLER, was developed by A. Sali and is available at http://salilab.org/modeller/modeller.html [8.12]; pre-computed results of the application of MODELLER to proteins of known sequence are available through MODBASE at http://modbase.compbio.ucsf.edu/modbase-cgi/search_form.cgi [8.13].

2.4.3 Secondary-structure prediction

The goal of secondary-structure prediction is to identify those residues within the sequence that will form helices and strands of β-sheet in the native structure, independent of their spatial arrangement in the tertiary structure. One of the original motivations for secondary-structure prediction was the belief that this should be substantially easier than – but a step towards – the prediction of tertiary structure. Moreover, both experimental evidence from copolymers of amino acids, and statistical evidence from the observed residue compositions of helices and β-sheets in solved protein structures, suggested that there are preferences among the residues for forming (or breaking) helices.

Early work on secondary-structure prediction made *a priori* predictions, based on tables of residue preferences. Methods were tested according to a three-state model in which each residue in the prediction and in the experimental structure was assigned to the classes 'helix', 'sheet', and 'other', and the percentage of residues assigned to the correct class was defined as a measure of success called Q_3. Such methods achieved typical accuracies corresponding to Q_3 of ~55% (10).

Progress depended on the recognition that tables of many aligned sequences contained consensus information that could improve the prediction accuracy. The idea is to apply pattern-recognition algorithms to a set of aligned sequences homologous to the sequence of unknown structure for which the secondary structure is to be predicted. The patterns are based on the distribution of residues at the positions in the alignment table.

The most powerful pattern recognition algorithms now being applied to secondary-structure prediction include neural networks and hidden Markov models. The basic idea is to develop a network of implications containing a large set of adjustable parameters governing the assignment of each residue to the three classes: helix, sheet, or other. The systems are quite general, and the parameters must be adjusted by 'training' on known sequences and structures. The best current methods claim average accuracies of Q_3 of ~75%.

Secondary-structure prediction used to be evaluated by CASP, but this category has been removed. Instead, a project called EVA performs continuous testing of methods based on web servers, using the short delay between posting of the sequences of structures about to be released by the PDB and the release of the structures themselves. The EVA team includes but is not limited to groups led by B. Rost at Columbia University in New York, USA, A. Sali in San Francisco, USA, and A. Valencia in Madrid, Spain.

Many methods for secondary-structure prediction are available through http://cubic.bioc.columbia.edu [8.14]. A list of secondary-structure prediction servers can be found at: http://abs.cit.nih.gov/main/otherservers.html [8.15]. *Protocol 3* describes the use of one such server.

Protocol 3

Using the PredictProtein server to predict secondary structure

1. Go to the PredictProtein server (http://cubic.bioc.columbia.edu/predictprotein/[8.16]) and click the link to submit a protein sequence for prediction.

2. On the resulting page, enter an e-mail address for reporting results, an optional title for the query, and the sequence itself (see *Fig. 9*), then click the **Submit/run prediction** button. The sequence used in this example is available as the file Q2PR35.fasta[8.17] from the Protocol_3 folder for this chapter at the book's website.

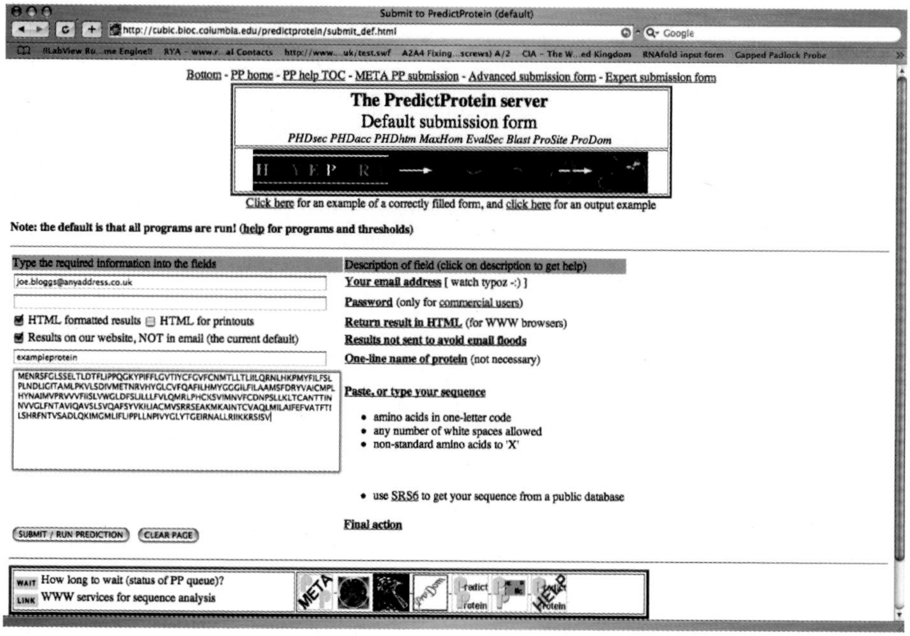

Figure 9. The server page of PredictProtein (http://cubic.bioc.columbia.edu)[8.14].

3. The results are returned by e-mail. For the example used here, the first part of the result is as follows:

```
              ....,....1....,....2....,....3....,....4....

AA            MENRSFGLSSELTLDTFLIPPQGKYPIFFLGVTIYCFGVFCNMT

PROF_sec         EEEEEEEEE     HHHHHHHHHHHHHHHHHHHHHH

Rel_sec       97666421125666405687122344778888888887730300


              ,....5....,....6....,....7....,....8....,....

AA            LLTLIILQRNLHKPMYFILFSLPLNDLIGITAMLPKVLSDIVME

P PROF_sec    EEEEEEE       HH HHHHH   HHHHHHHH HHHHHHHHHH

Rel_sec       13467504324652677874110121334200037888874236
```

Each block of the results contains three lines: the first ('AA') is the submitted amino acid sequence. The second ('PROF_sec') is the predicted secondary structure, indicating regions predicted to lie in helices ('H') or sheet ('E', standing for 'extended'); a blank signifies a prediction of a nonsecondary-structured region. The third line ('Rel_sec') is the reliability index: the number from 0 to 9 is a confidence estimate (0 = low confidence, 9 = high confidence) for that part of the prediction. Notice that the confidence values in the predictions in the right part of the first helical region are higher than those for the preceding sheet region.

2.4.4 ROSETTA

ROSETTA is a program by D. Baker and colleagues that predicts protein structure from amino acid sequence by assimilating information from known structures. At several recent CASP programs, ROSETTA showed the most consistent success on targets in both the Novel Fold and Fold Recognition categories (11).

ROSETTA predicts a protein structure by first generating structures of fragments using known structures and then combining them. For each contiguous region of three and nine residues, instances of that sequence and related sequences are identified in proteins of known structure. For fragments this small, there is no assumption of homology to the target protein. The distribution of conformations of the fragments in the proteins of known structure models the distribution of possible conformations of the corresponding fragments of the target structure.

ROSETTA explores the possible combinations of fragment conformations, evaluating compactness, paired β-sheets, and burial of hydrophobic residues. The structures that result from these simulations are separated into clusters, and the centers of the largest clusters are presented as predictions of the target structure. The idea is that a structure that emerges many times from independent simulations it is likely to have favorable features. There is a general belief that the work of Baker and colleagues represents a major breakthrough in the field of protein structure prediction.

Robetta (http://robetta.bakerlab.org [8.18]) is a web server designed to integrate and implement the best of the protein structure prediction tools (12). The central pipeline of the software first involves the parsing of a submitted amino acid sequence of a protein of unknown structure into putative domains. Then homology modeling techniques are applied to those domains for which suitable parents of known structure exist, and the *de novo* methods developed by Baker and coworkers to other domains. In addition, the user will receive the results of other prediction methods based on software developed outside the Robetta group. These include, for example, predictions of secondary structure, coiled-coils, and transmembrane helices.

2.5 Specialized methods for particular types of structure

The methods discussed so far are intended to be entirely general, at least where soluble, globular proteins are concerned. In contrast, specialized prediction methods have been developed for several types of protein of particular interest.

2.5.1 Prediction of antibody structures

Antibodies have some unusual features that suggest the need for specialized prediction methods. In most families of proteins, the binding site is the most strongly conserved part of the molecule. Other regions that do not play as direct a role in function diverge more quickly in evolution. In contrast, the antigen-binding sites of antibodies are the most variable part. The supporting region, the framework, is highly conserved.

The structural variation in antibody-combining sites is extensive, even within a single species, and the divergence is created not by the normal processes of evolution but by mechanisms designed specifically to generate structural diversity. These mechanisms create a repertoire of antibodies within a given vertebrate individual that contains molecules that recognize the entire organic world. Other mechanisms allow antibodies to 'mature' by somatic mutation to tune their specificity and affinity to specific threats such as infection by a specific viral strain.

A typical antibody (of the IgG class) contains four polypeptide chains, two identical light chains and two identical heavy chains (see *Fig. 10*). The light chains contain two domains and the heavy chains (of IgGs) contain four domains. The N-terminal domains of each chain are called the variable (V) domains, whilst the others are called the constant (C) domains. V domains of heavy and light chains interact; indeed such $V_L–V_H$ dimers, called F_V fragments, contain a complete antigen-binding site. A full immunoglobulin, comprising two light and two heavy chains, contains two equivalent antigen-binding sites. Other classes of antibody contain different numbers of constant domains and different quaternary structures. Some antibodies from the camel and related species contain only heavy chains and produce antigen-binding sites from V_H domains alone.

Each antibody domain contains a double-β-sheet framework (see *Fig. 11*). The frameworks of corresponding domains of antibodies from mammalian species are all quite similar. The tertiary structure is also common to other molecules of the immunoglobulin superfamily, including MHC proteins, and many molecules involved in cell–cell recognition.

Except for the unusual camel-type antibodies, the antigen-binding site is created by six loops, three from the V_L and three from the V_H domain (see *Fig. 11*).

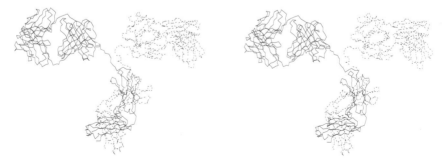

Figure 10. An intact antibody containing two light chains and two heavy chains.
One light–heavy chain pair is shown in solid lines; the other in broken lines. There are two antigen-combining sites, at the 'wing-tips' of the structure. (Stereo pair.)

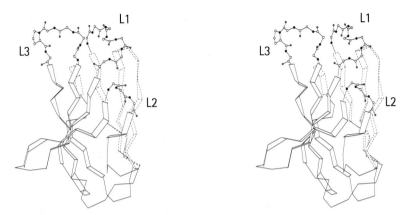

Figure 11. The Vκ domain of the Bence Jones protein REI.
Like other domains characteristic of the immunoglobulin superfamily, the tertiary structure includes two β-sheets (solid and broken ribbons, respectively) packed face-to-face and linked by a disulphide bridge. Three loops, L1, L2, and L3, contribute to the antigen-combining site. In a typical complete antibody, three homologous loops from the V_H domain would also contribute to the combining site. (Stereo pair.)

These loops are called the complementarity-determining regions (CDRs) and are numbered in order of appearance in the amino acid sequence: CDR1, CDR2, and CDR3 of light and heavy chains. Alternatively, they are denoted L1, L2, and L3 (the CDRs of the light chain) and H1, H2, and H3 (the CDRs of the heavy chain). L2, L3, H2, and H3 are hairpins, relatively short loops connecting consecutive strands in the same β-sheet. L1 and H1 bridge two different β-sheets. The seminal work of E.A. Kabat and T.T. Wu first noted the larger sequence variation in the V domains over the C domains, and, within the V domains, islands of sequence hypervariability that they presciently (before any antibody crystal structure had been solved) recognized as the regions responsible for the different specificities of different antibodies (13, 14).

From the point of view of prediction, the challenge in trying to rationalize the structural basis of antibody specificity is to understand the sequence–structure relationships in the variable domains. Within these domains, the framework is usually well within the range of applicability of standard homology modeling methods, so the difficulty is to predict the structures of the antigen-binding loops. It has been observed that the structures of short β-hairpins follow rules based on their local amino acid sequences (15). These rules are useful in predicting the conformations of some antigen-binding loops. A more general approach applies the canonical-structure model of C. Chothia and A. M. Lesk (16, 17). This model is based on the following observations:

- Five of the six antigen-binding loops of immunoglobulin structures (L1, L2, L3, H1, and H2) have only a small, discrete repertoire of main-chain conformations, called canonical structures.
- These conformations are determined by a few particular residues within the loop, or outside the loop but interacting with it. Among corresponding loops

of the same length, only these residues need to be conserved to maintain the conformation of the loop.

- Other residues in the sequences of the loops are thus left free to vary, to modulate the surface topography and charge distribution of the antigen-binding site.
- The signature patterns of the different loop conformations in the sequence provide useful methods for prediction of structure of antibody-combining sites.

An example of a canonical structure is the L3 loop from Vκ REI, a β-hairpin, containing a *cis*-proline at position 95, and stabilized by hydrogen bonds between the side chain of the residue at position 90, just N-terminal to the loop, and main-chain atoms of residues in the loop (see *Fig. 12*). The side chain at position 90 is a glutamine in REI (the residue at this position can also be an asparagine or a histidine in other Vκ chains). The combination of loop length, one of these polar side chains at position 90, and the proline at position 95 constitute the 'signature' of this conformation of this loop.

H3, the third hypervariable region of the heavy chain, is far more variable in length, sequence, and structure than the other antigen-binding loops. Several mechanisms contribute to generation of its diversity, including combinatorial choice of V_H, D, and J gene segments, and alternative splicing patterns at the junctions. Because of its greater variability, H3 does not follow the canonical structure model that governs the conformations of the other loops. However, sequence patterns determine alternative conformations of the proximal portions of the loop (18–20).

In expressed antibodies, H3 appears prominently at the center of the antigen-binding site. H3 – in contrast to the other five antigen-binding loops – has a conformation that depends strongly on its molecular environment.

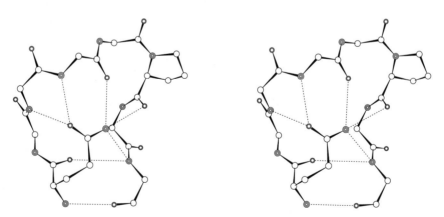

Figure 12. The most common canonical structure of Vκ L3 loops.
This is a six-residue hairpin, with a *cis*-proline at the fifth position of the loop, in a conformation stabilized by hydrogen bonding between a side chain at the residue just N-terminal to the loop and inward-pointing main-chain atoms. The two horizontal hydrogen bonds at the bottom of the structure connect the last two residues within the β-sheet. (Stereo pair.)

2.5.2 Prediction of transmembrane proteins and signal peptides

Many proteins are designed to sit within membranes. Membrane proteins mediate the exchange of matter, energy, and information between cell interiors and their surroundings. Examples of membrane protein functions include energy transduction via the generation or release of concentration gradients across cell or organelle membranes, and signal reception and transmission. It is estimated that, in the human genome, approximately 30% of genes encode membrane proteins. Approximately 70% of known targets of drugs are membrane proteins.

Given that membrane proteins are so common, it is important to have reliable tools for their identification. Relatively few membrane protein structures have been determined experimentally. This places a greater burden on computational tools for sequence analysis, to identify and characterize them.

Among their adaptations, membrane proteins contain regions of mostly nonpolar residues that interact with the organic layer. Many membrane proteins contain a set of seven consecutive α-helices that traverse the membrane, oriented approximately perpendicular to the plane of the membrane. These helices are connected by loops that protrude into the aqueous surroundings. A second class of membrane protein structures contains a β-barrel. Transmembrane helices are typically 15–30 residues long. Although enriched in hydrophobic residues, they contain some polar side chains, usually in interfaces between helices packed together in the structure. A useful clue to the orientation of the helices across the membrane is the 'positive– inside rule'. The loops between helices lie either entirely inside or entirely outside the cell or organelle, and those inside contain a preponderance of positively charged residues.

A simple approach to prediction of membrane proteins involves looking for amino acid segments of 15–30 residues that are rich in hydrophobic residues. However, signal peptides also contain hydrophobic helices: the signal sequence typically comprises a positively charged n-region, followed by a helical hydrophobic h-region, followed by a polar c-region. Methods for recognizing transmembrane helices tend to pick up the h-regions of signal peptides as false positives, whilst methods for recognizing signal peptides tend to pick up transmembrane helices as false positives.

L.A. Käll, A. Krogh and E.L. Sonnhammer trained hidden Markov models to test simultaneously for transmembrane helices and signal peptides (21). The goals are to find both at the same time, to discriminate between them in the results, and to predict not only the positions of the transmembrane helices but the locations – cytoplasmic or interior – of the loops. The method is called PHOBIUS, and is the most successful algorithm currently available for recognizing signal peptides and helical transmembrane proteins and for predicting the orientation of the transmembrane segments. PHOBIUS is capable of distinguishing h-domains of signal peptides from transmembrane helices: the number of false classifications of signal peptides was 3.9% and the number of false classifications of transmembrane helices was 7.7%, representing a great improvement over previous methods. It is interesting that addressing the two problems at once proved to be more successful than treating them separately.

Protocol 4

Using PHOBIUS to predict transmembrane helical segments and signal sequences

1. Go to the PHOBIUS web server at http://phobius.cgb.ki.se[8.19]. Choose **PolyPhobius** (link towards the top of the screen) to make use of information from homologous sequences in the prediction.

2. For this example, we will use the same protein sequence as was used in *Protocol 3* (Q2PR35. fasta[8.17]); either paste this sequence into the text window, or upload the file. Leave the other options at their default settings and click **Submit**.

3. The results returned by PHOBIUS are shown in *Fig. 13*, also available in the color section. At the top is a feature table showing the residue ranges predicted as transmembrane (TRANSMEM) or loops, the latter being assigned to cytoplasmic or noncytoplasmic locations.

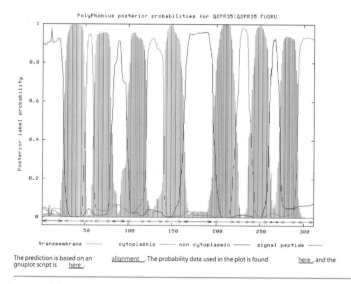

PolyPhobius prediction

```
Prediction of Q2PR35|Q2PR35_FUGRU

ID    Q2PR35|Q2PR35_FUGRU
FT    TOPO_DOM      1     25      NON CYTOPLASMIC.
FT    TRANSMEM     26     50
FT    TOPO_DOM     51     59      CYTOPLASMIC.
FT    TRANSMEM     60     80
FT    TOPO_DOM     81     97      NON CYTOPLASMIC.
FT    TRANSMEM     98    120
FT    TOPO_DOM    121    140      CYTOPLASMIC.
FT    TRANSMEM    141    162
FT    TOPO_DOM    163    196      NON CYTOPLASMIC.
FT    TRANSMEM    197    222
FT    TOPO_DOM    223    238      CYTOPLASMIC.
FT    TRANSMEM    239    260
FT    TOPO_DOM    261    272      NON CYTOPLASMIC.
FT    TRANSMEM    273    292
FT    TOPO_DOM    293    312      CYTOPLASMIC.
//
```

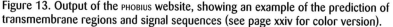

The prediction is based on an alignment . The probability data used in the plot is found here , and the gnuplot script is here .

Figure 13. Output of the PHOBIUS website, showing an example of the prediction of transmembrane regions and signal sequences (see page xxiv for color version).

4. The bottom part of the results page shows the graph giving, for each residue, the probability that the residue is a transmembrane region (densely spaced vertical bars) or an interhelical loop (solid lines). For example, the segment 197–222 is predicted to be a transmembrane region. It corresponds to a peak in the probability distribution (derived from the hidden Markov model) that is as high as 1.0 (= certainty) and over 0.9 for most of the central range of this region.

2.5.3 Coiled-coil regions

Proteins containing coiled coils are known among structural proteins, such as α-keratin, and also in a variety of globular proteins associated with a number of functions, prominently including transcription regulation. *Fig. 14(a)*, showing the structure of the jun–DNA complex, is a typical example. Such coiled-coil domains contain a signature pattern in their amino acid sequences. They show heptad repeats – seven-residue patterns – containing positions denoted a, b, c, d, e, f, and g, of which the first and fourth positions – a and g – are usually hydrophobic (see *Fig. 14b*).

Programs for predicting coiled coils include COILS (22) with a web server at http://www.ch.embnet.org/software/COILS_form.html[8.20] and PAIRCOIL (23) with a server at http://paircoil.lcs.mit.edu/cgi-bin/paircoil[8.21]. *Protocol 5* illustrates the use of the first of these programs.

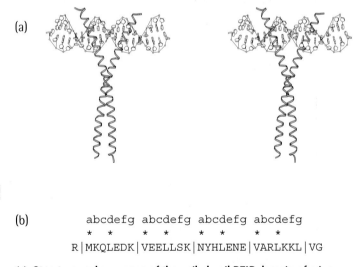

(a)

(b)   abcdefg abcdefg abcdefg abcdefg
             *   *   *   *   *   *   *   *
        R|MKQLEDK|VEELLSK|NYHLENE|VARLKKL|VG

Figure 14. Structure and sequence of the coiled-coil BZIP domain of c-jun.
(a) The coiled-coil BZIP domain from proto-oncogene c-jun bound to the cAMP response element DNA target. (b) Amino acid sequence of the coiled-coil region, split by vertical bars into heptads; asterisks indicate the hydrophobic residues at the 'a' and 'd' positions in each heptad.

Protocol 5

Prediction of coiled-coil regions using COILS

1. Go to the COILS server at http://www.ch.embnet.org/software/COILS_form.html [8.20].

2. For this example, we will use the c-fos protein from chicken. This is provided in the Protocol_5 folder for this chapter as a FASTA file, P11939.fasta [8.22].

3. Paste the complete FASTA file into the query sequence window and type an optional title into the **Query title** window. Leave the other settings at their default values and click the **Run Coils** button.

4. *Fig. 15* (also available in the color section) shows the prediction. The probabilities of the structure being a coiled coil are shown for analyses performed with sliding windows of 14, 21,

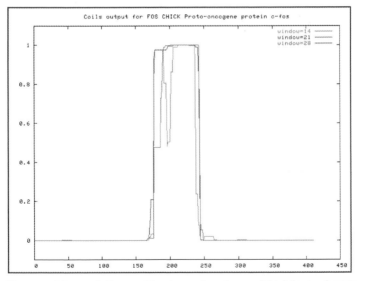

Coils output for FOS_CHICK Proto-oncogene protein c-fos

[ISREC-Server] Date: Mon Nov 27 23:53:01 Europe/Zurich 2006

```
coils -def -in=../wwwtmp/.COILS.29000.7081.seq
-out=../wwwtmp/.COILS.29000.7081.out -mat=2

# COILS version 2.1
# using MTIDK matrix
# no weights
# Input file is ../wwwtmp/.COILS.29000.7081.seq
#>FOS_CHICK Proto-oncogene protein c-fos, 419 bases, 4DDEA701 checksum.
```

You can get the prediction graphics shown above in one of the following formats:

- GIF-format
- Postscript-format
- numerical format (window 14, 21, 28)

Back to ISREC home page

Figure 15. Output of the COILS website, showing an example of the rediction of coiled-coiled regions (see page xxv for color version).

and 28 residues. In each case, there is a clear region where the probability is 1.0 (certainty), giving us confidence that there is a coiled region of around 60 residues near the middle of the sequence. We can be less confident of the exact boundaries of the coil-coiled region. The COILS documentation states that the results for the 'window = 21' are likely to be the most precise; on this basis we would expect the region 177–243 to form a coiled coil.

5. The results can also be downloaded in various graphic formats and as numerical data, using the links provided at the bottom of the results page.

3. REFERENCES

★ 1. **Branden C & Tooze J** (2005) *Introduction to Protein Structure*, 2nd ed. New York: Garland. – *A fine introduction to the entire field, setting it in context.*

2. **Lesk AM** (2002) *Introduction to Protein Architecture: The Structural Biology of Proteins.* Oxford University Press, Oxford, UK.

3. **Lesk AM** (2004) *Introduction to Protein Science: Architecture, Function and Genomics.* Oxford: Oxford University Press.

4. **Levitt M & Chothia C** (1976) *Nature*, **261**, 552–558.

5. **Lo Conte L, Brenner SE, Hubbard TJ, Chothia C & Murzin AG** (2002) *Nucleic Acids Res.* **30**, 264–267.

6. **Wixon J** (2001) *Comp. Funct. Genomics*, **2**, 103–113.

★ 7. **Tramontano A** (2006) *Protein Structure Prediction: Concepts and Applications.* Wiley-VCH, Weinheim. – *An overview of the current state of the art, explaining and discussing methods, and presenting results.*

★ 8. **Pieper U, Eswar N, Davis FP, et al.** (2006) *Nucleic Acids Res.* **34**, D291–D295. – *A description of one of the major homology modeling projects, which has both created software and collected large-scale results of predictions.*

★ 9. **Schwede T, Kopp J, Guex N & Peitsch MC** (2003) *Nucleic Acids Res.* **31**, 3381–3385. – *A description of another major homology modeling project, also providing software and collected results.*

10. **Kabsch W & Sander C** (1983) *FEBS Lett.* **155**, 179–182.

11. **Schueler-Furman O, Wang C, Bradley P, Misura K & Baker D** (2005) *Science*, **310**, 638–642.

★ 12. **Kim DE, Chivian D & Baker D** (2004) *Nucleic Acids Res.* **32**, W526–W531. – *Description of the most successful program for* de novo *prediction.*

13. **Kabat EA** (1978) *Adv. Protein Chem.* **32**, 1–75.

14. **Kabat EA, Wu TT & Bilofsky H** (1977) *J. Biol. Chem.* **252**, 6609–6616.

15. **Sibanda BL & Thornton JM** (1985) *Nature*, **316**, 170–174.

16. **Al-Lazikani B, Lesk AM & Chothia C** (1997) *J. Mol. Biol.* **273**, 927–948.

17. **Chothia C & Lesk AM** (1987) *J. Mol. Biol.* **196**, 901–917.

18. **Morea V, Tramontano A, Rustici M, Chothia C & Lesk AM** (1997) *Biophys. Chem.* **68**, 9–16.

19. **Morea V, Tramontano A, Rustici M, Chothia C & Lesk AM** (1998) *J. Mol. Biol.* **275**, 269–294.

20. **Shirai H, Kidera A & Nakamura H** (1996) *FEBS Lett.* **399**, 1–8.

21. **Kall L, Krogh A & Sonnhammer EL** (2004) *J. Mol. Biol.* **338**, 1027–1036.

22. **Lupas A, Van Dyke M & Stock J** (1991) *Science*, **252**, 1162–1164.

23. **Berger B, Wilson DB, Wolf E, Tonchev T, Milla M & Kim PS** (1995) *Proc. Natl. Acad. Sci. U.S.A.* **92**, 8259–8263.

CHAPTER 9

Gene ontology

Vineet Sangar

1. INTRODUCTION

Proteins carry out the individual operations that, suitably organized, constitute life. As a first step towards understanding the relationships among them, it is useful to be able to classify protein functions. Gene ontology, or GO (1–3), is a widely used system for such classification.

Without an effective system of classification, the value of the vast and expanding hoard of nucleotide sequence data is compromised by the inconsistent terminologies used by different researchers. For example, 'transcription' might have 'mRNA synthesis', 'polynucleotide synthesis', and 'mRNA generation' as synonyms. Biologists have no difficulty in equating these synonyms in the literature; however, computer-assisted searches will treat these terms as distinct, so that the result is sure to be incomplete. Classification systems have the potential to standardize nomenclature, supplanting the many synonyms that handicap computers and thereby making a wealth of data more accessible to analysis.

Many protein classification systems have been proposed, and we will return to some of these in more detail later in this chapter. For example, the Enzyme Commission (EC) classification began with a decision by the General Assembly of the International Union of Biochemistry (IUB), in consultation with the International Union of Pure and Applied Chemistry (IUPAC), in 1955, to establish an International Commission on Enzymes. The EC emphasized that '...enzyme nomenclature is primarily a matter of naming reactions catalyzed, not the structures of the proteins that catalyze them' (http://www.chem.qmul.ac.uk/iupac/jcbn/[9.1]), a philosophy adopted by GO. The EC classification scheme (4) (http://www.chem.qmul.ac.uk/iubmb/enzyme/[9.2]) consists of numbers arranged in trees that are three nodes deep. For example, EC 1.1.1.1 stands for alcohol dehydrogenase, and has been assigned to 182 alcohol dehydrogenases from different organisms, tissues, and organs. Parsing the number, the rightmost '1' stands for the class (oxidoreductases), the second '1' tells us which bond is cleaved (in this case CH-OH), the third '1' tells us about the acceptor (in this case, NAD or NADP), whilst the final '1' completes the specification (alcohol dehydrogenases). The EC nomenclature classifies enzymes according to the specific reaction catalyzed, but provides no means for classification of nonenzyme proteins. Nor does it attempt to classify according to the broader biological processes in which proteins participate, or to their physical locations in a cell.

1.1 Gene ontology

The GO Consortium was set up in 1998 with the goal of developing systems to classify protein functions, overcoming some of the restrictions of systems such as the EC classification. Ontologies 'provide controlled, consistent vocabularies to describe concepts and relationships, thereby enabling knowledge sharing' (5). The stated objectives of GO (http://www.geneontology.org/[9.3]) include the following:

- To design structured vocabularies describing aspects of molecular biology.
- To support annotation of gene products using vocabulary terms.
- To provide database access via these common terms to gene product annotations and associated sequences.

It is important to understand what GO is *not*, in order to make clearer what it *is*. In particular, GO is not a catalog of gene products, nor is it a complete database of all information – for example, evolutionary relationships and protein domain structures do not fall within its purview. Nor is GO a dictated standard for organizing data, or a 'compendium' database that unifies all other databases. It is not a mandatory or standard vocabulary, but a database for voluntary participation and usage by groups and individuals to achieve controlled and structured vocabularies. In other words, GO offers users the opportunity to share a common controlled vocabulary. Finally, it should be re-emphasized that GO is a system for describing the *functions* of gene products, not the products themselves (just as the EC system is a classification of enzyme *activities*, not enzymes).

Initially, GO was integrated with just three databases; Flybase (*Drosophila*), the *Saccharomyces* Genome Database (SGD) and the Mouse Genome Database (MGD). Since then, other plant, animal, and microbial databases have been integrated with GO in a two-way process: GO provides the standardized vocabulary needed to accommodate new types of protein in each database it embraces and, conversely, the respective databases use (and provide links to) the GO vocabulary in their annotations. Now, almost all major sequence databases are integrated with GO. Moreover, relationships have been set up to link GO to other 'classification-oriented' databases such as the EC database of enzymes, or the Pfam protein family database, a topic to which we will return later.

1.2 Structure of the GO database

GO provides controlled and carefully defined vocabularies, and a set of inter-relationships among the terms, called ontologies. The relationships are arranged in the form of 'directed acyclic graphs' (DAGs). A graph (in this context) is simply a network of connections, and a DAG is a network organized as parent–child relationships, in which a more specialized term can be thought of a 'child' of a more generalized 'parental' term. For example, the more general term 'membrane' is a parent of a more specialized term such as 'mitochondrial membrane'. A DAG differs from the more-familiar tree structure because, in a DAG, there may be alternative pathways from an ancestor (parent) to a descendent (child). However, the interconnections in a DAG are directed so that, although there may be paths from an ancestor to a descendent, there is no path leading from a descendent back to the ancestor (see *Fig. 1*).

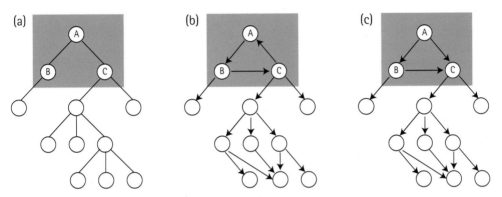

(a) (b) (c)

Figure 1. Graphs showing parent–child relationships between terms.
In each case, the nodes (circles) represent terms and the connecting lines show the relationships between them. (*a*) A conventional 'tree' structure, in which each parent (for example, A) can have one or more children (B and C), but each child has only one parent; hence, there is only one route leading from parent to child. (*b*) In this structure, each child can have two or more parents, but in some cases there are circular relationships (shaded box). (*c*) A directed acyclic graph (DAG) as used by GO; each child can have multiple parents (and vice versa), but the direction of the relationships (arrows) is restricted to prevent circular paths (shaded box). In this example, C is a child of both B and A; and B is also a child of A.

Fig. 2 shows a specific example, emphasizing the point that there can be more than one route from a parental term (such as 'cell') to child or grandchild terms (such as 'chloroplast'). GO graphs have two distinct types of relationship between the terms (nodes). The first relationship is 'is_a', meaning that the child term is one example of the more general class specified by the parental term. The second relationship is 'part_of', meaning that the child term is one component of the parental term. Thus, elaborating the relationships from *Fig. 2*, 'membrane' (child) is 'part of' 'cell' (parent); 'mitochondrial membrane' (child) 'is a' 'membrane' (parent). 'Chloroplast membrane' is linked to one parent, 'membrane', by an 'is_a' relationship (as the chloroplast membrane *is a* membrane) and to its other parent, 'chloroplast', by a 'part_of' relationship (as the chloroplast membrane is *part of* the chloroplast).

Another important characteristic of the relationships between GO terms is that they should obey the 'true path' rule. This means that if a child represents something, then all its parents should *also* encompass that same thing. In the example (*Fig. 2*), 'chloroplast membrane' is a type of 'membrane', which in turn is a part of 'cell'; and it is also true to say that 'chloroplast membrane' is a part of 'cell'. Thus, as we move downwards in GO DAGs, the information becomes more and more specific but never contradicts the information in a higher node.

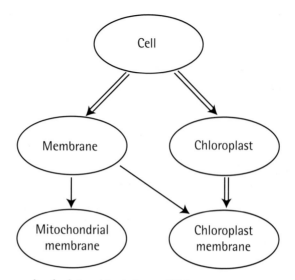

Figure 2. An example of relationships between GO terms.
Arrows point from the parental term to the child term. Single-lined arrows show
the relationship 'is_a' and double-lined arrows show the relationship 'part_of'. For
instance, 'chloroplast membrane' is a 'part_of' 'chloroplast' and 'is_a' 'membrane'. Both
'chloroplast' and 'membrane' are 'part_of' 'cell'.

These few simple relationships and rules, combined with the controlled definition
of the terms they connect, provide the controlled but flexible ontology of GO.

1.3 The three GO ontologies

GO consists of not one ontology but three, describing different aspects of a gene
product's role, and a protein may be described under all three ontologies. The three
GO ontologies are as follows (definitions taken from http://www.geneontology.
org/GO.doc.shtml [9.4]):

- **Molecular Function.** This ontology represents the activities at a molecular level.
  They do not specify where or when, or in what context, the action takes place.
  Molecular functions generally correspond to activities that can be performed
  by individual gene products, but some activities are performed by assembled
  complexes of gene products. (However, remember that GO is a classification of
  functions, not an assignment of functions to proteins or complexes.)
- **Biological Process.** A biological process is series of events accomplished by one
  or more ordered assemblies of molecular functions.
- **Cellular Component.** A cellular component is just that, a component of a cell
  but with the proviso that it is part of some larger object, which may be an
  anatomical structure (e.g. rough endoplasmic reticulum or nucleus) or a gene
  product group (e.g. ribosome, proteasome, or a protein dimer).

The distinction between these ontologies (especially between 'biological process'
and 'molecular function') can be confusing. A simple analogy would be with car

manufacture. If a bolt is tightened to fix the gear box to the engine, then the 'molecular function' would be 'tightening' (the specific action involved; many other instances of 'tightening' might occur in other contexts and places on the car); the 'biological process' would be 'transmission assembly' (the overall process of which this particular action is a part; the process will involve many other actions); and the 'cellular component' might be 'engine bay'. Similarly, hydrolysis (molecular function) of carbohydrates is one step in fermentation (biological process), which occurs in the cytosol (cellular component).

1.4 GO terms

We can now examine a typical GO term to understand its structure. The following is the definition of one such term (taken from http://www.geneontology.org/ontology/gene_ontology.obo [9.5]):

```
[Term]
id: GO:0000019
name: regulation of mitotic recombination
name_space: biological_process
def: "Any process that modulates the frequency, rate or extent ↻
    of DNA recombination during mitosis." [GOC:go_curators]
narrow_synonym: "regulation of recombination within rDNA ↻
    repeats" []
is_a: GO:0000018 ! regulation of DNA recombination
relationship: part_of GO:0006312 ! mitotic recombination
```

This format is as follows: 'id' is the identifier (GO:0000019) for this GO term; 'name_space' is the ontology (biological process, in this case) to which this term belongs; 'name' and 'def' (definition) are the crux of the GO database: the controlled vocabulary; 'synonym' represents all of the synonyms that are used in the databases; and 'is_a' and 'part_of' show the relationship of this GO id (GO:0000019) to other ids (GO:0000018 and GO:0006312): 'regulation of mitotic recombination' 'is_a' (i.e. is an instance of) 'regulation of DNA recombination' and is a 'part_of' 'mitotic recombination'.

1.5 Evidence codes

So far, we have seen a system that can be used to classify a gene product under any one of three different ontologies, and which shows how the terms within any one ontology are related to one another.

However, it is also important to consider the reliability with which a GO term is assigned to a given gene product. This is especially important as every node is connected to others in the GO database. Thus, if a protein is incorrectly assigned the GO term 'GO:0000019' ('regulation of mitotic recombination'), it will also be incorrectly classified as being involved in regulation of DNA recombination and as being part of the process of mitotic recombination, as these GO terms are parents of GO:0000019.

GO annotations therefore follow a cardinal rule that every annotation – every assignment of a GO term to a protein – must be accompanied by one or more *evidence codes*, which represent the broad categories of experimental or other evidence for that annotation. As with GO terms, evidence codes are standardized, and each is indicated by a code (see *Table 1*). The meanings are partly self-explanatory, but full definitions can be found at http://www.geneontology.org/GO.evidence.shtml[9.6].

Table 1. Evidence codes used in GO

Code	Meaning
IEA	Inferred from Electronic Annotation
IDA	Inferred from Direct Assay
IEP	Inferred from Expression Pattern
IGI	Inferred from Genetic Interaction
IMP	Inferred from Mutant Phenotype
IPI	Inferred from Physical Interaction
ISS	Inferred from Sequence Similarity
TAS	Traceable Author Statement
NAS	Nontraceable Author Statement
IC	Inferred from Curator
RCA	Inferred from Reviewed Computational Analysis
ND	No Data

2. METHODS AND APPROACHES

The controlled vocabularies of GO have opened a whole new set of possibilities for extracting information. GO can be used as an index to most of the available databases and can form the basis of efforts to integrate them.

A wide range of tools has been developed to use GO, and the following sections examine a few of these tools.

2.1 GO browsers

Browsers have been developed by curators of many different databases to link the information in their databases with GO. Some such browsers are listed at the end of this chapter (see section 3), and further information on many is provided by the GO consortium at http://www.geneontology.org/GO.tools.browsers.shtml[9.7].

These browsers have been developed according to the objectives of the projects being pursued by the investigators. As a first example, we will use the CGAP GO Browser (see *Protocol 1*) as a simple way to browse GO and to find human and mouse cancer-related genes assigned to specific terms – in this example, mouse genes having peroxidase activity.

Protocol 1

The CGAP GO Browser

1. Go to the CGAP GO Browser at http://cgap.nci.nih.gov/Genes/GOBrowser[9.8]. The introductory page will, initially, just display the three GO ontologies (biological_process, cellular_component, and molecular_function). To the right of each of these are listed the numbers of human (**Hs**) and mouse (**Mm**) proteins that have been assigned GO terms under the respective ontology.

2. To the left of each ontology is a gray button containing a '+'. Click on the '+' to the left of molecular_function.

3. This opens the molecular function ontology, displaying (at the time of writing) 17 GO terms that are at the highest level in the molecular function ontology. Again, links to the right of each term indicate how many human and mouse proteins have been assigned to each term. Some of the terms have an 'l' symbol to their left, indicating that there are no children of that term. Most, however, have a '+' symbol; click on the '+' to the left of antioxidant activity.

4. This opens a further set of terms (children of 'antioxidant activity'), of which the third member is peroxidase activity. The '+' symbol to its left indicates that there are children of this term (specific types of peroxidase activity). However, in this case, we will not look at all the children of 'peroxidase activity'. Instead, click on the blue link to the right of peroxidase activity, which indicates the number of mouse proteins that have been assigned this term. (At the time of writing, the link was **Mm:54** indicating 54 proteins, but this number is liable to change.)

5. A new window opens, displaying a table of the mouse cancer-implicated genes that have been annotated with the GO molecular function 'peroxidase activity'. At the time of writing, the seventh gene in the table is 'Aminoadipate-semialdehyde synthase'; click on the **Gene Info** link (on the right of the screen) for this entry.

6. The resulting page displays information and links for this gene. Towards the bottom of the page, under the heading 'Gene Ontology', you will see all of the GO terms assigned to this gene. In this case, **peroxidase activity** is only one of six GO terms assigned under molecular function (MF) ontology. There are also GO terms for the biological process (BP; in this example, **electron transport** and **response to oxidative stress**) and cellular compartment (CC; in this case, **mitochondrion**) ontologies.

7. Use your browser's 'back' button to return to the previous page showing the table of mouse genes with peroxidase activity. The box towards the upper right allows you to select and display certain types of information for all of the genes listed. Ensure that the **Ontology** box is ticked and click the **View** button beneath it.

8. The table now shows all of the GO terms assigned (under all three ontologies) to each of the genes in the list.

Next, we will look at AmiGO, a browser maintained by the GO consortium. This browser offers many options to search the ontologies and to query data from many linked databases on the basis of GO annotations. The following protocol is a brief tour to introduce some of its features. As an arbitrary example, we will suppose that a researcher is interested in S-adenosylmethionine transport.

Protocol 2

A brief tour of AmiGO

1. Go to http://www.geneontology.org/[9.3] and click the link **Browse the Gene Ontology with AmiGO**. The page will then show, underneath the box headed 'Filter Tree View', the 'top' of the GO hierarchy. The topmost item in the hierarchy is **all:all**, corresponding to all gene products that are annotated with any GO term, in any of the linked databases. The number in square brackets (192 662 at the time of writing, but sure to increase) indicates the number of GO-annotated gene products; click the small pie-chart icon to its right.

2. This opens a new window that shows, as a pie-chart, the number of GO annotations (in all of the databases linked to AmiGO) under each of the three ontologies (molecular function, biological process, and cellular component). Use your browser's 'back' button to return to the previous page.

3. We will now search for GO terms that relate to S-adenosylmethionine transport. Type 'adenosyl' into the text field next to **Search GO** towards the top of the page, ensure that the adjacent **Terms** button (not the **Gene Symbol/Name** button) is selected, and click **Submit**. The result is a list of GO terms and associated information about each one.

4. Scrolling down, you should find 'S-adenosylmethionine transport' and also 'S-adenosylmethionine transporter activity'. To the right of each term, you will find links to show the definition (**show def**), the GO ID, and, on the extreme right, an indication of which ontology the term belongs to. In this case, 'S-adenosylmethionine transport' is a biological process, whilst 'S-adenosylmethionine transporter activity' is a molecular function. Clearly, most proteins annotated with one term would be likely also to be annotated with the other, but the distinction is important (for example, it is conceivable that some proteins might be involved in the process of S-adenosylmethionine transport, without themselves having transporter activity. Assume for the moment that we are interested in only those gene products with transporter activity (GO:0000095). Click on the GO term itself (**S-adenosylmethionine transporter activity**).

5. This opens a new window, which displays the ontology of this term: it is a treasure-chest of information (see *Fig. 3*). Under the heading 'Term Lineage' (and beneath the box headed 'Filter tree view'), the text is arranged in a hierarchy from left to right on the screen, showing the ancestors (parents, grandparents, etc.) of our chosen term. The highest (left-most) level, is 'all' (indicating all GO terms), which is a parent of '**molecular function**', which is a parent of '**transporter activity**', and so on through successively narrower terms until we reach '**S-adenosylmethionine transporter activity**'. As before, the numbers in brackets after each term show how many gene products are annotated with that term.

6. Note that **S-adenosylmethionine transporter activity** appears at several points in the hierarchy, in lineages starting from **amine transporter activity**, **cofactor transporter activity**, and **organic acid transporter activity**.

7. Note also that **S-adenosylmethionine transporter activity** (wherever it appears in the hierarchy) has a '+' symbol to its left. This means that there are further levels of the hierarchy below this term. Click on the '+' and the more specialized children will appear. (In this case, there is only one child – **S-adenosylmethionine permease activity**.) Click on the '−' next to **S-adenosylmethionine transporter activity** to close this term again.

8. You can also view the siblings (rather than the ancestors) of the GO term. Click on the **Term parents, siblings and children** button above the term lineage, then on the **Set filters** button. The screen will refresh and will now show all of the siblings of 'S-adenosylmethionine transporter activity' at each point in the hierarchy where it occurs. For example,

'S-adenosylmethionine transporter activity' is a child of 'cofactor transporter activity', and so all of the other children of this term will be displayed. Return to the 'Term ancestors' view by clicking **Term ancestors** and **set filters**.

9. Click on the **Graphical view** link towards the top right of the term lineage. This will display the same data in a graphical form. Note that, perhaps counterintuitively, the more specialized term (S-adenosylmethionine transporter activity) is shown at the top of the screen, with arrows pointing towards its parents. This view clearly shows the way in which three hierarchies (diverging at 'transporter activity') each lead to 'S-adenosylmethionine transporter activity'. Use your browser's 'back' button to return to the text representation of the hierarchy.

10. Click on the **Term associations** link at the top of the page to see all of the proteins (in this case, only three) from the linked databases that have been annotated with the GO term GO:0000095 (S-adenosylmethionine transporter activity), or with any of its children. For example, PET8, from the SGD, has been assigned this GO term. You can also check the evidence on which this assignment was made (in this instance, from three types of evidence, all experimental), and there are links to the original database entry, literature citations, etc.

Figure 3. Ontology of the GO term 'S-adenosylmthionine transporter activity' (GO ID GO:0000095), as displayed in AmiGO.

11. Click on the gene name **PET8**. The updated page will show more detailed information on this protein including, particularly, a list of all of the other GO terms assigned to it. For example, you will see that this protein has been assigned the cellular component terms 'mitochondrion' and 'mitochondrial inner membrane', as well as biological process and molecular function terms reflecting its role in S-adenosylmethionine transport. Use your browser's 'back' button to return to the previous screen.

12. The menus just above the list of gene products allow you to select and filter the results (for example, you might want to view only those 'S-adenosylmethionine transporter activity' gene products that come from SGD and that have been annotated with this function on the basis of a biological assay (IDA, IGI, IEP, IMP, or IPI), or just by the sequence similarity (IEA, ISS, or RCA) (see *Table 1* for definitions).

13. Now click on the link **S-adenosylmethionine transporter activity ; GO:0000095** to return to the term lineage. Perhaps we want to broaden our search to include the more general 'sulfur amino acid transporter activity', which, as you will see, is a parent of S-adenosylmethionine transporter activity. Click on **sulfur amino acid transporter activity.**

14. The new window displays the term lineage for 'sulfur amino acid transporter activity'. As before, you can use the **Term associations** link towards the top of the page to explore further.

2.2 GO annotation tools

Many tools are available to help researchers annotate genes with appropriate GO terms. In many cases, of course, this process is equivalent to making a prediction of the function of a novel protein. This topic is covered in Chapter 10, along with illustrations of several programs (including GOFIGURE and GOPET), which attempt to assign appropriate GO terms based mainly on similarities with other GO-annotated proteins. Additional annotation tools are listed in section 3, and a further list (along with direct links to the respective tools) is available at http://www.geneontology.org/GO.tools.annotation.shtml [9.9].

GOTCHA (6) associates GO terms with a sequence of interest based on the sequence similarity. These are ranked by probability and displayed graphically on a subtree of GO. The main page accepts input in two ways: (i) pasting the sequence of interest in the provided window, and (ii) uploading the file with multiple sequences of interest. The page provides the option of searching many databases, which can be selected and deselected. Similarly, there is an option to filter the data based on the evidence codes for the genes. The output can be filtered further depending on the P value (suggested P value is 10–20%). For an explanation of the methodology for calculating the P value, see (6).

After selecting the sequence, database, evidence code, and the P value, you click the icon **Get GOing**. After submitting the job, the database provides the option of viewing the results online or downloading as an archive. The results are presented in a tabular form and also as a subtree in all three ontologies (molecular function, cellular component, and biological process). The table ranks the GO IDs that could be associated with the sequence with the score and estimated likelihood (P value), as well as the description.

2.3 Gene expression tools

Many research groups have developed tools to make better use of high-throughput gene expression assays such as microarrays in various organisms by using annotations from the GO database. The studies vary from pathological and pharmacological studies (GARBAN) to implementing the Kolmogorov–Smirnov continuous statistics approach to analyze Affymetrix microarray data (GODIST), to visualization of molecular interaction tools (BINGO). Some of the GO-related tools for gene expression analysis are listed in section 3, and a list of links to these and further tools is maintained at http://www.geneontology.org/GO.tools.microarray.shtml [9.10].

As an example, we will look more closely at one such tool – FUNCASSOCIATE (7). This program accepts a list of genes (typically a set of genes over- or underexpressed in a microarray-based expression assay) and finds all of the GO terms with which they have been annotated. It then identifies GO terms that are significantly over- or under-represented in the gene set, taking into account the frequency of that term in the complete gene set of the organism in question (or, optionally, in the subset of genes examined in the experiment).

Protocol 3

Use of FUNCASSOCIATE to find trends in expression data

1. For this example, we will use a short list of human genes (UniProt identifiers), which can be downloaded from the Protocol_3 folder for this chapter; the file is called GeneList.txt [9.11]. This dataset was created specifically for this example and represents a set of genes that are overexpressed in a hypothetical microarray expression assay, with the most overexpressed genes listed first.

2. Go to http://llama.med.harvard.edu/cgi/func/funcassociate [9.12]. In the **Select organism** pull-down menu, select **H. sapiens**.

3. Copy and paste the list of UniProt identifiers into the top-most of the two text windows. Click the button to the right of this window, which specifies **My query list of genes is ranked in descending order...** (This option places more weight on the first genes in the list – those that are most overexpressed in our example dataset; if your dataset was not ranked in this way, you would use the default option instead.)

4. Leave the lower text window empty. (If our microarray experiment had only examined a small subset of genes, then we could list these genes here so that our list of overexpressed genes would be analyzed in the context of this subset, rather than in comparison with the full set of UniProt human entries).

5. Tick the boxes to find both over- and under-represented attributes. Leave the P value at 0.05 (setting this to a lower value, such as 0.001, would ensure that only very significant representation biases were reported).

6. Click **Functionate!** and await the results.

7. The results screen will, towards the top, indicate the total number of human genes with which your query set of 184 genes was compared in evaluating over- or under-representation of GO terms. (At the time of writing, this number was 26 505.)

8. Below this is a brief explanation of the output and, below this, a ranked list of the attributes (GO terms) that were significantly over-represented in our gene set. The most important columns of the table are the adjusted probability value (P-adj), which reflects the statistical significance of the result, and the GO attribute.

9. In this example, most of the top hits in the list relate to olfactory perception; several other aspects of biology also seem to be well represented. Scrolling down, you will note that no attributes are listed as significantly under-represented; this is not surprising for such a small gene set: most GO attributes will not be represented at all.

10. Clicking on the button **Show gene attributes table** opens a new window showing which of the genes in the list contain which of the over-represented GO terms. (Check boxes on the list of attributes can be used to select only a subset of the attributes, if you prefer.)

11. Although this was an artificial example, a real dataset (from yeast) is available and can be examined as a demonstration, from the starting page of FUNCASSOCIATE. Many other microarray datasets can be downloaded (for example, from http://genome-www.stanford.edu/cgi-bin/webminer/mkjavascript[9.13]); a little editing will usually be needed to extract a list of gene identifiers from such data before pasting it into FUNCASSOCIATE.

2.4 Integration of GO with other classification systems

GO is one of the most widely used systems for classifying gene functions and provides a widely used 'common language' among databases. Nevertheless, many other classification systems exist, overlapping to various degrees with GO. The EC classification system was mentioned in section 1 and is approximately parallel to the 'molecular function' ontology of GO. Other systems focus on particular molecule types, organisms, or biological processes; or look at proteins from different perspectives (such as conserved sequence or structure motifs).

The GO Consortium has made a concerted effort to integrate GO annotations with these other databases. In particular, indices showing the relationship between GO terms and their equivalents in the other databases have been established and are updated continuously. These indices are available at http://www.geneontology.org/GO.indices.shtml?all[9.14], and each index consists of a list of terms from the relevant database (such as Pfam) and their equivalent GO terms and GO IDs. The equivalence is not always perfect (whether because of imprecise definitions of terms in one database or the other; because of a lack of one-to-one correspondence; or because the indexed database looks at biomolecules with a different perspective from GO), but is a useful guide. Moreover, the majority of these other databases provide direct links to GO so that, for example, a Pfam entry has links to the corresponding GO terms.

3. ADDITIONAL WEB RESOURCES

GO browsers

- CGAP GO : http://cgap.nci.nih.gov/Genes/GOBrowser[9.15]
- COBrA: http://www.xspan.org/cobra/index.html[9.16]

- Dr ZooView: http://genome4.ars.usda.gov/cgi-bin/dr.zoo.view.3.1/index.pl [9.17]
- EP GO: http://ep.ebi.ac.uk/EP/GO/ [9.18]
- GeneInfoViz: http://genenet.org/geneinfoviz/search.php [9.19]
- GoFish: http://llama.med.harvard.edu/~berriz/GoFishWelcome.html [9.20]
- PANDORA: http://www.pandora.cs.huji.ac.il/ [9.21]
- GenNav: http://mor.nlm.nih.gov/perl/gennav.pl [9.22]
- Goblet: http://goblet.molgen.mpg.de/ [9.23]

GO annotation tools

- GeneTools: http://www.genetools.no/ [9.24]
- GoAnnotator: http://xldb.fc.ul.pt/rebil/tools/goa/ [9.25]
- GOTCHA: http://www.compbio.dundee.ac.uk/Software/GOtcha/gotcha.html [9.26]
- HT-GO-FAT: http://199.133.147.108/mainbioinformatics.html [9.27]
- InGOt: http://www.inpharmatica.co.uk/ingot/ [9.28]

GO gene expression tools

- FUNCASSOCIATE: http://llama.med.harvard.edu/cgi/func/funcassociate [9.12]
- FUNCEXPRESSION: http://www.barleybase.org/funcexpression.php [9.29]
- GARBAN: http://www.garban.org/garban/home.php [9.30]
- GODIST: http://basalganglia.huji.ac.il/links.htm [9.31]

4. REFERENCES

★ 1. **Gene Ontology Consortium** (2001) *Genome Res.* **11**, 1425–1433. – *This paper gives an outline of the aims and methodologies behind GO.*
★ 2. **Gene Ontology Consortium** (2006) *Nucleic Acids Res.* **34**, D322–D326. – *This paper discusses updates to the GO vocabularies and tools, and other ontologies including SO – the Sequence Ontology*
★ 3. **Harris MA, Clark J, Ireland A, et al.** (2004) *Nucleic Acids Res.* **32**, D258–D261. – *This paper discusses the structure of GO and associated resources.*
   4. **NC-ICBMB** (1992) *Enzyme Nomenclature 1992: Recommendations of the Nomenclature Committee of the International Union of Biochemistry and Molecular Biology.* Edited by the NC-ICBMB and EC Webb. Academic Press, San Diego, CA.
   5. **Gruber TR** (1993) *Knowledge Acquisition,* **5**, 199–220.
   6. **Martin DM, Berriman M & Barton GJ** (2004) *BMC Bioinformatics,* **5**, 178.
   7. **Berriz GF, King OD, Bryant B, Sander C & Roth FP** (2003) *Bioinformatics,* **19**, 2502–2504.

CHAPTER 10

Prediction of protein function

Rodrigo Lopez

1. INTRODUCTION

In protein function prediction (sometimes termed 'gene function prediction'), the aim is to identify and characterize the physicochemical properties of the product of a transcript. Depending on the field of interest, these properties can be expressed in terms of biophysical, biophysiological, biomedical, and pure biological events. These in turn can tell us if the molecule acts as a protein kinase, is a tumor suppressor, or may be involved in development or cell death.

Whole-genome sequencing projects are the major source of unknown proteins at the current time and assignment of functions to gene products is, in most cases, done on the basis of the amino acid sequence alone. Protein structures, when available, often enable functional predictions to be made where sequence-based methods alone fail: this is one of the driving factors behind structural genomics projects. This is mainly because evolution will retain the folding pattern, and thus the function, long after sequence similarity becomes undetectable. However, it has to be noted that most of the structural genomic projects producing hundreds of protein structures are also generating many with unknown function. Inference of function based on sequence methods is tenuous: the methods do provide reasonable guesses at function, but they are not foolproof, as will be discussed in this chapter.

In this chapter, the reader will be introduced to many traditional sequence-based methods of predicting protein function. Most of the necessary software is available as open source and can be installed to run locally on any average modern computer. Advanced methods that use combinations of several traditional analyses – and make it easier for the user to carry out these simultaneously using web servers – will also be introduced. Moreover, state-of-the-art methodologies that employ machine-learning (i.e. artificial intelligence) techniques and further refine the analysis of protein sequences will be reviewed and compared towards the end of the chapter.

In addition to large-scale genome projects, the more traditional sources of novel protein sequences include peptide degradation by chemical means (Edman chemistry), direct peptide sequencing, mass spectroscopy, and lastly, but

Bioinformatics: *Methods Express* (Paul H. Dear, ed.)
© Scion Publishing Limited, 2007

most importantly, sequences obtained from crystallographic experiments that represent the one-dimensional structural chains in proteins and that are closely tied to secondary, tertiary, and quaternary conformations that give rise to specific biological activities. These technologies do not produce sequences en masse, but they do produce very accurate ones and these are often used as evidence that a catalytic residue is where it is expected or predicted; that a specific function attributed to a domain is backed by evidence; or that a fold, turn, or coil is in fact where it is expected.

2. METHODS AND APPROACHES

2.1 Required tools

There are many application suites for bioinformatics worth considering for the prediction of protein function. The most important commercial one is possibly GCG (Genetics Computer Group), known previously as the Wisconsin Package and now available from Accelrys (see http://www.accelrys.com/products/gcg/ [10.1]). The main open-source package is EMBOSS (1) (European Molecular Biology Open Software Suite) available from http://www.emboss.org/ [10.2]. These application suites complement and overlap each other in a variety of ways. *Tables 1* and *2* list the most commonly used tools for protein annotation and analysis in the EMBOSS and GCG suites, respectively. GCG may be the most consistent, but EMBOSS is the most functional and complete of the two. Furthermore, EMBOSS is distributed as source code and can be modified by the user to suit a particular requirement. Installing EMBOSS is straightforward if you have administration privileges on your computer. It will run well on a fairly average computer running Linux or one running CygWin (see http://www.cygwin.com [10.3]) under Microsoft Windows. EMBOSS is widespread and is available from various sites on the Internet. *Table 3* gives information on the portals providing EMBOSS in an unrestricted manner, as well as very comprehensive lists that show equivalences between EMBOSS and other application suites. Most of the EMBOSS programs can also be used on line via various web servers. The web sites that provide this service are many and include many of the EMBnet (European Molecular Biology network) nodes: a selection of EMBOSS web server sites is given in *Table 4*. Particularly noteworthy are the web services (2) from the European Bioinformatics Institute (EBI) – these work as if you had EMBOSS and the databases installed on your computer.

2.2 Prediction and determination of physicochemical properties of proteins

The first step in characterizing a new protein from its primary structure (sequence) is to establish whether it is in fact a good candidate on which to carry out further studies. It is especially worth establishing whether the protein shares any basic physicochemical properties with other proteins that have been studied experimentally. The most important use of these values is when qualitatively

Table 1. EMBOSS applications most commonly used in protein functional prediction and analysis

EMBOSS application name	Comment
CODERET	Extracts CDS, mRNA, and translations from feature tables
FUZZTRAN	Protein pattern search after translation
RECODE	Removes restriction sites but maintains the same translation
RECODER	Removes restriction sites but maintains the same translation
REMAP	Displays sequence with restriction sites, translation, etc.
SHOWORF	Pretty output of DNA translations
SHOWSEQ	Displays a sequence with features, translation, etc.
SIXPACK	Displays a DNA sequence with six-frame translation and ORFs
ANTIGENIC	Finds antigenic sites in proteins
BACKTRANSEQ	Back translates a protein sequence
CHARGE	Produces a protein charge plot
CHECKTRANS	Reports STOP codons and ORF statistics of a protein
DIGEST	Protein proteolytic enzyme or reagent cleavage digest
TRANALIGN	Aligns nucleic acid-coding regions given the aligned proteins
FUZZPRO	Protein pattern search
FUZZTRAN	Protein pattern search after translation
GARNIER	Predicts protein secondary structure
IEP	Calculates the isoelectric point of a peptide.
NEWCOILS	Predicts coiled protein secondary structure
OCTANOL	Displays protein hydropathy
ODDCOMP	Finds protein sequence regions with a biased composition
PATMATDB	Searches a protein sequence with a motif
PATMATMOTIFS	Searches the PROSITE motif database with a protein sequence
PEPNET	Displays proteins as a helical net
PEPSTATS	Protein statistics
PEPWHEEL	Shows protein sequences as helices
PEPWINDOW	Displays protein hydropathy
PREG	Regular expression search of a protein sequence
PSCAN	Scans proteins using PRINTS
SIGCLEAVE	Reports protein signal cleavage sites
TCODE	Fickett TESTCODE statistic to identify protein-coding DNA
TOPO	Draws an image of a transmembrane protein
TRANALIGN	Aligns nucleic acid-coding regions given the aligned proteins

assessing particular regions of a protein sequence that may share common physical characteristics with others in alignments with distantly related protein sequences. Typical parameters to measure and annotate are:

- Molecular weight
- Isoelectric point
- pK in neutral solution
- Overall charge and charge runs in the sequence
- Molar percentages
- Molar extinction coefficient (A_{280})
- Extinction coefficient at 1 mg/ml (A_{280})
- Residue composition
- Residue frequency
- Dayhoff statistic

Table 2. GCG applications most commonly used in protein functional prediction and analysis

GCG program	Description
MOTIFS	Looks for sequence motifs by searching through proteins for the patterns defined in the PROSITE Dictionary of Protein Sites and Patterns. Motifs can display an abstract of the current literature on each of the motifs it finds.
HMMERPFAM	Compares one or more sequences with a database of profile HMMs, such as the Pfam library, in order to identify known domains within the sequences.
PROFILESCAN	Uses a database of profiles to find structural and sequence motifs in protein sequences.
COILSCAN	Locates coiled-coil segments in protein sequences.
HTHSCAN	Scans protein sequences for the presence of HTH motifs, indicative of sequence-specific DNA-binding structures often associated with gene regulation.
SPSCAN	Scans protein sequences for the presence of secretory signal peptides.
PEPTIDESORT	Shows the peptide fragments from a digest of an amino acid sequence. It sorts the peptides by weight, position, and HPLC retention at pH 2.1, and shows the composition of each peptide. It also prints a summary of the composition of the whole protein.
ISOELECTRIC	Plots the charge as a function of pH for any peptide sequence.
PEPTIDEMAP	Creates a peptide map of an amino acid sequence.
PEPPLOT	Plots measures of protein secondary structure and hydrophobicity in parallel panels of the same plot.
PEPTIDESTRUCTURE	Makes secondary-structure predictions for a peptide sequence. The predictions include (in addition to α, β, coil, and turn) measures for antigenicity, flexibility, hydrophobicity, and surface probability. PLOTSTRUCTURE displays the predictions graphically.
PLOTSTRUCTURE	Plots the measures of protein secondary structure in the output file from PEPTIDESTRUCTURE. The measures can be shown on parallel panels of a graph or with a two-dimensional 'squiggly' representation.
MOMENT	Makes a contour plot of the helical hydrophobic moment of a peptide sequence.
HELICALWHEEL	Plots a peptide sequence as a helical wheel to help you recognize amphiphilic regions.
XNU	Replaces statistically significant tandem repeats in protein sequences with X characters. If a resulting protein sequence is used as a query for a BLAST search, the regions with X characters are ignored.
SEG	Replaces low-complexity regions in protein sequences with X characters. If a resulting protein sequence is used as a query for a BLAST search, the regions with X characters are ignored.

These values can be obtained using the EMBOSS PEPSTATS and PEPINFO programs. One particular statistic that requires some explaining is the Dayhoff statistic: this is the abundance of each amino acid in the protein compared with its abundance in a large number of well-characterized proteins. Clearly, no protein will have a truly 'average' composition (corresponding to a Dayhoff statistic of 1.000 for each amino acid), but large protein sequences in which many amino acids have extremely high or low Dayhoff statistics are compositionally biased and should be viewed with caution. *Protocols 1* and *2* illustrate the use of PEPSTATS and PEPINFO.

Table 3. Sites providing EMBOSS software and information

Web site address	Description
http://emboss.sourceforge.net/servers/ [10.57]	EMBOSS public servers
ftp://ftp.ncbi.nlm.nih.gov/repository/EMBOSS/ [10.58]	Table of equivalent programs at the NIH
http://www.ch.embnet.org/EMBOSS/index.html [10.59]	Swiss EMBnet node
http://bioportal.cgb.indiana.edu/tools/emboss/ [10.60]	Center for Genomics and Bioinformatics, Indiana University, USA
http://www.molgen.mpg.de/~beck/embossfaq.shtml [10.61]	Max-Plank Institute, Germany
http://www.csc.fi/molbio/progs/emboss/emboss.html [10.62]	Finish IT Centre for Science
http://emboss.imb.nrc.ca/ [10.63]	Canadian Bioinformatics Resource
http://www.molbiol.ox.ac.uk/analysis_tools/EMBOSS/ index.shtml [10.64]	CBRG at Oxford University, UK
http://www.psc.edu/general/software/packages/emboss/ [10.65]	Pittsburgh Supercomputing Center
http://bioinformatics.forsyth.org/gcgemboss/ [10.66]	The Forsyth Institute.
http://bioweb.pasteur.fr/seqanal/EMBOSS/ [10.67]	Institute Pasteur, France
http://csc-fserve.hh.med.imperial.ac.uk/emboss.html [10.68]	Imperial College, London, UK
http://zeon.well.ox.ac.uk/Pise/ [10.69]	Wellcome Trust Centre for Human Genetics, Oxford, UK
http://faculty.ucr.edu/~tgirke/Documents/EMBOSS/ EMBOSS_MANUAL.html [10.70]	EMBOSS manual
http://www.csc.fi/molbio/progs/emboss/emboss_qg.pdf [10.71]	EMBOSS quick guide

Table 4. EMBOSS web servers

Web site address	Description
http://srs.ebi.ac.uk [10.6]	Select the Tools tab, and then the link to EMBOSS. Many tools are available using a novel network technology called SOAP that allows the user access to applications and databases without depending on a WWW browser
http://proteas.uio.no/emboss/ [10.72]	WWW interface to EMBOSS running from the Norwegian EMBnet node
http://www.pasteur.fr/recherche/unites/sis/Pise/ [10.73]	Pise running from the Pasteur Institute in Paris, France
http://www.wemboss.org/ [10.74]	wEMBOSS from the Belgian EMBnet node
http://srs.ebi.ac.uk/ [10.6]	Access to EMBOSS over the WWW interface of SRS at the EBI
http://www.angis.org.au/ [10.75]	Biomanager at the Australian EMBnet node
http://embossgui.sourceforge.net/demo/ [10.76]	EMBOSS Explorer, supported by the National Research Council of Canada and by Genome Prairie

Protocol 1

Analysis of a protein sequence using PEPSTATS

1. We will first retrieve a protein sequence using the UniProt Knowledgebase (3). Visit http://www.uniprot.org/ [10.4]. Type the entry identifier 'ABCC9_HUMAN' into the text search window at the top of the screen and press the adjacent **Search** button. The record for the entry should appear.

2. We require the protein sequence in FASTA format. Locate and click the **Fasta** link (at the top of the main entry table) and the FASTA form of the sequence should appear. Copy and paste this sequence (including the header line beginning '>') into a text file and save this (being sure to save as 'Text only' if using Word.) A copy of the file that you should now have is provided in the Protocol_1 folder for this chapter, as ABCC9fasta.txt [10.5].

3. We will now run PEPSTATS on this sequence. This can be done locally if you have installed the software, or on any of the web servers detailed in *Table 4*. In this case, we will use the EBI web server. Go to http://srs.ebi.ac.uk/ [10.6], click on the **Tools** tab (the black-on-grey tab, not the white-on-black 'Tools' button above it), then choose **EMBOSS** from the list under 'Packages Information'.

4. Scroll down to find PEPSTATS in the alphabetical list and click on the adjacent **Launch** link.

5. On the resulting screen, paste the FASTA sequence into the text box and click **Launch**.

6. When the new screen appears, click on the **results** link (NOTE: this is the link in the text **Use Batch job status page** to view the **results**, *not* the 'Results' tab at the top of the screen). You will be taken to a page showing the status of all of your recent requests: this one should be the top-most or only one on the list. A tick in the **Status** column indicates your results are ready – if it shows an egg timer instead, wait and then click the refresh button.

7. When your results are ready, click on the job name to see the results.

8. *Fig. 1* shows the expected results for this sequence.

```
PEPSTATS of ACC9_HUMAN from 1 to 1549

Molecular weight = 174259.02              Residues = 1549
Average Residue Weight  = 112.498         Charge   = 15.0
Isoelectric Point = 7.3871
A280 Molar Extinction Coefficient  = 201130
A280 Extinction Coefficient 1mg/ml = 1.15
Improbability of expression in inclusion bodies = 0.636


Residue          Number                     Mole%       DayhoffStat
A = Ala          107                        6.908       0.803
B = Asx          0                          0.000       0.000
C = Cys          25                         1.614       0.557
D = Asp          66                         4.261       0.775
E = Glu          84                         5.423       0.904
F = Phe          75                         4.842       1.345
G = Gly          85                         5.487       0.653
H = His          34                         2.195       1.097
I = Ile          113                        7.295       1.621
K = Lys          79                         5.100       0.773
L = Leu          203                        13.105      1.771
M = Met          45                         2.905       1.709
N = Asn          75                         4.842       1.126
P = Pro          51                         3.292       0.633
Q = Gln          54                         3.486       0.894
R = Arg          69                         4.454       0.909
S = Ser          110                        7.101       1.014
T = Thr          105                        6.779       1.111
V = Val          98                         6.327       0.959
W = Trp          25                         1.614       1.241
X = Xaa          0                          0.000       0.000
Y = Tyr          46                         2.970       0.873
Z = Glx          0                          0.000       0.000


Property     Residues                        Number      Mole%
Tiny         (A+C+G+S+T)                     432         27.889
Small        (A+B+C+D+G+N+P+S+T+V)           722         46.611
Aliphatic    (I+L+V)                         414         26.727
Aromatic     (F+H+W+Y)                       180         11.620
Non-polar    (A+C+F+G+I+L+M+P+V+W+Y)         873         56.359
Polar        (D+E+H+K+N+Q+R+S+T+Z)           676         43.641
Charged      (B+D+E+H+K+R+Z)                 332         21.433
Basic        (H+K+R)                         182         11.750
Acidic       (B+D+E+Z)                       150          9.684
```

Figure 1. Output of EMBOSS PEPSTATS with ABCC9_HUMAN (ATP-binding cassette transporter).

Protocol 2

Analysis of a protein sequence using PEPINFO

1. Following the same steps as in *Protocol 1*, obtain the sequence of the mouse 5-hydroxytryptamine 2B receptor, which has entry identifier 5HT2B_MOUSE in FASTA format and save this as a text file. (A copy of the file you should have is given in the Protocol_2 folder as 5HT2Bfasta.txt [10.7]).

2. Run PEPINFO on this sequence. The steps are the same as in *Protocol 1*, but, when you get the alphabetic list of EMBOSS programs, find PEPINFO. Remember to click on the adjacent **Launch** link.

3. When your results are ready, clicking on the job name will produce a page showing several different analyses of the sequence. Click on the link above the first analysis (the link will be **PEPINFO:1_temp_job1** unless you chose a different name for the job in the preceding step).

4. On the resulting page, you will see links to two graphics files (for example, temp_job1.pepinfo.1.png and temp_job1.pepinfo.2.png). Clicking on the first of these will create a new window showing the distribution of different classes of amino acids (tiny, small, aliphatic, aromatic, etc.) in the sequence. Note the distribution, particularly the scarcity of polar and charged residues in some parts of the protein.

5. Close the graphics window and click on the second of the graphics file links. This will produce a page showing plots of hydropathy (using several different parameters). Note the peaks in hydropathy corresponding to the regions with fewer polar or charged residues. We will return to hydropathy plots later.

2.3 Determination of secondary structure from sequence

Inferring the secondary structure of a protein from its sequence is an abstract art. This is mostly due to the fact that most theoretical models yield only approximate results. It largely involves guessing what major structural elements (helices, α- and β-sheets, coils, and turns) a protein is likely to have. However, they do help because they provide an overall picture of the gross secondary structure expected to be observed if the tertiary structure was determined. Furthermore, when comparing secondary structure predictions from closely related protein sequences, they provide clear evidence of what core elements are shared between these. They also help determine and characterize protein sequences at the level of families and domains. There are basically two ways in which secondary structure is determined: (i) by the analysis of residue composition and residue-run order, and (ii) via alignments, in which a sequence with unknown structure is aligned with one or more for which the structure is known. Examples of tools that can predict secondary structure using the single sequence methods include EMBOSS's GARNIER (4) (one of the earliest methods) and GCG's PEPTIDESTRUCTURE and PLOTSTRUCTURE applications, which are based around Chou and Fasman predictions (5). In the following two protocols, we will apply both methods to a human protein sequence.

Protocol 3

Basic secondary structure prediction using GARNIER

1. Obtain the amino acid sequence of UniProt entry ABCC9_HUMAN, as described in *Protocol 1*.

2. In this example, we will use the EBI web server to run EMBOSS GARNIER. Go to http://srs.ebi. ac.uk/ [10.6], click on the **Tools** tab at the top of the window, then choose **EMBOSS** from the list under **Packages Information**.

3. Scroll down to find GARNIER in the alphabetical list and click on the adjacent **Launch** link.

4. Paste the FASTA sequence into the sequence window and click **Launch**.

5. When the new screen appears, click on the **results** link (NOTE: this is the link in the text **Use Batch job status page to view the results**, *not* the 'Results' tab at the top of the screen). You will be taken to a page showing the status of all of your recent requests: this one should be the top-most or only one on the list. A tick in the **Status** column indicates your results are ready – if it shows an egg timer instead, wait and then click the refresh button.

6. When your results are ready, click on the job name to see the results.

7. The result shows the complete amino acid sequence, with each residue assigned to helix, sheet, turn, or coil regions. A summary at the bottom of the screen gives the number and percentage of residues assigned to each type of structure.

8. Copy the text in the main result window, paste it into a text document, and save this. (A copy of the file that you should have is given in the Protocol_3 folder as Garnier_result.txt [10.8].)

Protocol 4

Basic secondary-structure prediction using CHOFAS

1. Obtain the amino acid sequence of UniProt entry ABCC9_HUMAN, as described in *Protocol 1*.

2. Go to http://fasta.bioch.virginia.edu/ [10.9], and click **Plot Kyte–Doolittle Hydropathy**.

3. In the new window, change the topmost menu from **Kyte–Doolittle Hydropathy Plot** to **Chou–Fasman Secondary Structure prediction**, and paste the sequence into the protein sequence window. Click **Submit Sequence**.

4. The result will show the assignment of the amino acids to helix, sheet, or turn structures, with a summary at the bottom indicating the number and percentage of amino acids in each category. Save the results to a text file. (A copy of the file that you should have is given in the Protocol_4 folder as CHOFAS_result.txt [10.10].)

The main difference between GARNIER and CHOFAS is in the specificity of the prediction. GARNIER reports four types of major structure: helices, sheets, turns, and coils. CHOFAS, on the other hand, reports only helices, sheets, and turns. Predictions of secondary structure using methods such as CHOFAS and GARNIER are about 60–70% accurate. If a comparison is made between the output of CHOFAS and GARNIER and the annotation of the ABCC9_HUMAN entry at the level of topo domains and transmembrane regions, this is easily illustrated. *Fig. 2*, obtained from the SRS server at the EBI, is a cartoon of the features annotated in the entry. This can be used for the comparison mentioned above.

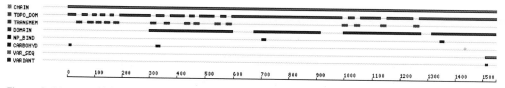

Figure 2. Diagram of the features of UniProt ABCC9_HUMAN.
Compare the TOPO domain and TRANSMEM regions with the output of CHOFAS and GARNIER.

GARNIER reports 40% of the sequence as being potentially involved in helices, whilst CHOFAS gives 77.1% helices. There are 15 transmembrane regions in the UniProt/Swiss-Prot annotation (see *Fig. 2*; some of these are predictions, whilst some are supported by experimental evidence) accounting for some 30% of the sequence in total. To examine this disparity, and to focus more specifically on transmembrane helices, we can use TMAP (6), an EMBOSS application designed specifically to determine transmembrane regions.

Protocol 5

Using EMBOSS TMAP to look for transmembrane regions

1. Obtain the amino acid sequence of UniProt entry ABCC9_HUMAN, as described in *Protocol 1*.

2. Use the EBI web server to run TMAP on this sequence. The process is identical to that used to run GARNIER in *Protocol 3*, except of course that you will choose TMAP instead of GARNIER from the alphabetic list of EMBOSS programs and click the adjacent **Launch** button.

3. The result shows the predicted transmembrane regions in a text format. Copy this information into a new text file. (A copy of the file that you should now have is given in the Protocol_5 folder as Tmap_result.txt [10,11].)

4. Beneath the text output will be a link to a graphics file. A copy of this image is shown in *Fig. 3*.

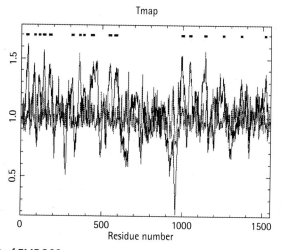

Figure 3. Graphical output of EMBOSS TMAP.
Note the marker line at the top that indicates where the potential transmembrane regions are located.

The output of TMAP can be used to compare the secondary-structure prediction of GARNIER and CHOFAS at the level of helices alone in order to distinguish between helices involved in transmembrane regions and those that are not. TMAP predicts 17 transmembrane domains for this sequence, most of which are in good agreement with the annotation found in the UniProt/Swiss-Prot entry.

Although the older sequence-based methods seem to overpredict (for example, CHOFAS reports 25 helices), the user must realize that these are predicting gross secondary structure and do not take into consideration that in order for a helix to be involved in a transmembrane region, it must be a part of a structure that crosses the plasma membrane going outward, then turning and going inward through the membrane again. At least 19 residues are required to accomplish this.

The shorter helices predicted by the older methods should not be dismissed: they could be exposed at the surface of the protein. In this case, one would be quite likely to consider this to be a globular protein, perhaps involved in the transport of ions (e.g. transferrin), in regulation of biological processes (insulin), or as an enzyme (which are generally spherical in shape). Amongst the most commonly known globular proteins are hemoglobin and immunoglobulins.

CHOFAS, TMAP and GARNIER are 'old-fashioned' algorithms that lack the sophistication of methods based on machine-learning techniques for pattern and profile matching such as hidden Markov models (HMMs) (7). HMMs are algorithms that are specifically designed to identify patterns in irregular sequences. Unlike the preceding sequence-based methods, TMHMM (8) uses HMMs to predict the membrane topology of the protein. HMMs do this by generating a pattern or grammar that is specific to the overall structure of transmembrane regions – alternating segments of membrane and nonmembrane segments with a well-defined organization of positively charged residues. The following protocol illustrates the use of TMHMM.

Protocol 6

Using TMHMM to identify transmembrane regions

1. Obtain the amino acid sequence of UniProt entry ABCC9_HUMAN, as described in *Protocol 1*.

2. Go to http://www.cbs.dtu.dk/services/TMHMM/ [10.12] and paste the FASTA sequence into the large window. Leave the other options at their default settings and click **Submit**.

3. The first part of the results is a text output, listing the parts of the sequence predicted to lie inside and outside the cell, and the transmembrane helices between them. Copy this text into a new text document. A copy of the document you should now have is given in the Protocol_6 folder as TMHMM_result.txt [10.13].

4. There is also a graphical output representing the same information. This is shown in *Fig. 4* (also available in the color section).

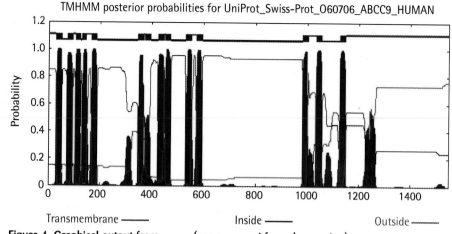

TMHMM posterior probabilities for UniProt_Swiss-Prot_O60706_ABCC9_HUMAN

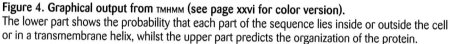

Transmembrane ——————   Inside ——————   Outside ——————

Figure 4. Graphical output from TMHMM (see page xxvi for color version).
The lower part shows the probability that each part of the sequence lies inside or outside the cell or in a transmembrane helix, whilst the upper part predicts the organization of the protein.

TMHMM 2.0 predicts 14 transmembrane regions. In fact, none of the prediction methods accurately coincide with each other. However, they all provide predictions that overlap in subtle ways. It is important to consider what training sets were used by the artificial intelligence-based methods, as their outputs will naturally be biased towards these. TMHMM 2.0 is particularly well suited to the analysis of eukaryotic protein sequences. The user needs to be careful about the presence of the signal peptide in the sequence, as this may decrease the accuracy of the prediction.

Without supporting experimental evidence that the transmembrane regions are actually located where they have been placed by the annotators or by the prediction methods, it is in cases like this that the user is warned to be a little wary of making assumptions about the accuracy of results. An important factor that may sometimes explain differences in the results of the same application but in different sites is the program configuration, which may have been tweaked to fit a particular requirement.

Transmembrane regions can also be detected and predicted accurately by using hydropathy indices. The most efficient way in which these are used is in the generation of graphs or hydropathy profiles, also know as Kyte–Doolittle (9) graphs. We have already seen examples of such graphs produced by PEPINFO (see *Protocol 2*). There are several indices of hydropathy in existence, but the most commonly used is that proposed by Nakai *et al.* in 1992 (10). These graphs are mainly used to predict whether a protein contains transmembrane regions that indicate it may be a fibrous protein, or surface regions that may indicate that it is of a globular nature. By choosing the window size over which hydrophobicity is plotted, different features can be emphasized. In the following protocol, we will use PEPWINDOW to look at hydropathy using different window sizes.

Protocol 7

Using EMBOSS PEPWINDOW to plot hydropathy over different window sizes

1. Obtain the amino acid sequence of UniProt entry ABCC9_HUMAN, as described in *Protocol 1*.

2. Go to http://srs.ebi.ac.uk/ [10.6] and select the **Tools** tab at the top of the page and then the link to EMBOSS on the left. Scroll down to find PEPWINDOW in the alphabetic list of programs and then click the adjacent **Launch** link.

3. Paste the FASTA amino acid sequence into the sequence window.

4. Click the small '+' in the **Advanced Options** box to show the available options. Change the window size to 19, then click **Launch**.

5. Retrieve your results (click the **results** link – not the 'Results' tab – and then click on the job name shown). You should get a graphics file similar to that shown in *Fig. 5(a)*, plotting hydropathy over a sliding 19-residue window.

6. Note the good correlation between the highest peaks on this plot and the 15 transmembrane regions of the annotation (see *Fig. 2*). Most of the highest peaks correspond to annotated transmembrane portions, with the exception of some peaks towards the right.

7. Repeat the analysis, but this time use a window size of nine residues. *Fig. 5(b)* shows the expected output. Note the many additional smaller peaks and troughs compared with those seen with a window of 19 residues. These additional features indicate regions that may be exposed on the surface of the protein.

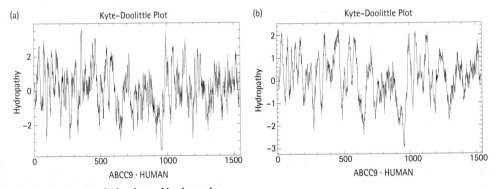

Figure 5. Kyte–Doolittle plots of hydropathy.
(*a*) With a window size of 19 residues, short potential helices are suppressed and only longer helices – particularly those likely to be transmembrane regions – are apparent as peaks. (*b*) With a window size of nine residues, smaller features become apparent – these indicate regions of the protein likely to be exposed on the surface.

Determining the secondary structure of a protein from its sequence helps in identifying the major structural elements and, in doing so, in inferring some of the core structural and functional elements. In the previous example, a very simple technique has been shown that helps determine whether a protein is potentially involved in carrier functions (although it gives us no clue as to what is transported) or whether it may be involved in more complex biological pathways, such as the transport of substances across the cell membrane.

Transcription factors provide a further example of proteins where *de novo* analysis of the sequence can be informative. Transcription factors bind DNA in specific ways at specific places – promoter or enhancer regions – to modulate transcription. Determining whether a protein has DNA-binding functionality is often done by the identification of certain residue patterns within the sequence. These patterns (or 'motifs') often occur only within constrained structures in a protein. For example, leucine zippers are an important group of DNA-binding motifs and are characterized by having four or more leucine residues spaced at seven-residue intervals and arranged along one side of a helical secondary structure. Two such motifs can interdigitate like the teeth of a zipper to form a stable interaction between two protein helices.

The following example uses the sequence of a well-known oncogene class called the FOS family. The *c-fos* proto-oncogene has been found in all mammalian model organisms and has also been identified in other vertebrates such as birds and fish. It is a relatively simple protein characterized by a helical structure known as a coiled coil around the middle of the sequence, which contains the DNA-binding domain and a leucine zipper. The following protocol uses the 2ZIP server (11) to find leucine zippers and related features.

Protocol 8

Using 2ZIP to identify leucine zipper motifs

1. Go to http://www.uniprot.org/ [10.4] and retrieve the entry for FOS_HUMAN in FASTA format (see steps 1 and 2 of *Protocol 1*). Copy and paste the sequence into a text document. A copy of the document you should now have is given in the Protocol_8 folder as FOSfasta.txt [10.14].

2. Go to the 2ZIP server at http://2zip.molgen.mpg.de/ [10.15], paste the FASTA sequence into the sequence window, and press **Submit Zipper Query**.

3. The main part of the output is shown in *Fig. 6* (the complete output is given in the Protocol_8 folder as 2zip_prediction.txt [10.16]). A leucine zipper region has been predicted, roughly in the middle of the protein.

```
1) number of potential LEUCINE ZIPPERS: 1

1---------11--------21--------31--------41--------51-------
MMFSGFNADYEASSSRCSSASPAGDSLSYYHSPADSFSSMGSPVNAQDFCTDLAVSSANF




61--------71--------81--------91--------101-------111------
IPTVTAISTSPDLQWLVQPALVSSVAPSQTRAPHPFGVPAPSAGAYSRAGVVKTMTGGRA




121-------131-------141-------151-------161-------171------
QSIGRRGKVEQLSPEEEEKRRIRRERNKMAAAKCRNRRRELTDTLQAETDQLEDEKSALQ
                     CCCCCCCCCCCCCCCCCCCCCCCCCCCCCCCCCC
                     L------L------L-
                     LZLZLZLZLZLZLZLZLZ

181-------191-------201-------211-------221-------231------
TEIANLLKEKEKLEFILAAHRPACKIPDDLGFPEEMSVASLDLTGGLPEVATPESEEAFT
CCCCCCCCCCCCCCCCC
-----L------L
LZLZLZLZLZLZLZL

241-------251-------261-------271-------281-------291------
LPLLNDPEPKPSVEPVKSISSMELKTEPFDDFLFPASSRPSGSETARSVPDMDLSGSFYA




301-------311-------321-------331-------341-------351------
ADWEPLHSGSLGMGPMATELEPLCTPVVTCTPSCTAYTSSFVFTYPEADSFPSCAAAHRK




361-------371-------
GSSSNEPSSDSLSSPTLLAL
```

Figure 6. Prediction from the 2ZIP server of the protein FOS_HUMAN.
Underneath the amino acid sequence, a predicted coiled-coil region is indicated by 'CCC...'. A predicted leucine zipper is indicated by 'LZLZ...' and the leucine residues within the zipper are indicated above by 'L'.

2ZIP can provide information on where in the sequence this particular pattern is located. However, it does not say whether it is located in a particular type of secondary structure. The output of TMAP (see *Protocol 5*) is useful in predicting that it is located within a helix and that perhaps the leucine residues are arranged in the outside of the coil structure. A more direct way of viewing this is to use the graph of the helical topology of the protein generated by the EMBOSS application PEPNET. PEPNET simply displays the amino acid sequence of the protein 'wrapped' into a helix of seven residues per turn. This does not, of course, mean that the entire protein is helical, but it does make it easy to spot motifs that would run down one face of helical regions. This is illustrated in *Protocol 9*.

Protocol 9

Using EMBOSS PEPNET to visualize a leucine zipper motif

1. Obtain the amino acid sequence of UniProt entry FOS_HUMAN, as described in *Protocol 8*.

2. Go to http://srs.ebi.ac.uk/ [10.6] and select the **Tools** tab at the top of the page and then the link to EMBOSS on the left. Scroll down to find PEPNET in the alphabetic list of programs and then click the adjacent **Launch** link.

3. Paste the FASTA amino acid sequence into the sequence window.

4. Leaving the options at their default values, click **Launch**.

5. The result will be presented in two graphics files (the protein is too large to show in a single file). The first of these is shown in *Fig. 7* (also available in the color section) – note the leucine zipper motif at residues 165–193, presenting five leucine residues along one side of the helix. (The fact that the line of leucines happens to be on the 'top' of the helical plot is fortuitous – they could equally have been in a line along the middle or bottom of the graph.)

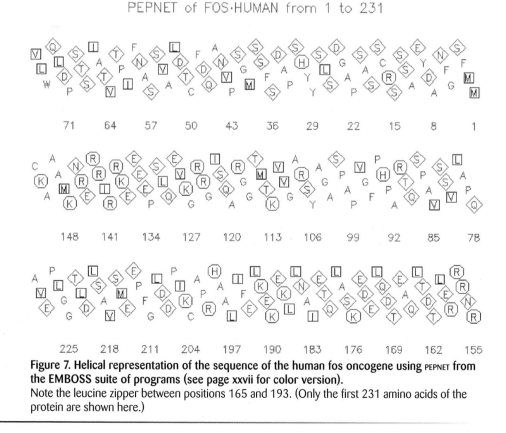

PEPNET of FOS·HUMAN from 1 to 231

Figure 7. Helical representation of the sequence of the human fos oncogene using PEPNET from the EMBOSS suite of programs (see page xxvii for color version).
Note the leucine zipper between positions 165 and 193. (Only the first 231 amino acids of the protein are shown here.)

So far, we have seen how to make predictions of some easily recognizable secondary structures and motifs, which can help in inferring function for many proteins. As an exercise, the reader is encouraged to run GARNIER and CHOFAS (see Protocols 3 and 4) on the mouse 5-hydroxytryptamine 2B receptor (see Protocol 2) and compare the results. Then try running PEPWINDOW (see Protocol 7) on the same sequence, varying the window size between nine and 19 residues; is there ground to assume that there are any transmembrane regions or short helical regions in the sequence?

2.4 Determination of functional domains using pattern-matching methods

Another approach to identifying functional elements, or regions that are associated with function, is pattern matching to reveal motifs shared with other proteins. Often it is not possible to recognize a relationship between a known and an unknown protein sequence from a pairwise alignment alone. Nevertheless, a relatively short pattern or motif may exist that, if highly conserved, will establish a specific functionality within the unknown sequence. Many experts have contributed to the creation of databases dedicated to storing and annotating such signatures, patterns, and motifs. They have also created the tools with which to search these databases. Often, the results of searches against these databases will provide insights into the putative function of a protein. In some cases, these results also yield valuable information about the structure and about functional mechanisms. The most commonly known databases that deal with signatures, patterns, and motifs include Prosite (12), PRINTS (13), Blocks (14), and Pfam (15). The following protocol illustrates the use of pattern-matching databases such as Prosite, using the ScanProsite searching tool from ExPASy.

Protocol 10

Using ScanProsite to search for motifs

1. Download the example file USERSEQ1_fasta.txt [10.17] from the Protocol_10 folder.

2. Go to http://www.expasy.org/ [10.18], click on the **PROSITE** link (under 'Databases'), and, on the resulting page, click on the **ScanProsite** link.

3. On the resulting page, paste the contents of the FASTA file into the window under **Protein(s) to be scanned**. Leave the settings at their default values and click on **START THE SCAN** at the bottom of the page.

4. The result (at the time of writing) is shown in *Fig. 8*. There are two hits (PS00036 and PS50217), one to a profile and another to a pattern. The annotation for these can be viewed below, and although they predict the same DNA-binding structure, the leucine zipper, they do so using two different matching methodologies: a search using consensus patterns and another using a profile derived from a matrix, respectively.

5. Clicking on either the link to the profile (**PS50217**) or to the corresponding matrix (**PS00036**) will take you to the same page, giving descriptive information about both the pattern and the corresponding matrix (see *Fig. 9*).

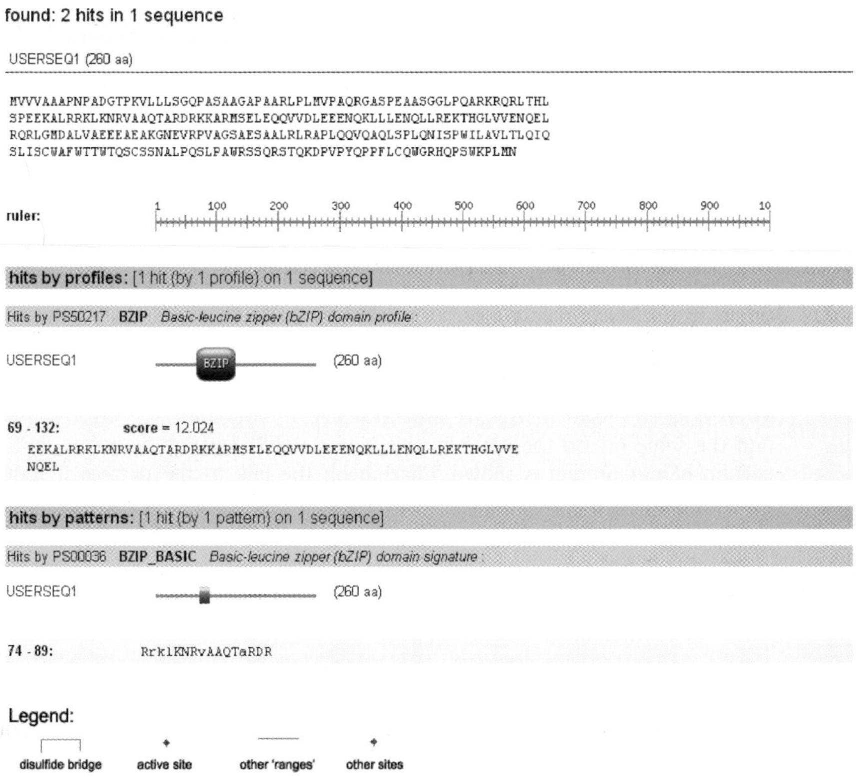

found: 2 hits in 1 sequence

USERSEQ1 (260 aa)

MVVVAAAPNPADGTPKVLLLSGQPASAAGAPAARLPLMVPAQRGASPEAASGGLPQARKRQRLTHL
SPEEKALRRKLKNRVAAQTARDRKKARMSELEQQVVDLEEENQKLLLENQLLREKTHGLVVENQEL
RQRLGMDALVAEEEAEAKGNEVRPVAGSAESAALRLRAPLQQVQAQLSPLQNISPWILAVLTLQIQ
SLISCWAFWTTWTQSCSSNALPQSLPAWRSSQRSTQKDPVPYQPPFLCQWGRHQPSWKPLMN

ruler:

hits by profiles: [1 hit (by 1 profile) on 1 sequence]

Hits by PS50217 **BZIP** *Basic-leucine zipper (bZIP) domain profile* :

USERSEQ1   ——————[BZIP]——————   (260 aa)

69 - 132:   **score** = 12.024

EEKALRRKLKNRVAAQTARDRKKARMSELEQQVVDLEEENQKLLLENQLLREKTHGLVVE
NQEL

hits by patterns: [1 hit (by 1 pattern) on 1 sequence]

Hits by PS00036 **BZIP_BASIC** *Basic-leucine zipper (bZIP) domain signature* :

USERSEQ1   —————[■]—————   (260 aa)

74 - 89:   RrklKNRvAAQTaRDR

Legend:

disulfide bridge   active site   other 'ranges'   other sites

Figure 8. Expected result of using ScanProsite to look for motifs in the protein sequence 'USERSEQ1_fasta.txt [10.20]'.

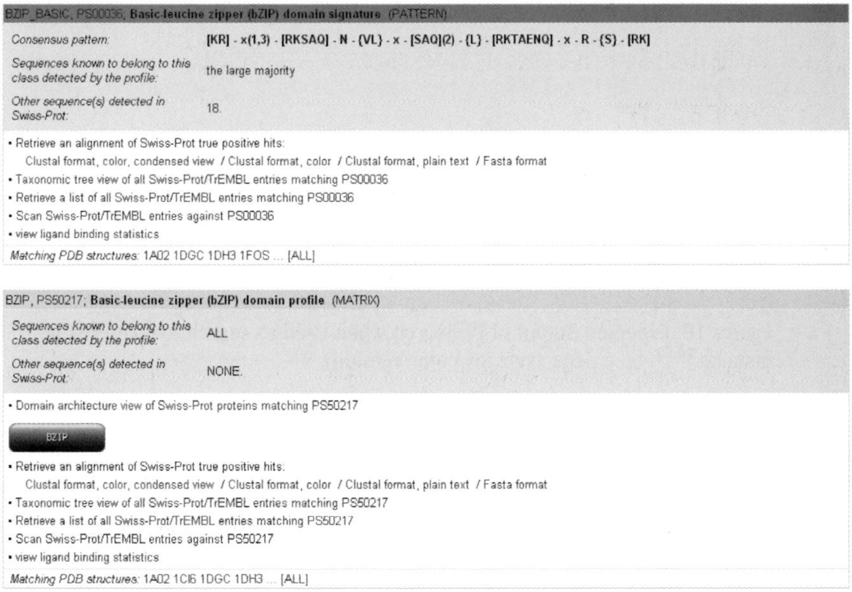

BZIP_BASIC, PS00036, Basic-leucine zipper (bZIP) domain signature (PATTERN)	
Consensus pattern:	[KR] - x(1,3) - [RKSAQ] - N - {VL} - x - [SAQ](2) - {L} - [RKTAENQ] - x - R - {S} - [RK]
Sequences known to belong to this class detected by the profile:	the large majority
Other sequence(s) detected in Swiss-Prot:	18.

- Retrieve an alignment of Swiss-Prot true positive hits:
     Clustal format, color, condensed view / Clustal format, color / Clustal format, plain text / Fasta format
- Taxonomic tree view of all Swiss-Prot/TrEMBL entries matching PS00036
- Retrieve a list of all Swiss-Prot/TrEMBL entries matching PS00036
- Scan Swiss-Prot/TrEMBL entries against PS00036
- view ligand binding statistics

Matching PDB structures: 1A02 1DGC 1DH3 1FOS ... [ALL]

BZIP, PS50217; Basic-leucine zipper (bZIP) domain profile (MATRIX)	
Sequences known to belong to this class detected by the profile:	ALL
Other sequence(s) detected in Swiss-Prot:	NONE.

- Domain architecture view of Swiss-Prot proteins matching PS50217

[BZIP]

- Retrieve an alignment of Swiss-Prot true positive hits:
     Clustal format, color, condensed view / Clustal format, color / Clustal format, plain text / Fasta format
- Taxonomic tree view of all Swiss-Prot/TrEMBL entries matching PS50217
- Retrieve a list of all Swiss-Prot/TrEMBL entries matching PS50217
- Scan Swiss-Prot/TrEMBL entries against PS50217
- view ligand binding statistics

Matching PDB structures: 1A02 1CI6 1DGC 1DH3 ... [ALL]

Figure 9. Description of the bZIP domain pattern (upper) and the corresponding matrix (lower) that produced the positive hits seen in *Fig. 8*.

We can also use another tool to search the Prosite database – PPSearch – which illustrates more clearly the actual region of our query sequence that matches the pattern. This is shown in *Protocol 11*.

Protocol 11

Using PPSearch to search for motifs

1. Go to http://www.ebi.ac.uk/ppsearch/ [10.19] and paste the contents of the file USERSEQ1_ fasta.txt [10.20] (see *Protocol 10*) into the sequence window.

2. Leaving other options at their default settings, click **Run**. The expected output (at the time of writing) is shown in *Fig. 10* (also available in the color section). As you can see, we find the same hit (to the basic leucine zipper (bZIP) pattern) as before, but the matching portion of our protein is shown. Clicking on the link to the pattern (**PS00036**) will take you to a page giving details of the pattern and of the many sequences from which it was derived.

PPSearch Output

```
------------------------------------------------------------
|            ppsearch (c) 1994 EMBL Data Library           |
|          based on MacPattern (c) 1990-1994 R. Fuchs      |
------------------------------------------------------------

PROSITE pattern search started: Thu Jun 15 12:56:22 2006

Sequence file: /ebi/extserv/old-work/ppsearch-20060615-12562181829685.input

-----------------------------------------
Sequence /ebi/extserv/old-work/ppsearch-20060615-12562181829685.input (260
residues):

Matching pattern PS00036 BZIP_BASIC:
    74: RRKLKNRVAAQTARDR
Total matches: 1

Total no of hits in this sequence: 1


==========================================

1311 pattern(s) searched in 1 sequence(s), 260 residues.
Total no of hits in all sequences: 1.
Search time: 00:00 min
```

Figure 10. Expected output of PPSearch when used to search for patterns in 'USERSEQ1_ fasta.txt [10.20]' (see page xxviii for color version).

The user needs to consider carefully what his or her data are being compared against when searching databases of any sort (patterns or profiles). The databases distinguish frequent motifs and ambiguous profiles, and options are available to include or exclude these when searching. These frequent patterns are often short signals associated with phosphorylation or myristylation sites. These are almost always detected around serine, threonine, and arginine residues and often are considered false-positive predictions. (Try repeating *Protocol 11*, but change the 'Include abundant patterns' option from **No** to **Yes**.)

Another thing to consider is that results, even excellent predictions, need to be reviewed in terms of the origin and context of the sequence. There are several motifs that are found only in prokaryotes and such findings in proteins of nonprokaryotic origin need to be filtered out. One typical example would be the prokaryotic membrane lipid attachment site. Conversely, features normally confined to eukaryotes may be predicted in prokaryotic proteins, where they should be carefully reviewed; examples might include homeodomains and helix–loop–helix (also known as helix–turn–helix or HTH), zinc finger, leucine zipper (in the presence or absence of basic DNA-binding domains), and winged helix motifs.

The important thing to remember before assigning function based on these types of prediction is that an ambiguous pattern, although identified in the sequence, cannot be depended upon in the absence of further supporting evidence. In certain cases, their presence must be viewed in the evolutionary context (for instance, it may be what remains of an earlier function of a protein or part of it). When it comes to transcription factors, it is important to remember that not all transcription factors bind DNA. Some just bind other transcription factors or other molecules in an allosteric fashion. Conversely, not all transcription factors that do bind DNA actually activate transcription.

Other examples of transcription factor motifs that can be determined by pattern-matching analysis of the protein sequence are the HTH (16) and zinc-finger domains. The HTH motif is found in proteins involved in transcription regulation, usually as repressors, and is composed of 20–25 residues arranged in two α-helices that bind DNA at the major groove. Consider the sequence of SHOX_HUMAN, a 292-residue protein involved in fundamental aspects of growth and development, which is expressed in a wide variety of tissues, including skeletal muscle, placenta, pancreas, heart muscle, and bone marrow. It contains a homeobox domain spanning residues 117–176. Within this region, there is an HTH motif that can help us determine what kind of regulation this protein is involved in. Analysis of SHOX_HUMAN with the EMBOSS program HELIXTURNHELIX yields a positive prediction for this motif within the homeobox domain spanning residues 145–166. (HELIXTURNHELIX is used in essentially the same way as the other EMBOSS programs.)

Zinc fingers usually occur in groups that form a 'glove' around a stretch of DNA. They consist of two antiparallel β-strands and one α-helix. Zinc fingers are responsible for binding zinc ions using a pair of cysteine residues located on the β-sheets and a pair of histidine residues on the α-helix. This structure is highly conserved and is represented by the Prosite entry ZINC_FINGER_C2H2. Analysis of SHOX_HUMAN using PPSearch (see *Protocol 11*; the sequence of SHOX_HUMAN is given in the folder Other_Files) suggests 12 consecutive zinc fingers (see *Fig. 11*).

```
Sequence ZN492_HUMAN (574 residues):

Matching pattern PS00028 ZINC_FINGER_C2H2_1:
   186: CKECEKSFCMLSHLAQHKRIH
   214: CKECGKAYNETSNLSTHKRIH
   242: CEECGKAFNRLSHLTTHKIIH
   270: CEECGKAFNQSANLTTHKRIH
   298: CEECGRAFSQSSTLTAHKIIH
   326: CEECGKAFSQSSTLTTHKIIH
   354: CEECGKAFSQLSHLTTHKRIH
   382: CEECGKAFKQSSTLTTHKRIH
   410: CEVCSKAFSRFSHLTTHKRIH
   438: CEECGKAFNLSSQLTTHKIIH
   466: CEECGKAFNQSSTLSKHKVIH
   522: CEECGKAFNNSSILNRHKMIH
Total matches: 12


Total no of hits in this sequence: 12
```

Figure 11. Zinc-finger motifs found by PPSearch in the sequence of SHOX_HUMAN.
Note the conserved cysteines towards the left of each finger and the histidines towards
the right.

Finally, we will take one more example of using motif-searching approaches.
Consider the following sequence:

```
MRLAVGALLVCAVLGLCLAVPDKTVRWCAVSEHEATKCQSFRDHMKSVIPSDGPSVACVK
KASYLDCIRAIAANEADAVTLDAGLVYDAYLAPNNLKPVVAEFYGSKEDPQTFYYAVAVV
KKDSGFQMNQLRGKKSCHTGLGRSAGWNIPIGLLYCDLPEPRKPLEKAVANFFSGSCAPC
ADGTDFPQLCQLCPGCGCSTLNQYFGYSGAFKCLKDGAGDVAFVKHSTIFENLANKADRD
QYELLCLDNTRKPVDEYKDCHLAQVPSHTVVARSMGGKEDLIWELLNQAQEHFGKDKSKE
FQLFSSPHGKDLLFKDSAHGFLKVPPRMDAKMYLGYEYVTAIRNLREGTCPEAPTDECKP
VKWCALSHHERLKCDEWSVNSVGKIECVSAETTEDCIAKIMNGEADAMSLDGGFVYIAGK
CGLVPVLAENYNKSDNCEDTPEAGYFAVAVVKKSASDLTWDNLKGKKSCHTAVGRTAGWN
IPMGLLYNKINHCRFDEFFSEGCAPGSKKDSSLCKLCMGSGLNLCEPNNKEGYYGYTGAF
RCLVEKGDVAFVKHQTVPQNTGGKNPDPWAKNLNEKDYELLCLDGTRKPVEEYANCHLAR
APNHAVVTRKDKEACVHKILRQQQHLFGSNVTDCSGNFCLFRSETKDLLFRDDTVCLAKL
HDRNTYEKYLGEEYVKAVGNLRKCSTSSLLEACTFRRP
```

Sequence database searches performed with this sequence indicate that it is
closely related to a family of globular proteins known as transferrins – iron-binding
transport proteins that are expressed in the liver and secreted in plasma. They
are characterized by containing two peptidase domains (known as S60 domains),
which are nonfunctional as the catalytic residues are missing. In light of this
information, the next step is to try to identify these metal-binding domains using
pattern-matching approaches that hopefully will yield predictions that support
the sequence database search observations.

A search of Prosite using PPSearch (see *Protocol 11*; the protein sequence can
be found in the Protocol_12 folder) returns the result shown in *Fig. 12*. Note that
essentially the same motif has been reported twice, at positions 114 and 115:
this is because the definition of the motif is ambiguous. Apart from this, three
different motifs have each been found at two distinct locations. In fact, each of
the three motifs is a conserved part of the larger P60 domain, and the protein
contains two such domains.

Next, we will analyze the same sequence using FINGERPRINTSCAN, a program that
compares proteins against the PRINTS database. PRINTS is a database of protein

'fingerprints' – groups of motifs characteristic of specific protein families. FINGERPRINTSCAN, therefore, looks not just for isolated sequence motifs, but for the presence (and relative positioning) of a suite of motifs.

```
Sequence PROBABLETRANSFERRIN (698 residues):

•  Matching pattern PS00205 TRANSFERRIN_1:
   114: YYAVAVVKKD
   115: YAVAVVKKD
   445: YFAVAVVKKS
Total matches: 3


•  Matching pattern PS00206 TRANSFERRIN_2:
   207: YSGAFKCLKDGAGDVAF
   536: YTGAFRCLVEKGDVAF
Total matches: 2


•  Matching pattern PS00207 TRANSFERRIN_3:
   241: QYELLCLDNTRKPVDEYKDCHLAQVPSHTVV
   577: DYELLCLDGTRKPVEEYANCHLARAPNHAVV
Total matches: 2


Total no of hits in this sequence: 7
Total no of Patterns matched for this sequence: 3
```

Figure 12. Prosite matches against a probable transferrin protein reported by PPSearch.

Protocol 12

Using FINGERPRINTSCAN to compare a protein against the PRINTS database

1. Download the protein sequence file PROBABLETRANSFERRIN_fasta.txt [10.21] from the Protocol_12 folder.

2. Go to http://www.ebi.ac.uk/printsscan/ [10.22] and paste the complete FASTA sequence into the sequence window. Leave other settings at their default values and click **Run**.

3. The first part of the results is shown in *Fig. 13*. There are matches to four different PRINT entries, but only the first (transferrin) is significant (note the low *E* value). Of ten motifs forming the transferrin fingerprint, the program has found all ten. The row of ten 'I's simply indicates that all ten transferrin motifs were found with >35% identity compared with the respective consensus sequence for that motif.

```
Sn; PROBABLETRANSFERRIN
, 698 bases, 62BD8E34 checksum.
Si; Fasta sequence
2TBS
2TBT FingerPrint    No.Motifs SumId    AveId    ProfScore  Ppvalue    Evalue     GraphScan
2TBH TRANSFERRIN    10 of  10 670.31   67.03    6548       9.9e-98    7.9e-93    IIIIIIIIII
2TBN 5HT2BRECEPTR    2 of   8  61.06   30.53    421        0.00016    13         i....I..
2TBN OPSINBLUE       2 of   6  62.68   31.34    379        0.0016     1.3e+02    .I...i
2TBN LVIRUSORF2      2 of   9  43.48   21.74    326        0.0052     4.1e+02    i......i.
```

Figure 13. Results of FINGERPRINTSCAN applied to a probable transferrin protein.

4. The rest of the results (not shown here) show the alignment between the query sequence and the complete set of motifs making up the matching fingerprint. (For a full explanation of the output, follow the **FingerPRINTScan Help** link on the left of the web page.)

As a third line of attack, we will search the Pfam database – a collection of multiple sequence alignments (and models that reflect the consensus features of those alignments) for many protein families and domains.

Protocol 13

Searching Pfam

1. Go to http://www.sanger.ac.uk/Software/Pfam/ [10.23] and, under the **Search by** tab at the top of the page, select **Protein name or sequence**.

2. Paste the probable transferrin sequence (see *Protocol 12*) into the large sequence window. Leave other settings at their default values and click **Search Pfam**.

3. Follow the prompts to retrieve the results. You should find that two segments of the query sequence match the full-length transferrin domain with high scores. Click on the **align** link (under **Alignment** in the table of results) to see the quality of the alignment in each case.

All three searches indicate that our 'probable transferrin' is indeed a transferrin, but the details of the supporting evidence differ in each case. Naturally, a truly novel protein sequence is likely to give less-clear-cut results: relying on a single method, even the most advanced one, is not a good approach. As a rule, the user should always consult at least two independent methodologies.

2.5 Advanced methods combining several protein function prediction algorithms

The availability of large families of homologous sequences has revolutionized secondary-structure prediction and several web-based services have emerged that attain levels of accuracy (when compared with the real tertiary and quaternary structures) of 70% or more. Some of these later methods include PredictProtein (17) at http://www.predictprotein.org/ [10.24], and JPRED (18) using Jnet (19) at http://www.compbio.dundee.ac.uk/~www-jpred/ [10.25].

What characterizes these methods is that they combine several different approaches into one. This is obviously very useful as it means that the user does not need to run the sequence independently through a myriad of programs or web sites. That said, one should always be aware that these systems are dependent on the databases that they use being up to date. Although the above methods are more accurate, they present limitations with protein sequences or alignments that may contain multiple domains or exceed a certain length. The lists below show which methods are employed for PredictProtein and JPRED.

PredictProtein uses the methods:

- Prosite search
- Low-complexity region filtering
- ProDom (20) domain search
- PSI-BLAST (21) against the UniProt Knowledgebase
- *MaxHom* (22), multiple sequence alignment based on PSI-BLAST highest scores
- Coils prediction
- Disulfate bridge prediction
- Nuclear localization signals to predict whether the protein binds DNA
- Prediction of membrane-bound helices
- Prediction of α- and β-sheets
- Prediction of protein globularity and solvent accessibility
- HMMPFAM
- Reliability

PredictProtein can produce quite a substantial output. In the 'Other_files' folder, you will find two files – XBP1_HUMAN.txt [10.26] containing a human protein sequence and PredProtXBP [10.27] containing the output of PredictProtein (using default settings) for this sequence.

JPRED uses the following methods:

- Low-complexity region filtering
- PSI-BLAST against the UniProt Knowledgebase
- Pair-wise comparison of all matches in order to generate percentage identities
- Clustering of the sequences on the basis of percentage identity
- Removes clusters with >75% identity (these are redundant)
- Removes gaps from the target sequence and correct alignments
- Extracts PSI-BLAST and HMMPSSM profiles and passes them to the Jnet neural network

JPRED is a particularly fast and accurate method. The best way of demonstrating this server is by using it. The user should try these methods on his own and frequently read the instructions, as sequence format, sequence length, and other requirements can change without notice. The following protocol shows the main steps required to submit data to the Jnet server.

Protocol 14

Using JPRED

1. Upload the file ExampleSeq.txt [10.28] from the Protocol_14 folder. (This is an amino acid sequence in raw format, with no spaces or header.)

2. Go to http://www.compbio.dundee.ac.uk/~www-jpred/ [10.29] and click on the **Prediction** link.

3. Enter your e-mail address as requested and paste the stripped sequence into the sequence window.

4. Leave the other options at their default settings and click **Run secondary structure predictions**. You will see a page with a job identifier that will be used to generate unique results pages for your analysis.

5. When the job is completed you will receive an e-mail or see a link in the browser to the results page. Links will take you to various representations of the output, including:

- A graphical representation of the alignment of your sequence with the closest matches in a similarity search of the UniProt Knowledgebase using JALVIEW (23)
- HTML versions of the prediction to review using your browser
- PostScript output
- The profiles generated during the analysis
- A gzipped file of all results for downloading

Many of these representations display the same information in various formats. *Fig. 14* (also available in the color section) shows the JALVIEW representation; only the central part of the protein is shown in this view.

6. Click on the link to view your results in HTML format and use the links (top left of page) to look at the annotations of some of the closely similar proteins.

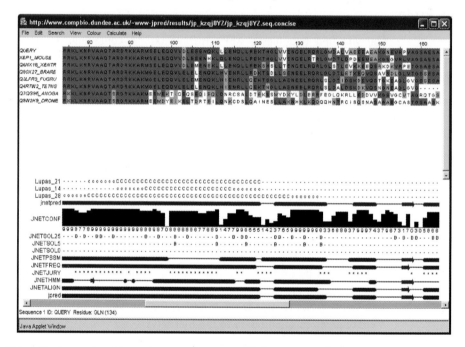

Figure 14. JPRED output shown in JALVIEW (see page xxix for color version).
The upper part of the screen shows the alignment of the query protein with others of significant similarity (shading indicates conservation) and contains a screen image of the results obtained when analyzing the sequence above using JALVIEW. Below this, the lines beginning 'Lupas' indicate predicted coiled-coil regions (at three different window sizes). The 'JNETSOL' lines indicate which residues are likely to be accessible to solvent ('B' indicates a buried residue – one that is less than 25%, less than 5% or 0% exposed to solvent in the three tracks). The remainder of the tracks all relate to secondary structure predictions: 'JNETPSSM', 'JNETFREQ', 'JNETHMM', and 'JNETALIGN' show predictions made by various methods (red tubes are α-helices; green arrows are β-sheets); asterisks in the 'JNETJURY' track show where these predictions disagreed and had to be resolved, whilst the 'jnetpred' track shows the consensus secondary-structure prediction. Finally, the histogram (and the corresponding numerical values beneath it) show the reliability of the prediction shown in 'jnetpred': for example, the large helix on the left includes one region towards its right end where the prediction is less certain, approximately at residues 113–116.

In this example, we are fortunate in that the protein is closely similar to many well-annotated proteins in other species; the unknown protein is clearly an X-box-binding protein. However, even without these close similarities (or without the similar proteins being themselves well annotated), we could make several inferences from the secondary-structure predictions. For example, the protein is probably globular: it contains many short helices interspersed with β-sheet regions and is peppered with buried residues inconsistent with a fibrous protein. It contains a strongly predicted coiled-coil region, which might suggest a possible leucine zipper, and closer inspection of the sequence shows a series of leucines, spaced at seven-residue intervals, from position 97 to 132 – strong evidence for a leucine zipper.

2.6 Protein function prediction by transfer of annotation

So far, we have looked mainly at inferences based on secondary-structure predictions, on the prediction of certain easily recognized motifs such as leucine zippers, and on the identification of motifs that have been found to recur in certain types of protein. In the following, the concept of inferring protein function by comparing similarities and homologies – a process also known as transfer of annotation – will be discussed.

In the previous example (*Protocol 14* and *Fig. 14*), the user will note the presence of a multiple sequence alignment on the top of the display, generated from the top matches found using a sequence similarity search. There are some 250 000 proteins known to date for which much information is available. If the result of a sequence similarity search yields results with very high percentage identity and good expectation (*E*) values (see Chapter 3), it is worth realigning these best hits with a program such as CLUSTALW (24), T-COFFEE (25), or MUSCLE (26). These programs are specifically designed to align common regions among the input sequences and help the user to borrow annotation from the known ones and apply it to the unknown.

Fig. 15 (also available in the color section) shows the results of a sequence similarity search using WU-BLAST2 (27) on the example sequence used in *Protocol 14*, searching against the UniProt/Swiss-Prot database. This yielded three matches with identities in the range 82–99%. These three sequences were subsequently aligned using DBCLUSTAL (28), using the services at http://www.ebi.ac.uk/blast2 [10.30]. The matches indicate a high degree of similarity between the query sequence and X-box-binding protein 1 (XBP-1), a transcription factor essential for hepatocyte growth, the differentiation of plasma cells, and immunoglobulin secretion. The XBP-1 proteins are part of a family of proteins known as the bZIP family and contain a basic motif DNA-binding domain as well as a leucine zipper in positions 69–119 of the query sequence. From the alignment, it is obvious that there is a strong degree of similarity among the four sequences. This implies that they may have the same overall structure, but, due to lack of crystallographic experimental evidence in the annotation, it is unsafe to assume that this is the case.

```
CLUSTAL (1.0) multiple sequence alignment


Sequence      MVVVAAAPNPADGTPKVLLLSGQPASAAGAPAAR-LPLMVPAQRGASPEAASGGLPQARK 59
XBP1_HUMAN    MVVVAAAPNPADGTPKVLLLSGQPASAAGAPAGQALPLMVPAQRGASPEAASGGLPQARK 60
XBP1_RAT      MVVVAAAPSAASAAPKVLLLSGQPASGG-----RALPLMVPGPRAAGSEAS--GTPQARK 53
XBP1_MOUSE    MVVVAAAPSAATAAPKVLLLSGQPASGG-----RALPLMVPGPRAAGSEAS--GTPQARK 53
              ********..* .:*************..     : ******. *.*..**:  * *****


Sequence      RQRLTHLSPEEKALRRKLKNRVAAQTARDRKKARMSELEQQVVDLEEENQKLLLENQLLR 119
XBP1_HUMAN    RQRLTHLSPEEKALRRKLKNRVAAQTARDRKKARMSELEQQVVDLEEENQKLLLENQLLR 120
XBP1_RAT      RQRLTHLSPEEKALRRKLKNRVAAQTARDRKKARMSELEQQVVDLEEENQKLQLENQLLR 113
XBP1_MOUSE    RQRLTHLSPEEKALRRKLKNRVAAQTARDRKKARMSELEQQVVDLEEENHKLQLENQLLR 113
              ************************************************.:** *******


Sequence      EKTHGLVVENQELRQRLGMDALVAEE--EAEAKGNEVRPVAGSAESAALRLRAPLQQVQA 177
XBP1_HUMAN    EKTHGLVVENQELRQRLGMDALVAEE--EAEAKGNEVRPVAGSAESAALRLRAPLQQVQA 178
XBP1_RAT      EKTHGLVIENQELRTRLGMNALVTEEVSEAESKGNGVRLVAGSAESAALRLRAPLQQVQA 173
XBP1_MOUSE    EKTHGLVVENQELRTRLGMDTLDPDEVPEVEAKGSGVRLVAGSAESAALRLCAPLQQVQA 173
              *******:****** ****::* ..:* *.*:**. ** ************ ********


Sequence      QLSPLQNISPWILAVLTLQIQSLISCWAFWTTWTQSCSSNALPQSLPAWRSSQRSTQKDP 237
XBP1_HUMAN    QLSPLQNISPWILAVLTLQIQSLISCWAFWTTWTQSCSSNALPQSLPAWRSSQRSTQKDP 238
XBP1_RAT      QLSPPQNIFPWILTLLPLQILSLISFWAFWTSWTLSCFSNVLPQSLLIWRNSQRSTQKDL 233
XBP1_MOUSE    QLSPPQNIFPWTLTLLPLQILSLISFWAFWTSWTLSCFSNVLPQSLLVWRNSQRSTQKDL 233
              **** *** ** *::*.*** **** *****:** ** **.***** **.********


Sequence      VPYQPPFLCQWGRHQPSWKPLMN----------- 260
XBP1_HUMAN    VPYQPPFLCQWGRHQPSWKPLMN----------- 261
XBP1_RAT      VPYQPPFLCQWGPHQPSWKPLMNSFVLTMYTPSL 267
XBP1_MOUSE    VPYQPPFLCQWGPHQPSWKPLMNSFVLTMYTPSL 267
              ************ **********
```

Figure 15. A multiple alignment using DBCLUSTAL (see page xxx for color version).

2.7 Multiple sequence alignments and secondary databases

The reader will have noted that the advanced methods rely on producing a multiple sequence alignment (MSA) in order to generate as accurate a prediction of function as possible. MSAs are used to generate signatures, patterns, and statistical profiles (also known as HMMs). MSAs will typically show where a set of sequences have common elements, called domains and architectures, and where they are fundamentally different from each other. Producing good MSAs has always been a challenge and the choice of algorithm will depend on suitability, availability, and application support and maintenance. The most accurate and most commonly used tools available today and their most important web-based services are listed in *Table 5*.

Starting from an MSA, there are several ways in which a model can be built that describes relationships among the aligned sequences. These models may be represented as consensus patterns, as signatures, or as statistically inferred profiles. In its most basic form, a consensus pattern can be derived that reflects the various levels of conservation of each column in the alignment. A signature, which also reflects content, will present a set of expressions that identify all possible compositionally constrained positions. On the other hand, a profile is a statically derived expression obtained by analyzing the distribution of amino acids in a training set of related proteins. Furthermore, an HMM can be deduced

Table 5. Programs and web sites providing multiple protein sequence alignment

Program	Comment	URL
CLUSTALW 1.8 (DNA/protein)	Global progressive	http://www.ebi.ac.uk/clustalw/ [10.77]
		http://searchlauncher.bcm.tmc.edu/ [10.78]
MAP (DNA/protein)	Global progressive in linear space	http://searchlauncher.bcm.tmc.edu/ [10.78]
PIMA 1.4 (protein only)	Pattern-Induced (local) Multiple Alignment	http://searchlauncher.bcm.tmc.edu/ [10.78]
MSA 2.1 (protein only)	Near-optimal sum-of-pairs global	http://xylian.igh.cnrs.fr/msa/msa.html [10.79]
		http://searchlauncher.bcm.tmc.edu/ [10.78]
BLOCK MAKER (protein only)	Finds conserved blocks in sequence sets	http://xylian.igh.cnrs.fr/msa/msa.html [10.79]
MULTALIGN (DNA/protein)		http://prodes.toulouse.inra.fr/multalin/ multalin.html [10.80]
T-COFFEE (DNA/protein)	Very strict and accurate	http://www.ch.embnet.org/software/ TCoffee.html [10.81]
		http://www.ebi.ac.uk/t-coffee/ [10.82]
MUSCLE (protein and DNA)	Alignment by log-expectation	http://www.drive5.com/muscle/ [10.83]
		http://www.ebi.ac.uk/muscle/ [10.84]
		http://phylogenomics.berkeley.edu/cgi-bin/ muscle/input_muscle.py [10.85]

that uses all known position states in an alignment and determines the statistical probability of change having taken place in each. HMMs are today a very common technique for generating sets of data and secondary databases that can be used to predict function by the statistical relatedness in a set of sequences.

2.8 An overview of InterPro and CDD

The relationships described above are the foundation for a large set of database projects and tools that capture and classify sequence relationships according to formalized methodologies. Each in turn presents a slightly different characterization of what are protein families, domains, and biologically functional sites. The beauty of all of these databases is that each provides the user with the ability to predict function and they often help determine what the structure of novel proteins may be. There are two mainstream services at present: InterPro (29) (http://www.ebi. ac.uk/interpro/ [10.31]) and the National Center for Biotechnology Information (NCBI) CDD (http://www.ncbi.nlm.nih.gov/Structure/cdd/cdd.shtml [10.32]). Important off-spins of these are the COGs database (30), which describes clusters of orthologous groups of proteins in completed genomes, and Integr8 (31), which uses data integrated from a large number of data resources including Genome Reviews (31), the UniProt Knowledgebase, InterPro, CluSTr (32), and GOA (33). Integr8 provides access to general information about each species, statistical analyses, information about orthologs and synteny, and links to the source data for more information.

InterPro and CDD each have services that allow a user to analyze sequence data and identify functional domains, families, and architectures. InterPro currently uses data from 11 distinct members of the InterPro consortium and uses a tool called

INTERPROSCAN (34) for sequence searching. NCBI's CDD uses a special version of NCBI's BLAST called reverse position-specific BLAST (RPS–BLAST) – a derivative of PSI–BLAST.

INTERPROSCAN is, in fact, a programmed workflow that searches each of the databases in the consortium, using over a dozen methods that fall into broad types. The first type are sequence based and make use of sequence and sequence motifs to predict function. The second type are based on HMMs and differ from one another in the way in which the classification and detection of profiles is measured.

One difference between INTERPROSCAN and CDD is the ability of INTERPROSCAN to work with DNA sequences (i.e. contigs). The user must choose an appropriate translation table in order for the service to translate DNA into protein correctly. Noteworthy also is the fact that INTERPROSCAN can be downloaded from the EBI's ftp server (ftp:// ftp.ebi.ac.uk/pub/software/unix/iprscan/[10.33]) and installed locally. This means that users can integrate their own data into the system. The web-based system can also be used programmatically via a simple object access protocol (SOAP)-based web service from the EBI (http://www.ebi.ac.uk/Tools/webservices[10.34]).

CDD uses sequence data derived from SMART, Pfam, and COG. RPS–BLAST is one of the most important tools developed by the NCBI and offers ease of use but lacks the completeness of INTERPROSCAN. It has been suggested that one of the most serious challenges that both INTERPROSCAN and CDD face is keeping up to date with the 'donor' databases, but this must be viewed in relation to the existence of experimental evidence for each prediction and consolidation among the different databases. However, it is important to note that these databases are updated at irregular intervals and can change dramatically from release to release. It is possible that positive matches observed at one time will not be reproduced at another. Furthermore, positives observed at any time may need constant reviewing until experimental evidence establishes with certainty the existence (or not) of a particular domain in a sequence subclass or species.

Another general-purpose version of BLAST, specifically designed to characterize protein function using the traditional protein sequence databases, is position-specific iterative BLAST (PSI–BLAST). PSI–BLAST refers to a feature of BLAST 2.0 in which a profile (or position-specific scoring matrix, PSSM) is automatically constructed from a multiple alignment of the highest-scoring hits in an initial BLAST search. The PSSM is generated by calculating specific scores for each position in the alignment. Highly conserved positions receive high scores and weakly conserved positions receive scores near zero. The profile is used to perform further rounds of the BLAST search and the results of each 'iteration' are used to refine the profile. This iterative searching strategy results in increased sensitivity, at the cost of increased running times. Increased sensitivity can also be obtained with WU–BLAST as can be seen on the EBI's WU–BLAST server. This has settings that control sensitivity in very much the same way as in RPS–BLAST/PSI–BLAST.

2.8.1 Using INTERPROSCAN

As described earlier, INTERPROSCAN is a wrapper-type application that encompasses all of the databases and analysis methods of the InterPro consortium databases.

In terms of predicting function in proteins, INTERPROSCAN is perhaps one of the most complete applications. INTERPROSCAN exists in two related versions: one is a Perl-based version that can be downloaded from the EBI's ftp server and run locally. However, the application makes high demands on computation power, and it is advisable to run it on a cluster of computers rather than on a single CPU. The second version of INTERPROSCAN is the service from the EBI that includes access via the web, mail, or via SOAP-based web services. *Protocol 15* illustrates the use of the EBI web-server version.

Protocol 15

Using INTERPROSCAN

1. Obtain the sequence of UniProt entry RXRA_HUMAN in FASTA format. (See *Protocol 1*; a copy of this sequence is given in the Protocol_15 folder as 'RXRAfasta.txt [10.35].')

2. Go to http://www.ebi.ac.uk/InterProScan [10.36], and paste the FASTA sequence into the large sequence window.

3. Note that all of the applications currently used by INTERPROSCAN are listed as options that can be included or omitted (the default has them all included). Clicking on the name of any one of these options (**BlastProDom, FPrintScan, HMMPIR**, etc.) will open the web page for that resource. (If you try this, navigate back to the INTERPROSCAN page.)

4. Leave the other options at their default settings and click **Submit job.**

5. *Fig. 16* (also available in the color section) shows part of the results page viewed in the default mode (at the time of writing). For each match, the left-hand column has the accession

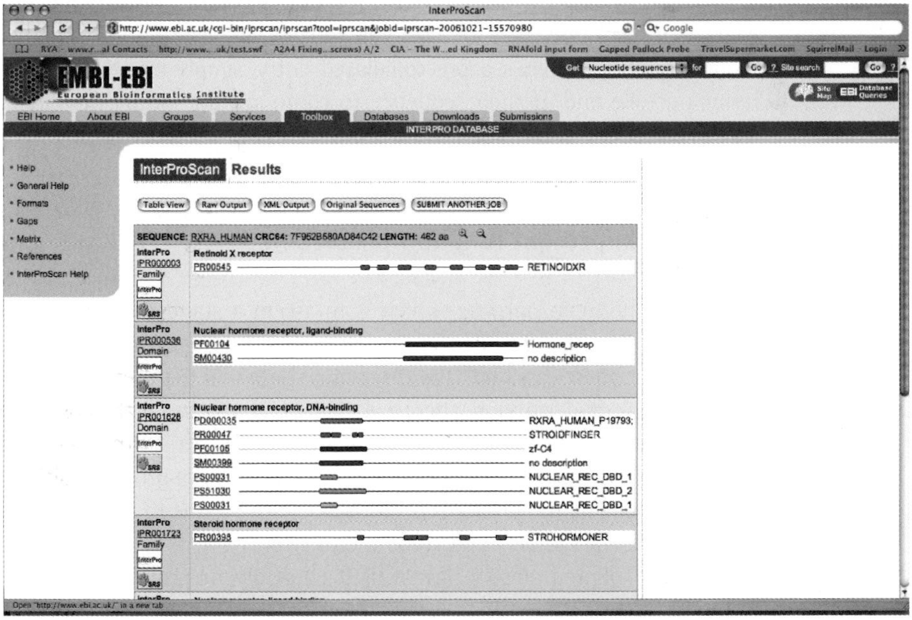

Figure 16. Part of the results of INTERPROSCAN for the protein RXRA_HUMAN (see page xxxi for color version).

number of the InterPro domain or family to which the match belongs. The second column contains the accession numbers of the match in each of the member databases that InterPro has examined, followed by a cartoon of the signature superimposed on the virtual sequence. The accession numbers are hyperlinked and will take the user to the official web site hosting the database. Note that you can use the mouse to hover over the colored regions of the cartoon for each match: a box will pop up that shows the name and range of the match.

6. Click on the **Table View** button (upper left of main part of window) to view the same data in text-only format. This view contains InterPro-specific parent/child relationships and the entries in InterPro where these occur. Most importantly, this view contains gene ontology (GO) annotation and thus provides access to the control vocabularies that should help the user to choose the correct and accepted terminology. It also highlights which are the most common semantics used by the member databases to describe the function, family, or domain of the matches found.

In order to reduce computation time, INTERPROSCAN uses a database called InterPro Matches (or IPRMATCHES for short) that contains the pre-calculated results of INTERPROSCAN runs for all UniProt sequences. This database is produced quarterly in parallel with the InterPro database and is available from the EBI ftp server (ftp:// ftp.ebi.ac.uk/pub/databases/interpro/ [10.37]). When a sequence is submitted to INTERPROSCAN, its CRC64 (a 'checksum', which is effectively a numerical fingerprint unique to that sequence) is calculated and compared with the CRC64s of each entry in the IPRMATCHES database. If one is found, the pre-calculated results are displayed – this will have been the case for the example used in *Protocol 15* (the CRC64 of the query sequence is shown towards the top of the results window). If no match to the CRC64 is found (i.e. if the query sequence is not identical to a UniProt sequence with a pre-calculated result), the results are calculated 'live'. However, the user can force a 'live' calculation, even when the input sequence matches a pre-calculated entry, simply by adding an arbitrary residue at one end of their sequence to confound the CRC64 matching. In this way, INTERPROSCAN always provides up-to-date results irrespective of the quarterly releases of InterPro.

2.9 Recent advances in protein function prediction

GO serves one main purpose, which is to assign a standardized terminology to proteins by use of controlled vocabularies (see Chapter 9). As more and more biological databases assign GO terms to their data and these cross-reference each other, the task of identifying where a novel protein may be located in the cell, what function it may have, and in which biological processes it may be involved may become as simple as running a similarity search. This is a gross oversimplification of the problem, as the methods used to assign GO terms to biological entities in the databases vary widely and are prone to error.

A number of programs are now available that, given a protein sequence, attempt to assign GO terms describing its function, location, and biological process. These programs rely largely on similarity searches across a range of databases, coupled with a variety of approaches to interpret these similarities to make intelligent predictions about the unknown protein. In this section, we will use the human retinoic acid receptor as our 'unknown' protein and see how well four of these automated GO-assignment programs perform. No explicit protocols are given here, as the use of the web servers for these programs is fairly self explanatory. The protein sequence file to be used as the input in each case is RXRAfasta.txt [10.35] in the Protocol_15 folder.

The first site is GOPET (35) (http://genius.embnet.dkfz-heidelberg.de/menu/cgi-bin/w2h-open/w2h.open/w2h.startthis [10.38]). The default settings can be used throughout, but, when choosing the options, ask for predictions on both molecular process and biological function (the site does not, at the time of writing, offer predictions in the 'cellular compartment' aspect). The result is a table listing the most likely GO terms (in both the 'function' and 'process' aspects) for the protein, along with a confidence level for the prediction. The output of GOPET for this example is given in the 'Other_files' folder as RXRA_GOPET.html [10.39].

The second site is the ProtFun server (36) (http://www.cbs.dtu.dk/services/ProtFun/ [10.40]). This server attempts to classify the protein not only by GO category, but also under several other classification schemes. Within each scheme, the output gives probabilities and odds for the protein belonging to each category. The output for ProtFun on the human retinoid receptor is given in the 'Other_files' folder as RXRA_PROTFUN.html [10.41].

The third site, PFP (37) (http://dragon.bio.purdue.edu/pfp/ [10.42]), starts by using iterations of PSI-BLAST to identify matches against the UniProt database and extracts an initial set of GO terms from these matches. It then relies on the known associations between various functions to identify further GO terms that may be applicable to the protein, but that could not have been predicted solely from the sequence similarity. Results from this server are returned by e-mail, and the output in this case is given in the 'Other_files' folder as RFRA_PFP.txt [10.43].

Finally, a fourth site is GOFigure (38) (http://udgenome.ags.udel.edu/gofigure/ [10.44]). This program gives an extensive output, based again on similarity of the query protein to GO-annotated sequences in a number of databases. (Note that the results returned by GOFigure are displayed, rather confusingly, with the caption 'GO_Annotator'; however, there is another unrelated program called GOANNOTATOR, which tabulates literature citations supporting the GO annotation of sequences). The complete analysis can be downloaded as a single 'tar' file, a copy of which is given in the 'Other_files' folder as RXRA_GOFIGURE.tar [10.45]. After unpacking this, open the file index.html in the resulting folder with your browser and follow the links to show the classification of the protein under three ontologies. There is also a simple text output, which is given separately in the 'Other_files' folder as RXRA_GOFIGURE.txt [10.46]. In this example, the results are unambiguous because our 'unknown' protein is of course a perfect match to the well-annotated human retinoid receptor protein!

You should now look over the results from these four sites and compare their predictions. All sites are in broad agreement, but they differ in the details of their predictions and in some of the finer details of functional classification. For this reason, it is always important to compare the predictions of as many such sites (and indeed the results of as many different types of tool) as possible. In this case, of course, our 'unknown' protein is already well characterized and so we can compare the predictions not only with each other, but with the 'official' annotation. The UniProt/Swiss-Prot annotation for the protein includes the following keywords: DNA-binding, Host–virus interaction, Metal-binding, Nuclear protein, Polymorphism, Receptor, Transcription, Transcription regulations, Zinc, and Zinc-finger; its GO annotation is shown in *Table 6*.

Table 6. GO annotation for RXRA_HUMAN from the UniProt website

The output of the prediction sites described in the text should be compared with this.

Ontology	Term	GO ID	Evidence
Molecular function	Protein binding	GO:0005515	Inferred from physical interaction
	Retinoid-X receptor activity	GO:0004886	Traceable author statement
	Transcription coactivator activity	GO:0003713	Traceable author statement
	Transcription factor activity	GO:0003700	Inferred from direct assay
Biological process	Positive regulation of transcription from R...	GO:0045944	Inferred from direct assay
	Signal transduction	GO:0007165	Non-traceable author statement
	Vitamin metabolism	GO:0006766	Traceable author statement

Just as INTERPROSCAN acts as an 'umbrella' for searching many protein family and domain databases, the JAFA (39) server at http://jafa.burnham.org [10.47] compiles results from many of the services that use the GO database to assign function to unknown proteins. JAFA makes remote calls to a variety of services that presently include GOTCHA (40), INTERPROSCAN, GOBLET (41), and GOFIGURE, and provides an output that makes it easy to compare their results and see consensus predictions. The server returns results as a text summary showing the most likely GO assignments and, more importantly, a summary of the results indicating which of the underlying servers agree with each assignment. The simple text part of the result for the human retinoic acid receptor, run using the default settings, is given in the 'Other_files' folder as RXRA_JAFA.txt [10.48], but you should examine the full output online.

Disagreement between the servers may indicate overprediction or misannotation of a protein in one of the server's databases. It may also indicate a new annotation that has not yet been taken into consideration on all of the servers: GO-based services need to keep up to date with the master production copy and also, to a lesser degree, with each other. At the level of protein function, the JAFA results largely agree with the UniProt/Swiss-Prot annotation of the entry: the protein is involved in DNA binding, has retinoid-X receptor activity, and thus has transcription coactivator activity and therefore has transcription factor activity. Most of the servers report that the protein is located in the nucleus; the GO annotation of the entry does not indicate this, but the keywords do so. This would indicate that there is a lag in annotation, which may indicate that, at the level of GO annotation, the entry itself is not a good reference. At the level of biological processes, there is agreement, but the user will notice that two of the GO terms are different. This is not necessarily an error but preference from the annotator or the developers of the JAFA service. This brings to light the fact that, whilst databases such as GO strive for standardization of the vocabularies they maintain, there is still some degree of subjectivity.

The important point to note from the above is that choosing a reference for assigning function to a protein sequence must be done with care. Predictions need to be backed up wherever possible with experimental evidence or, at the minimum, support from several up-to-date databases and computational methods.

2.10 Concluding remarks

Protein function is complex association of experimentally determined and computationally predicted attributes that describe biochemical, physiological, physicochemical, cellular, and developmental properties. Transfer of functional annotation based on sequence homology is an expert art, and inferring function from sequence searches alone should be seen as providing the basis for further analysis. Beyond transfer of annotation, the most successful approaches to date combine tools from artificial intelligence (HMMs, neural networks, etc.) with multiple sequence alignments, which are rich in evolutionary information.

The transfer of annotation between proteins in different organisms can only be done reliably in the presence of evidence that the respective proteins are indeed performing the same, or perhaps related, biological processes. To infer function without taking into consideration the few critical residues responsible for that function, and without considering the potential evolutionary relationships, is a recipe for disaster.

In this chapter, we have necessarily considered examples that are relatively straightforward. Large, multi-functional, multi-domain proteins are best analyzed in parts, tackling each domain in turn.

3. ADDITIONAL WEB RESOURCES

Analytical resources

- ProtFun server: http://www.cbs.dtu.dk/services/ProtFun/ [10.40]. Protein function prediction server.
- PFP: http://dragon.bio.purdue.edu/pfp/ [10.42]. Protein function prediction.
- Automated Function Prediction group: http://BioFunctionPrediction.org [10.50]. Home page of the Automated Function Prediction Special Interest group, with links to tools and other resources.
- PredictProtein: http://www.predictprotein.org/ [10.24]. Protein function and structure prediction.
- Expasy: http://www.expasy.org/tools/ [10.51]. Expasy proteomic tools.
- Pratt: http://www.ebi.ac.uk/pratt/ [10.52]. Pattern recognition and matching in protein sequences.

Educational resources

- 2Can: http://www.ebi.ac.uk/2can/ [10.53]. EBI's bioinformatics support and training portal covering a vast range of bioinformatic topics.
- NCBI Education: http://www.ncbi.nlm.nih.gov/Education/ [10.54]. A starting point for access to NCBI education and training resources.
- Bioinformatics Net: http://www.bioinformatics.vg/biolinks/bioinformatics/ Training%2520and%2520Education.shtml [10.55]. Training and education resources.
- EMBnet: http://www.embnet.org [10.56]. Home page of EMBnet, a worldwide bioinformatics network.

4. REFERENCES

★ 1. **Rice P, Longden I & Bleasby A** (2000) *Trends Genet.* **16**, 276–277. *– An overview of the EMBOSS suite and a valuable introduction to the wide range of available tools.*

★ 2. **Labarga A, Pilai S, Valentin M, Anderson M & Lopez R** (2005) *EMBnet News,* **11**, 18–23. *– For potential developers of protein function prediction servers: several of the methods described in this chapter use these services.*

★ 3. **Wu CH, Apweiler R, Bairoch A, et al.** (2006) *Nucleic Acids Res.* **34**, D187–D191. *– UniProt is the largest repository of protein sequences and keeping up to date with its development is essential reading.*

4. **Garnier J, Gibrat JF & Robson B** (1996) *Methods Enzymol.* **266**, 540–553.

5. **Chou PY & Fasman GD** (1974) *Biochemistry,* **13**, 222–245.

6. **Persson B & Argos P** (1994) *J. Mol. Biol.* **237**, 182–192.

7. **Durbin R, Eddy S, Krogh A & Mitchison G** (1998) *The Theory Behind Profile HMMs.* Cambridge University Press, Cambridge.

8. **Krogh A, Larsson B, von Heijne G & Sonnhammer EL** (2001) *J. Mol. Biol.* **305**, 567–580.

9. **Kyte J & Doolittle RF** (1982) *J. Mol. Biol.* **157**, 105–132.

10. **Nakai K & Kanehisa M** (1992) *Genomics,* **14**, 897–911.

11. **Bornberg-Bauer E, Rivals E & Vingron M** (1998) *Nucleic Acids Res.* **26**, 2740–2746.

12. **Hulo N, Bairoch A, Bulliard V, et al.** (2006) *Nucleic Acids Res.* **34**, D227–D230.

13. **Attwood TK, Bradley P, Gaulton A, Maudling N, Mitchell A & Moulton G** (2004) In *Encyclopaedia of Genetics, Genomics, Proteomics and Bioinformatics.* Edited by M Dunn, L Jorde, P Little & A Subramaniam. John Wiley & Sons.

14. **Henikoff JG, Greene EA, Pietrokovski S & Henikoff S** (2000) *Nucleic Acids Res.* **28**, 228–230.

15. **Finn RD, Mistry J, Schuster-Bockler B, et al.** (2006) *Nucleic Acids Res.* **34**, D247–D251.

16. **Brennan RG & Matthews BW** (1989) *J. Biol. Chem.* **264**, 1903–1906.

17. **Rost B & Liu J** (2003) *Nucleic Acids Res.* **31**, 3300–3304.

18. **Cuff JA, Clamp ME, Siddiqui AS, Finlay M & Barton GJ** (1998) *Bioinformatics,* **14**, 892–893.

19. **Cuff JA & Barton GJ** (2000) *Proteins,* **40**, 502–511.

20. **Bru C, Courcelle E, Carrere S, Beausse Y, Dalmar S & Kahn D** (2005) *Nucleic Acids Res.* **33**, D212–D215.

★ 21. **Altschul SF, Madden TL, Schaffer AA, et al.** (1997) *Nucleic Acids Res.* **25**, 3389–3402. *– Sequence similarity searches are largely done using* BLAST. *This is important reading for anyone wanting to use this tool.*

22. **Sander C & Schneider R** (1991) *Proteins,* **9**, 56–68.

23. **Clamp M, Cuff J, Searle SM & Barton GJ** (2004) *Bioinformatics,* **20**, 426–427.

★ 24. **Thompson JD, Higgins DG & Gibson TJ** (1994) *Nucleic Acids Res.* **22**, 4673–4680. *–* CLUSTALW *is the most ubiquitous multiple sequence alignment tool. Understanding its strengths and limitations is important, in particular when used as part of functional prediction workflow.*

25. **Notredame C, Higgins DG & Heringa J** (2000) *J. Mol. Biol.* **302**, 205–217.
26. **Edgar RC** (2004) *Nucleic Acids Res.* **32**, 1792–1797.
27. **Lopez R, Silventoinen V, Robinson S, Kibria A & Gish W** (2003) *Nucleic Acids Res.* **31**, 3795–3798.
28. **Thompson JD, Plewniak F, Thierry J & Poch O** (2000) *Nucleic Acids Res.* **28**, 2919–2926.
★ 29. **Mulder NJ, Apweiler R, Attwood TK, *et al.*** (2005) *Nucleic Acids Res.* **33**, D201–D205.
    – *The InterPro consortium produces a unified database product that is invaluable for protein function prediction studies.*
30. **Tatusov RL, Koonin EV & Lipman DJ** (1997) *Science,* **278**, 631–637.
31. **Kersey P, Bower L, Morris L, *et al.*** (2005) *Nucleic Acids Res.* **33**, D297–D302.
32. **Petryszak R, Kretschmann E, Wieser D & Apweiler R** (2005) *Bioinformatics,* **21**, 3604–3609.
33. **Camon EB, Barrell DG, Dimmer EC, *et al.*** (2005) *BMC Bioinformatics,* **6** (Suppl. 1), S17.
★ 34. **Quevillon E, Silventoinen V, Pillai S, *et al.*** (2005) *Nucleic Acids Res.* **33**, W116–W120.
    – *A hands-on explanation of the power of InterProScan.*
35. **Vinayagam A, Konig R, Moormann J, *et al.*** (2004) *BMC Bioinformatics,* **5**, 116.
★ 36. **Jensen LJ, Gupta R, Staerfeldt HH & Brunak S** (2003) *Bioinformatics,* **19**, 635–642. – *An excellent example of how to use GO.*
37. **Hawkins T, Luban S & Kihara D** (2006) *Protein Sci.* **15**, 1550–1556.
38. **Khan S, Situ G, Decker K & Schmidt CJ** (2003) *Bioinformatics,* **19**, 2484–2485.
39. **Friedberg I, Harder T & Godzik A** (2006) *Nucleic Acids Res.* **34**, W379–W381.
40. **Martin DM, Berriman M & Barton GJ** (2004) *BMC Bioinformatics,* **5**, 178.
41. **Hennig S, Groth D & Lehrach H** (2003) *Nucleic Acids Res.* **31**, 3712–3715.

CHAPTER 11

Multiple sequence alignment

Burkhard Morgenstern

1. INTRODUCTION

Sequence alignment is of crucial importance for all aspects of biological sequence analysis. Virtually all methods of nucleic acid and protein sequence analysis rely directly or indirectly on alignments, so the output of these methods depends on the quality of the underlying alignments. Applications of sequence alignment include genome sequence annotation and gene prediction, phylogeny reconstruction, RNA and protein structure analysis, and functional classification of proteins, to mention just a few important applications. An alignment of two sequences is called a *pairwise* alignment, an alignment of three or more sequences is called a *multiple* alignment. Pairwise alignment is used mainly for database searching; here the major challenge is to combine program speed with sensitivity and specificity. An overview of methods for sequence database searching is given in Chapter 3; the focus of the present chapter is on multiple alignment. Multiple alignments contain much more information than pairwise alignments, but constructing multiple alignments involves additional challenges compared with pairwise alignment. A good overview of multiple-alignment methods is given by Notredame (1) and in the widely used textbook by Durbin *et al.* (2).

Sequence alignment can be carried out manually by experienced experts, and many researchers say that this is still the best possible alignment method. Nevertheless, most alignments are, of course, automatically calculated by software programs. Alignment algorithms are therefore a central area of research in bioinformatics. Developing and improving alignment methods involves a number of nontrivial algorithmic questions. The present chapter places the emphasis on methods, and is written for nonbioinformaticians who want use bioinformatics tools to analyse their data.

From this perspective, it may seem unnecessary to care about the algorithmic ideas behind these methods: all of the tools have corresponding web servers, which can, largely, be used without understanding the theoretical concepts, just as one can drive a car without knowing anything about car engines. So why should a molecular biologist care about algorithmic concepts of multiple-alignment programs? The trouble is that alignment programs – as with other bioinformatics

Bioinformatics: *Methods Express* (Paul H. Dear, ed.)
© Scion Publishing Limited, 2007

tools – are far from perfect and can produce total nonsense. What is worse, it is not at all obvious whether the result of an alignment program makes sense or whether it is completely meaningless. It is not difficult to tell whether a car works, but it is often not easy to find out whether the output of an alignment program is biologically meaningful. It is therefore necessary for a user of an alignment program to know at least some of the basic ideas behind the program in order to understand what the program is able to do and what its limitations are.

This chapter outlines the concepts behind some of the major software tools for multiple alignment in general terms and without unnecessary jargon. Technically, these methods are easy to use, even by nonexperts. All of the methods described below have user-friendly and well-documented web sites that are more or less self-explanatory. Therefore, the focus of this chapter is not on how to run these programs, but rather an outline of how these programs work, in which situations they are useful, and what their limitations are. The 'protocols' are, therefore, confined mainly to a few illustrative worked examples that highlight the different strengths of the programs. In addition, some approaches are mentioned that can be used to deal with the limitations of existing multiple-alignment software and to overcome some of their shortcomings.

2. METHODS AND APPROACHES

2.1 The alignment problem in computational biology

The goal of pairwise and multiple sequence alignment is to pinpoint homologies between two or several nucleic acid or protein sequences as shown in *Fig. 1*. Ideally, an alignment should arrange those parts of the sequences that are biologically related by common function, structure, or evolution on top of each other. In *Fig. 1*, for example, there are five structurally conserved regions that are present in

```
1csy  shekmpWFHGKISREESEQIVligskTNGKFLIRARD..nnGSYALCLLH
1gri  emkphpWFFGKIPRAKAEEML.skqrHDGAFLIRESEs.apGDFSLSVKF
1aya  ...mrrWFHPNITGVEAENLLltrg.VDGSFLARPSKs.npGDFTLSVRR
2pna  .lqdaeWYWGDISREEVNEKLrdt..ADGTFLVRDAStkmhGDYTLTLRK
1bfi  hhdektWNVGSSNRNKAENLLrgk..RDGTFLVRESS..kqGCYACSVVV

1csy  EGKVLHYRIdkdktgklsipegk.kFDTLWQLVEHYsyka......dgll
1gri  GNDVQHFKVlrdgagkyfl.wvv.kFNSLNELVDYHrsts.vsrnqqifl
1aya  NGAVTHIKIqn..tgdyydlyggekFATLAELQYYmehhgqlkekngdv
2pna  GGNNKLIKIfh.rdgkygfsdpl.tFNSVVELINHYrnes.laqynpkld
1bfi  DGEVKHCVInktatg.ygfaepynlYSSLKELVLHYqhts.lvqhndsln

1csy  rvl.tVPcqk
1gri  rdieqVPqq.
1aya  iel.kYPln.
2pna  vkl.lYPvs.
1bfi  vtl.aYPvya
```

Figure 1. Multiple alignment of a protein family taken from the BAliBASE database at http://bips.u-strasbg.fr/fr/Products/Databases/BAliBASE/ [11.1].
The alignment highlights five conserved structural motifs (shown in bold capital letters).

all members of a protein family; the alignment highlights these regions by writing them on top of each other. We call such an alignment a *biologically correct* alignment. A major question in the development of alignment methodology is how a computer program should decide what a 'biologically correct' alignment of a given set of input sequences is. A computer is a stupid machine and knows nothing about biology. The only input information for standard alignment programs is sequence data, i.e. linear strings of characters. To produce meaningful alignments from arbitrary sets of input sequences, a computer program needs a mathematically defined *scoring scheme* that can be used to assign a quality score to every possible alignment from a given input data set. Computer scientists call such a scoring scheme an *objective function*. The first and most important prerequisite component of an alignment software program is therefore an objective function that – hopefully – assigns numerically high scores to those alignments that are biologically meaningful and vice versa.

2.2 Pairwise sequence alignment

An alignment of two sequences can be interpreted as a hypothesis about the evolution of these sequences: by writing two residues in the same column of the alignment, one assumes that these residues go back to a common origin in evolution. A gap in one sequence corresponds either to a deletion in this sequence or to an insertion in the other sequence at the corresponding position. Based on these considerations, one can define a quality score for every possible alignment of a given pair of sequences. The idea is to give high scores to those alignments that represent a plausible hypothesis of the evolution of the aligned sequences. In general, two sets of parameters are used to define such a score. First, a so-called *substitution matrix* assigns a similarity score to every possible pair of aligned amino acid or nucleic acid residues. The higher the degree of similarity between two residues, the higher the corresponding score. Commonly used substitution matrices are PAM (3) and BLOSUM (4); they are based on probabilistic models of protein evolution derived from known protein families. The idea is that an amino acid in a protein is more likely to be replaced with one of a similar physicochemical character (for example, leucine with isoleucine) than with one of a very different character (for instance, leucine with tryptophan), so similar amino acids are more likely to be in the same column in a biologically correct alignment. Aligning leucine with isoleucine therefore gets a higher score than aligning, for example, leucine and tryptophan. Next, a *penalty* is defined for each gap in the alignment. The score of a pairwise alignment as used in standard alignment programs is then defined as the sum of similarity scores of aligned residue pairs minus the corresponding gap penalty for each gap introduced in the alignment (see *Fig. 2*).

Once an objective function is defined for pairwise sequence alignment, the next major question is how to find an optimal alignment according to this objective function. There is an efficient computational technique called *dynamic programming* that calculates an alignment with maximum score in the sense of the

```
seq1 YARDIRD
seq2 FA--LRE
```

Figure 2. Score of a pairwise alignment as calculated by standard alignment methods.
Each pair of amino acids x and y is assigned a similarity score $s(x,y)$ and each gap of
length l is assigned a penalty called $g(l)$. The score of a pairwise alignment is defined as
the sum of the similarity scores of the aligned residues minus a gap penalty for each gap
introduced in the alignment. Thus, in the above example, the alignment score is given
as the sum of the five similarity scores minus the penalty for a gap of length 2: $s(Y,F)$
$+ s(A,A) - g(2) + s(I,L) + s(R,R) + s(D,E)$. For any given input sequence set, standard
alignment algorithms return the alignment with maximum possible score.

above-defined scoring scheme for any pair of input sequences (5). It is important
to mention that an alignment with a *mathematically* optimal score is by no means
guaranteed to be *biologically* meaningful. As long as two sequences are *globally*
related, i.e. related over their entire length, a maximum-score alignment is *likely*
to be biologically correct, i.e. it is likely to align sections of the sequences that
are related by common ancestry or structure with each other. It is an interesting
experiment, however, to align two completely unrelated sequences, for example
two random sequences, using a standard alignment program. Most methods would
align these sequences from the beginning to the end, inserting just a few gaps
here and there. They do so in order to maximize the above-outlined mathematical
alignment score – but in this case such an alignment would make no sense at
all, as completely unrelated parts of the sequences would be aligned with each
other.

For situations where two input sequences are not globally related and share
only a conserved local domain, there is a version of the dynamic programming
method that detects and aligns these domains and ignores the remainder of the
sequences. This idea goes back to Smith and Waterman (6). Such 'local alignment'
methods are used mainly for database searching. The well-known search tool BLAST
(7) is the most widely used method for local alignment (in fact, it is the most
widely used bioinformatics tool of all). Database search tools such as BLAST do not
attempt to calculate mathematically optimal alignments, as this would be too
time-consuming if large databases are to be searched. Instead, they only try to
approximate optimal alignments to save computing time.

Finally, two fundamental limitations of most existing alignment programs
should be mentioned. First, these methods assume that homologous parts in two
or several sequences occur in the *same relative order* within these sequences. This
is, of course, a necessary condition that has to be satisfied if sequence homologies
are to be represented in a pairwise or multiple alignment; without this condition,
alignment programs cannot be reasonably applied. Secondly, most alignment
methods assume that similarity at the primary sequence level corresponds to true
homology, i.e. to common ancestry in evolution. However, this does not need to be
the case either. For distantly related sequences, primary sequence similarity may
be low and can be overshadowed by spurious random similarities. In this situation,
no sequence-based alignment program can reconstruct the true structural or
evolutionary relationship between sequences, so additional sources of information
have to be considered to compare distal sequences.

2.3 Multiple sequence alignment

The above-mentioned approach to pairwise alignment can be extended to multiple alignment. Again, the quality score of any given alignment can be defined by summing similarity values of aligned residues and subtracting a penalty for every gap in the alignment. One way of doing this is to do an all-against-all comparison of the residues aligned in one column of a multiple alignment and to add the corresponding similarity values. Usually, a weighting system is applied to the input sequences in order to down-weight groups of closely related sequences that could otherwise bias the resulting alignment. Theoretically, the dynamic programming technique that calculates pairwise alignments with maximum score can be generalized to produce maximal-scoring multiple alignments. Unfortunately, however, the running time and memory requirements of this approach would be excessive. Consequently, most methods for multiple alignment do not attempt to calculate an *exact* optimal alignment. Rather, they try to *approximate* an alignment with maximum score, usually without knowing how close the produced alignments come to the mathematically optimal alignment.

The most popular way of calculating global multiple alignments is the so-called *progressive approach* (8). Here, multiple alignments are calculated by performing a series of pairwise alignments of sequences and of already aligned groups of sequences until all sequences from the input data set are aligned in one resulting multiple alignment. In this approach, the order of the pairwise alignments follows the branching order of a so-called *guide tree*. Roughly speaking, this is an approximate phylogenetic tree calculated in a quick-and-dirty way from pairwise distances between the input sequences. The most widely used multiple-alignment program, CLUSTALW (9), is based on this progressive alignment approach.

Standard progressive methods such as CLUSTALW usually produce good alignments as long as the input sequences exhibit sufficient similarity over their entire length at the primary sequence level. However, they have a number of shortcomings. First, the outcome of progressive methods depends crucially on the order in which the pairwise alignments are performed, so they depend on the guide tree that the program uses. This tree, however, can only be a rough approximation to the true phylogenetic tree, as it is based on pairwise sequence distances only. This is particularly problematic if multiple alignment is used as a prerequisite for phylogeny reconstruction. All methods for sequence-based phylogeny reconstruction are based on multiple alignments. The trouble is that if a progressive method is used to calculate the input alignment for a phylogeny program, this alignment depends on the guide tree. As the output phylogenetic tree, in turn, depends on the multiple alignment, a quick-and-dirty guide tree that is only based on pairwise sequence distances can strongly affect the phylogenetic tree calculated by more advanced phylogeny methods.

Secondly, another shortcoming of standard progressive approaches is that misalignments in the beginning phase cannot be corrected later if more sequence information is taken into account. Thirdly, a general problem of standard approaches to pairwise and multiple alignment is their dependency on parameters for the gap penalties. These parameters crucially influence the resulting alignments, but there is no general applicable rule that tells the user how to adjust these parameters. Finally, standard progressive methods are limited to producing *global* alignments, so they fail to produce reasonable alignments if the input sequences share no detectable global similarity.

In addition to the above *global* approaches to multiple sequence alignment, there are a number of methods to calculate *local* multiple alignments, for example the Gibbs sampling approach (10), PIMA (11), MACAW (12), and MEME (13). Most of these local methods return conserved motifs that are present in *all* input sequences and ignore the remainder of the sequences. Such motif-finding programs are useful if conserved patterns are to be detected, for example regulatory signals in genomic sequences. The segment-based approach DIALIGN (14) combines global and local alignment features; this program works by constructing multiple alignments using pairwise local similarities as building blocks. It tries to align those parts of the sequences that are homologous, but does not align the rest of the sequences where no similarity is found. In this way, DIALIGN returns global alignments if sequences are homologous over their entire length, but local alignments if only local similarity is detectable. The main advantage of the program is its ability to align remotely related sequence sets that cannot be aligned by the traditional methods outlined above. However, if sequences are globally related but the degree of similarity among the sequences is low, DIALIGN may return spurious random similarities instead of a biologically meaningful global alignment. This is a major disadvantage of the program, and in such situations programs such as CLUSTALW or T-COFFEE (see below) do a better job.

Another major method for multiple alignment is T-COFFEE (15). Like CLUSTALW, T-COFFEE is a progressive alignment method, but it has two important advantages compared with earlier progressive methods such as CLUSTALW: first, T-COFFEE uses both global and local alignment methods to perform pairwise sequence comparisons and, secondly, before the multiple alignment is assembled based on these pairwise alignments, alignments of different pairs of input sequences are compared with each other in order to identify similarities that involve more than two sequences.

2.4 Benchmarking and evaluation of multiple-alignment software

During the last years, extensive studies have been carried out to systematically evaluate and compare the performance of protein multiple-alignment software, and several databases have been compiled for these studies. By far the most widely used benchmark database for multiple protein alignment is BAliBASE (16). BAliBASE was the first database designed to evaluate multiple protein aligners and its development was an important step for developers and users of multiple-alignment software. BAliBASE consists of protein families for which biologically meaningful reference alignments are available that are based on three-dimensional structure superposition. These reference alignments have been created automatically and later manually refined by experts. The first version of BAliBASE contained five major sets of reference alignments covering different types of alignment problems. Later additions cover sequence repeats and transmembrane proteins (17); the latest version of BAliBASE contains a total of 6255 protein sequences (18). Other databases used to evaluate multiple protein alignments are HOMSTRAD (19), SABmark (20), and PREFAB (21).

It should be mentioned that the focus of these benchmark databases is generally on *globally* related reference sets, so global alignment programs usually perform much better on these data than do local methods. In the first version of BAliBASE, the database developers artificially removed unrelated sections of the sequences for many of the benchmark alignments to make them globally related. Thus, test runs on these benchmark sequence sets are distorted, as global alignment programs perform better than they would perform on the corresponding real-world sequences. Fortunately, in the last version of BAliBASE, the original full-length sequences were made available, so it is now possible to do more realistic benchmarking studies on these protein families (18). An interesting study to evaluate the performance of multiple-alignment programs on locally related sequence families has been carried out by Lassmann and Sonnhammer (22). These authors implanted conserved motifs into unrelated random sequences to study the behaviour of multiple-alignment tools on distantly related sequence sets.

2.5 Visualization and comparison of multiple alignments

Multiple alignments of more than, say, three or four sequences may be hard to interpret without additional software tools. Therefore, a number of visualization tools have been developed that provide graphical representations of alignments. Well-known visualization tools for multiple alignment are, for example, MVIEW (23), JALVIEW (24), and QALIGN/PANTA RHEI (25, 26).

As no multiple-alignment tool is guaranteed to produce biologically meaningful alignments, it is common practice to compare the output of different programs. Regions where these different methods agree are generally thought to be more reliable than regions where different programs disagree. In particular, if alignment methods with different underlying concepts – for example a global and a local tool – agree on how sequences should be aligned, then one can have somewhat more confidence in these alignments. Alignments calculated by different methods from the same input data set are often compared manually, but recently some software tools have been developed to perform this type of analysis, for example ALTAVIST (27) and SINICVIEW (28).

2.6 Multiple alignment of large genomic sequences

Recently, alignment of long genomic sequences has become an important tool for genome sequence annotation. The idea is that, during evolution, functional parts of the genome such as protein-coding exons or gene-regulatory sites are more conserved than nonfunctional parts. Thus, if syntenic sequences at an appropriate phylogenetic distance are compared with each other, those functional parts often correspond to islands of local sequence similarity separated by regions of lower similarity. In this way, pairwise and multiple alignments of syntenic sequences can be used to discover functionally important sites in the genome.

Large-scale alignment of genomic sequences involves several new challenges. The sheer size of the input data requires programs that are computationally highly efficient. At the same time, these programs have to be sensitive enough to detect small functional elements in large sequences. Commonly used multiple-alignment programs for large genomic sequences include MUMMER (29), MGA (30), CHAOS/DIALIGN (31), MAVID (32), and MULTI-LAGAN (33); see Pollard *et al.* (34) for a systematic comparison and evaluation. These software tools have been used successfully in many studies, for example to detect regulatory signals (35–37), for gene prediction (38–41), and to detect pathogenic microorganisms (42, 43). In this chapter, we cannot cover software tools for large-scale genomic alignment. The interested reader is referred to the original papers and to the corresponding internet pages.

2.7 Software tools for multiple alignment

In this section, we briefly describe some widely used tools for multiple sequence alignment. All of these programs are available online through the Internet; their web sites are well documented and easy to use. To run the programs outlined below, all one needs to do is to go to the respective internet site, paste or upload a set of sequences – usually in FASTA format – and hit the 'run' button. As these steps are more or less self-explanatory, we restrict ourselves largely to providing their respective internet addresses, explaining their main strengths and weaknesses and mentioning some of the program parameters that can be adjusted by the user. A few worked examples are given, which allow the user to compare some of the tools and their results, and which draw attention to their respective strengths and weaknesses. The programs below are discussed in chronological order of their first publication.

2.7.1 CLUSTALW

CLUSTALW is by far the most widely used software program for multiple DNA and protein alignment; the method is an improvement of the original CLUSTAL program by Higgins and Sharp (44). The paper by Thompson *et al.* (9) that describes CLUSTALW is one of the most influential and most frequently cited publications in bioinformatics. CLUSTAL uses the *progressive* strategy described above for multiple alignment, where sequences are aligned following the branching order of a so-called *guide tree*; this guide tree is based on pairwise similarities among the input sequences. The main feature of CLUSTALW is sequence weighting based on pairwise sequence identities; furthermore, the program uses sophisticated methods to use substitution matrices and gap penalties depending on various parameters such as local sequence similarity and amino acid composition of protein sequences. CLUSTALW is a strictly global alignment program, i.e. it aligns the input sequences from the beginning to the end, even if they are only locally related.

CLUSTALW is available on line through many web sites, for example at the European Bioinformatics Institute (EBI) at http://www.ebi.ac.uk/clustalw/ [11.2]. At this server, an option for *fast* pairwise alignment is available to speed up the construction of the guide tree; this can be done by adjusting a number of parameters for the pairwise alignment procedure. In addition, there are parameters for gap opening and gap extension penalties, as well as specialized parameters for end gaps. Various substitution matrices are available, and finally a phylogenetic tree is constructed from the output alignment using a method that is specified by the user. (Phylogenetic analysis, including the use of CLUSTALW, is covered in more detail in Chapter 12). *Protocol 1* gives a simple example of the use of CLUSTALW.

Protocol 1

CLUSTALW

1. From the Protocol_1 folder for this chapter on the book's website, download the example file DHFR.fasta[11.3]. This file contains, in FASTA format, the protein sequences of dihydrofolate reductase from chicken, human, *Pneumocystis* (a fungus), and *Pseudomonas* (a bacterium).

2. Go to the CLUSTAL server at EBI at http://www.ebi.ac.uk/clustalw/ [11.2] and paste the contents of the DHFR.fasta[11.3] file into the large text box towards the bottom of the screen.

3. If you complete your e-mail address and select **email** from the pull-down **RESULTS** menu (upper part of the screen), you will be e-mailed when your results are ready. In this case, though, leave the **RESULTS** option set to **interactive**, to wait and view the results when they are ready. Leave all other options at their default settings and click **Run**.

4. After a few moments, the results page will appear. This page contains abundant information and links to numerous other files. The Scores Table gives a simple measure of the degree of alignment (as the percentage of identical amino acids in the alignment) between each pair of sequences; not surprisingly, it is greatest between chicken and human (74%), and lower between all other pairs. This is a crude reflection of the phylogenetic relationships, but see Chapter 12 for a detailed discussion of sequence-based phylogenetic analysis.

5. Further down the page is shown the alignment of the four sequences; this is reproduced in *Fig. 3*. The alignment of the chicken and human sequences to each other is clear; note the introduction of gaps in these sequences and that of *Pseudomonas*, primarily to align them with the longer *Pneumocystis* sequence. The alignment between the vertebrate and the other sequences is less strong than between the two vertebrates, but there are clear 'islands' where all sequences align, indicating the presence of conserved motifs and suggesting that the alignment is biologically meaningful.

6. The reader is encouraged to explore the many options (particularly for viewing and analysing the output) that are offered. In particular, the JALVIEW option (started by a button towards the top of the page) provides a useful view of the alignment, highlighting the quality of the alignment and the presence of conserved regions, and giving a consensus sequence.

7. As a cautionary tale, the reader is reminded that CLUSTALW will attempt to align all sequences globally, even if they are unrelated or have only local similarities – this is illustrated in a later protocol in this chapter.

```
CLUSTAL W (1.83) multiple sequence alignment


CHICKEN         ---VRSLNSIVAVCQNMGIGKDGNLPWPPLRNEYKYFQRMTSTSHVEG---KQNAVIMGK 54
HUMAN           ---VGSLNCIVAVSQNMGIGKNGDLPWPPLRNEFRYFQRMTTTSSVEG---KQNLVIMGK 54
PNEUMOCYSTIS    MNQQKSLTLIVALTTSYGIGRSNSLPWK-LKKEISYFKRVTSFVPTFDSFESMNVVLMGR 59
PSEUDOMONAS     MKTHLPLSLIAALGENRVIGVDNSMPWH-LPGDFKYFKATTLG---------KPIIMGR 49
                .*. *.*.*:   .   ** ...:**   *   :  **:  *      :  ::**:


CHICKEN         KTWFSIPEKNRPLKDRINIVLSRELKEAPKGAHYLSKSLDDALALLD---SPELKSKVDM 111
HUMAN           KTWFSIPEKNRPLKGRINLVLSRELKEPPQGAHFLSRSLDDALKLTE---QPELANKVDM 111
PNEUMOCYSTIS    KTWESIPLQFRPLKGRINVVITRNESLDLGNGIHSAKSLDHALELLYRTYGSESSVQINR 119
PSEUDOMONAS     KTWDSLG---RPLPGRLNLVVSRQTDLQLEGAEVFP-SLDAAVVRAE---QWAQEQGVDE 102
                *** *:    ***  .*.*:.*::*: .   ..   . *** *:         ::


CHICKEN         VWIVGGTAVYKAAMEKPINHRLFVTRILHEFESDTFFP----------EIDYKDFKLLTE 161
HUMAN           VWIVGGSSVYKEAMNHPGHLKLFVTRIMQDFESDTFFP----------EIDLEKYKLLPE 161
PNEUMOCYSTIS    IFVIGGAQLYKAAMDHPKLDRIMATIIYKDIHCDVFFPLKFRDKEWSSVWKKEKHSDLES 179
PSEUDOMONAS     VMLIGGAQLYAQGLAQ--ADRLYLTRVALSPEGDAWFP----------EFDTAQWALVSN 150
                : ::**: :*  .: :     :: * :  . .*.:**        .  . :.


CHICKEN         YPG--VPADIQEEDGIQYKFEVYQKSVLAQ 189
HUMAN           YPG--VLSDVQEEKGIKYKFEVYEKND--- 186
PNEUMOCYSTIS    WVGTKVPHGKINEDGFDYEFEMWTRDL--- 206
PSEUDOMONAS     AEN------AAVDEKPAYSFEVWERV---- 170
                 .   :      :.   *.**:: :   :
```

Figure 3. Global alignment of four dihydrofolate reductase sequences by CLUSTALW.
The symbols on the bottom line indicate regions of perfect (*), good (:), and moderate (.)
conservation.

2.7.2 The Gibbs motif sampler

Whilst CLUSTAL produces strictly global alignments, the Gibbs sampling approach
by Lawrence *et al.* (10) returns a purely local alignment of the input sequences.
The goal of the program is to identify one or several gap-free local alignments
of conserved motifs that are present in *all* of the input sequences; the length of
these motifs is to be specified by the user. To select motifs, the program uses a
quality criterion that is based on a stochastic model. Finding a mathematically
optimal motif according to this criterion is computationally demanding; the
Gibbs motif sampler therefore uses an iterative search algorithm that starts with
a randomly picked local multiple alignment, which is then iteratively improved.
This strategy is heuristic, i.e. it is not guaranteed to find an optimal solution,
but it is computationally efficient. The Gibbs motif sampler has been applied
successfully to many different situations, e.g. to detect protein domains and to
identify small regulatory regions in genomic DNA. A certain limitation of this
approach is its restriction to motifs that are contained in *all* sequences in a
given input data set. Thus, if a motif is present only in some but not in all of
the sequences under study, it is either not found at all or it is found in those
sequences where it is present but then aligned with unrelated segments from
the remaining sequences.

The homepage of the Gibbs sampling method is http://bayesweb.wadsworth.
org/gibbs/gibbs.html [11.4]. The program has various options, e.g. for motif length,
for the underlying probabilistic model, and for the optimization strategy that is
used.

2.7.3 PRRP

Like CLUSTALW, Gotoh's program PRRP is based on the progressive alignment strategy (45). However, to improve some of the shortcomings of the progressive approach, PRRP uses a randomized iterative refinement strategy, i.e. the program first calculates an initial multiple alignment using the progressive method and then improves this alignment step by step. In each iteration step, the current multiple alignment is split into two subalignments that are again realigned against each other. In this way, errors that occur early in the progressive alignment procedure can be corrected later if more sequence information is considered. The program simultaneously optimizes the multiple alignment, phylogenetic tree, and sequence weights; this is achieved by using a double nested iteration algorithm. This procedure is somewhat time-consuming, but several independent studies have shown that PRRP performs significantly better than more traditional progressive approaches. Like CLUSTALW, PRRP is a strictly global method that cannot be applied to local alignment problems.

PRRP is available online at http://prrn.ims.u-tokyo.ac.jp/ [11.5]. This server provides several expert options for the iterative refinement strategy. In addition, the commonly used gap penalties and substitution matrices can be adjusted by the user.

2.7.4 DIALIGN and DIALIGN-T

DIALIGN was introduced by Morgenstern *et al.* (14) as a method that combines local and global alignment features. The approach works by searching pairwise local ungapped similarities among the input sequences, so-called *fragments*. These fragments are then used to assemble a final multiple alignment. In this way, the program is able to identify local similarities shared by several or all of the input sequences, but it is not restricted to local alignment. If the input sequences are homologous over their entire length, the program will return a full global alignment similar to the output of CLUSTALW and other global methods. Recent improvements in the program include an *anchoring* option where expert knowledge can be used to guide the alignment procedure (46). This program feature is useful in situations where the user has some previous knowledge about (local) homologies among the input sequences. If the standard version of the program fails to align these homologies correctly, the anchoring option can be used to enforce a correct alignment of known homologies and the remainder of the sequences is then aligned automatically. A complete reimplementation of the program called DIALIGN-T has recently been published (47). This program version contains some algorithmic improvements compared with the original implementation; in particular, it is less susceptible to spurious random similarities.

Various versions of DIALIGN are available online through the Göttingen Bioinformatics Compute Server (GOBICS) at http://gobics.de/ [11.6]. A number of parameters can be adjusted, such as a *threshold* for the similarity of local fragments considered for alignment. For nucleic acid sequences, a *translation* option is available that translates local segments of the sequences into peptide segments according to the genetic code. Sequence similarity is then measured by comparing these implied peptide segments instead of the original nucleic acid segments.

Protocol 2 takes the user through a worked example comparing DIALIGN-T with CLUSTALW when dealing with sequences that have only local similarity.

Protocol 2

Comparison between DIALIGN-T and CLUSTALW with locally similar sequences

1. From the Protocol_2 folder for this chapter on the book's website, download the file motif. fasta[11.7]. The three sequences in this file all contain a conserved motif, but the remainder of each sequence is random; the motif is present at different locations in each of the three sequences.

2. First, use CLUSTALW (see *Protocol 1*) to attempt to produce a global alignment among these sequences. The result is shown in *Fig. 4(a)*. The alignment among the sequences is poor (less than 10% of residues between any two sequences), but an uncritical eye might spot pockets of alignment and infer that the result was valid. However, the conserved motif (highlighted in the figure, but not in the CLUSTALW output) has not been aligned among the three sequences – the result is biologically meaningless.

3. We will now align the same sequences using DIALIGN-T. Go to http://gobics.de/department/ software [11.8] and select the link **DIALIGN-T**. On the resulting page, under 'Online Submission', select the link for **web interface for dna sequences**.

4. Use the **Choose file** (or, depending on your browser, **Browse**) button to upload the motif.fasta[11.7] file. Leave the other options at their default values and click the **Run Dialign-T** button.

5. The expected results are shown in *Fig. 4(b)*. The conserved motif (again, highlighted in the figure) has been correctly aligned – only small pockets of the unrelated flanking sequence have been aligned (shown in uppercase) and these have not 'swamped' the correct alignment of the motif[a].

Note

[a]It should be mentioned that other global alignment programs such as PROBCONS (48, 49) and MAFFT (50, 51) performed better than CLUSTAL on these example sequences.

Figure 4 (opposite). Performance of different multiple-alignment programs on a sequence set that contains a locally conserved motif.
The three sequences have been artificially created by implanting a conserved motif (grey blocks) into unrelated random sequences. The motif is 25 amino acids and the average degree of sequence conservation among the three sequences is below 50% within the motif. The position of the implanted motif is at the amino terminus (sequence 1), in the middle (sequence 2) and at the carboxyl terminus (sequence 3) of the sequences, respectively. (a) CLUSTALW (9) produces a global alignment of the input sequences. Here, the conserved motif is not aligned correctly; instead, unrelated parts of the sequences are aligned with each other. (b) A local program (DIALIGN-T) (47) correctly aligns the conserved motif; only small parts of the unrelated random sequences are aligned. Note that in the DIALIGN output, only upper-case letters are considered to be aligned. Lower-case letters are not aligned, even if they appear in the same column. Thus, the program output is highly specific in that it aligns only small parts of the sequences.

(a) CLUSTALW alignment:

```
seq1    HVMSERKRREKLNEMFLVLKSLLPSVTWSKEKKHANSDCPPVQYCMFVQCRNRRQIGWMN
seq2    ----KNRHMVVYSGYYHLMPPVCQILDALWELMLETCVIEVPERVNICGGDFDAYTEEQA
seq3    -----AQNMQQSYHEPCWMTPQYDHKCLVDMIFFTPITHHMSMCKSECN--LQIHHDYYL


seq1    NNYACECIRWAPCKMNLHRYVTNGYLASFSSYLFVLYHNQAQFWPLEWVGFGAEAILIIC
seq2    N-IKHHCDRWHYCAEVYIGVFKVRYFDCGKRWSTGLKKRDGRCFDHSIAERLRRERIAER
seq3    DPGNPDCADIFYYQIMHDQYRDSCLHDYSYWWNTHLRFAKNPFCRGWVVGSYPKRAFFRC


seq1    GKCVTRHNACQAPVMNYFHCNYTEDDQ------TVHDCISGHAKTHMFKAMCLRYLPSMH
seq2    IRALQELVPTCSLTNPQMHDFWQWSQL------FIKCSVEGAWIPHSFIQEIITKP-QVE
seq3    LAGSFPMFAFCMNITPWQWDFIVPFDFGHLEFEQINWYEQSPRCIHCGDMMILTYN-KTD


seq1    MEHHWNPAYPDCPHFGNFCYAVHSKKLFAVWGRTPGLIVWPDQYWKFWNIN------
seq2    MEPNNYYFYNMAEPPFIFVRYVPFRIISYVVADTVPFEMGQYGCWCWIYYYPQSWHL
seq3    GSPLFAYRDPCCEGPAIPMWMIWPQKPPHIIAERKRREKLTQRFVALSALVPG----
```

(b) DIALIGN-T alignment:

```
seq1    ---------- ---------- ---------- ---------- ----------
seq2    k--------- ---------- ---------- --NRHMvvys g---------
seq3    aqnmqqsyhe pcwmtpqydh kclvdmifft piTHHMsmck secnlqihhd


seq1    ---------- ---------- ---------- ---------- ----------
seq2    ---------- ----YYHLMp ---------- ---------- ----PVCQil
seq3    yyldpgnpdc adifYYQIMh dqyrdsclhd ysywwnthlr faknPFCRgw


seq1    ---------- ---------- ---------- ---------- ----------
seq2    dal------- ---------- ---------- -WELMLetcv ievpervnic
seq3    vvgsypkraf frclagsfpm fafcmnitpw qWDFIVpf-- ----------


seq1    ---------- ---------- ---------- ---------- ----------
seq2    ggDFDAYTEE QANIKHHCDR WHYCAEVYIG VFKvryfdcg krwstglkKR
seq3    --DFGHLEFE QINWYEQSPR CIHCGDMMIL TYN------- --------KT


seq1    ---------- ---------- ---------H VMSERKRREK LNEMFLVLKS
seq2    DGrcfd---- ---------- ---------H SIAERLRRER IAERIRALQE
seq3    DGsplfayrd pccegpaipm wmiwpqkppH IIAERKRREK LTQRFVALSA


seq1    LLPSVTWSKE KKHansdcpp vqycMFVQCr nrrqiGWMNN NYACECIRWA
seq2    LVPTCSLTNP QMHdfwqwsq ----LFIKCs veg--AWIPH SFIQEIITKP
seq3    LVPg------ ---------- ---------- ---------- ----------


seq1    PCKMNLHRYv tngylasfss ylfvlYHNQA Qfwplewvgf gaeailiicg
seq2    QVEMEPNNYy ---------- -----FYNMA Eppfif---- ----------
seq3    ---------- ---------- ---------- ---------- ----------


seq1    kcvtrhnacq apvmnyfhcn yteddqtvhd cisghakthm fkamcLRYLP
seq2    ---------- ---------- ---------- ---------- -----VRYVP
seq3    ---------- ---------- ---------- ---------- ----------


seq1    smhmehhwnp aypdcphFGN FCYAVHSKKL FAVwgrtpGL IVWPDQYWKF
seq2    ---------- --*-----FRI ISYVVADTVP FEMgqy--GC WCWIYYYPQS
seq3    ---------- ---------- ---------- ---------- ----------


seq1    WNIn
seq2    WHL-
seq3    ----
```

DIALIGN also offers another option – the ability to provide *anchor* points between the sequences. In some situations, the user may know of one or more conserved motifs shared across several proteins, but these fail to be aligned correctly if they are weakly conserved and buried in larger, nonconserved sequences. The solution is to specify anchor points – one or more positions in the sequences that must be aligned with each other. In this way, the program is forced to align the anchored parts of the sequence and will align the flanking sequence as well as it can around these anchor points. DIALIGN allows multiple anchor points to be specified, as well as the priority for considering them in case they contradict each other and cannot be included in one single output alignment. The use of DIALIGN's anchor point option is covered in *Protocol 3*.

2.7.5 PRALINE

The program PRALINE developed by Heringa (52, 53) uses two new ideas to improve the quality of multiple alignments. First, it has been shown that the multiple alignment of a given set of input sequences is generally improved if additional sequences are added that are phylogenetically related to the original input data. PRALINE uses such additional sequences in the alignment procedure in order to improve the quality of the alignment result, but removes them from the final output alignment. To find phylogenetically related sequences, PRALINE runs the profile-based database search program PSI-BLAST (54) on the input sequences; it then replaces sequences by profiles obtained from these database searches. Secondly, for protein alignment, PRALINE uses several different programs for secondary-structure prediction and includes the resulting structural information for sequence alignment. A progressive alignment is carried out taking into account these additional sources of sequential and structural information.

PRALINE is available online via http://www.ibi.vu.nl/programs/ [11.9], and both a standard and an expert web interface are available. Output alignments can be colour coded.

2.7.6 T-COFFEE

After CLUSTALW, Notredame's program T-COFFEE (15) is currently the second most widely used method for multiple alignment of nucleic acid and protein sequences. It is sometimes seen as the successor of CLUSTAL; in fact the main developer of CLUSTAL is also a co-developer of T-COFFEE. Like DIALIGN, T-COFFEE combines local and global alignment features. The central idea is to use a so-called *library* of pairwise alignments of all possible pairs of input sequences. In the default version of the program, this library consists of 11 alignments for each possible pair of sequences, namely one global and ten nonoverlapping local alignments. The global pairwise alignments are calculated by CLUSTALW, whilst for local pairwise alignments the program SIM (55) is used. Next, these pairwise alignments are checked for local *consistency*. In the context of T-COFFEE, this means that the program searches for overlapping pairwise alignments that (indirectly) support each other. In this way, weak homologies can be identified that might have been overlooked in the respective initial pairwise alignments.

Based on these consistency calculations, pairwise sequence similarities are re-scored and, in a final step, a standard progressive procedure is applied to calculate a multiple alignment, similar to the procedure used in CLUSTAL.

T-COFFEE has two major advantages compared with traditional progressive methods: it incorporates local alignment features and it can correct mistakes in pairwise alignments by checking for consistency among these alignments. An attractive feature of the program is its ability to use *arbitrary* user-defined pairwise alignments in the initial library instead of the CLUSTAL and SIM alignments that are used by default. This makes it possible to use the consistency-based machinery of T-COFFEE to combine and integrate arbitrary pairwise alignments obtained from different sources of information into one resulting multiple alignment. Such an approach has been used successfully to construct multiple alignments based on RNA or protein structure information. Siebert and Backofen used T-COFFEE in their approach, where multiple alignments of RNA sequences are calculated based on local similarities of their predicted secondary structures (56). The authors of T-COFFEE developed a method called 3D COFFEE where protein sequence information is combined with three-dimensional structure information to create multiple protein alignments that are superior to alignments based on linear sequence information alone (57).

T-COFFEE and 3D COFFEE are both available from the program developer's home page at http://www.tcoffee.org [11.10].

2.7.7 MAFFT

MAFFT is a software tool for global multiple alignment developed by Katoh *et al.* (50, 51). Compared with other tools for global alignment such as CLUSTAL or T-COFFEE, MAFFT is extremely fast. Whilst most multiple-alignment tools are restricted to sequence sets of several hundred sequences, MAFFT can be applied to sequence families consisting of thousands of input sequences. Like many other methods for multiple alignment, MAFFT uses a progressive alignment approach. The high program speed is achieved by using a technique called *fast Fourier transform*. In addition, MAFFT has an option for fast pairwise sequence comparison by counting common 6-tuples (hexamers) instead of calculating full global alignments. The guide tree that determines the order of pairwise alignments in the multiple-alignment procedure is based on these 6-tuple counts. As the all-against-all pairwise alignment step is the most time-consuming step in the progressive alignment approach, this technique leads to a drastic reduction in program running time. According to the authors, the alignment quality is not statistically significantly affected by this heuristic.

To improve the outcome of the progressive method, MAFFT optionally uses an iterative refinement technique similar to the one used in PRRP (45). In addition, the developers of MAFFT implemented an idea that has been proposed previously by Simossis and Heringa (53). Input sequences provided by the user are searched against a database to retrieve close homologs. A multiple alignment is calculated for all sequences – input sequences plus hits from the database search – to improve the quality of the output alignment. Sequences from the database are then removed from the resulting multiple alignment. Test runs on various benchmark databases for global alignment have shown that the output quality of MAFFT is at least comparable to other existing global multiple-alignment programs.

MAFFT is available online at http://align.bmr.kyushu-u.ac.jp/mafft/online/server/ [11.11]. At the server, there is an option 'Mafft-homologs' to include the database searching option outlined above; homologs of the input sequences are retrieved by running BLAST (7) against the Swiss-Prot database (13).

2.7.8 PROBCONS

PROBCONS (48, 49) is one of the latest additions to the choice of multiple-alignment programs that are now available. The method combines the consistency approach used in T-COFFEE with a probabilistic approach based on so-called pair hidden Markov models. As in T-COFFEE, pairwise similarities among any two of the input sequences can be supported by similarities to other sequences from the input data set. However, whilst T-COFFEE uses standard alignment programs to create the initial 'library' of pairwise similarities, PROBCONS uses a rigorous probabilistic approach to score pairwise and multiple alignments. Based on these probabilistic consistency considerations, a multiple alignment is created using the standard progressive approach. The resulting alignment can be improved iteratively similar to MAFFT and PRRP. Systematic test runs on standard benchmark data such as BAliBASE have shown that PROBCONS outperforms competing alignment methods on these data in terms of alignment quality. The program is faster than T-COFFEE but slower than CLUSTAL and DIALIGN.

PROBCONS is available online at a WWW server at Stanford University where it has been developed: http://probcons.stanford.edu/ [11.12]. Parameters can be adjusted for consistency calculations, iterative refinement of the output alignment, and output formatting.

In the following protocol, we compare PROBCONS with DIALIGN (with and without the use of the *anchoring* option – see section 2.7.4). The sequences we will use are from the BAliBASE (16, 18) reference set 1r96. These four protein sequences are a challenging test example for multiple-alignment programs. Each sequence contains three functional motifs at similar respective positions within the sequences, but the similarity among the four sequences at the primary sequence level is low. Thus, global alignment programs that align sequences over their entire length usually perform better on these test sequences than local methods.

Protocol 3

Comparison of PROBCONS and DIALIGN with and without anchoring

1. From the Protocol_3 folder for this chapter, retrieve the file anchor.fasta[11.13], which contains four sequences.

2. Go to the PROBCONS server at http://probcons.stanford.edu/[11.12], enter your e-mail address, and upload the anchor.fasta[11.13] file following the prompts.

3. Leave the other settings at their default values and click **Run PROBCONS!**

4. After a few moments, you will receive an e-mail containing the alignment; the expected result is shown (with added shading) in *Fig. 5(a)*. Note the generally good alignment of the functional motifs.

5. We will now try the same alignment using DIALIGN in its default mode. Go to the DIALIGN home page at http://dialign.gobics.de/anchor/[11.14] and follow the link to the submission form.

6. Upload the anchor.fasta[11.13] file using the **Choose file** or **Browse** button.

7. If you want to receive the program results by e-mail – in addition to seeing them on the screen – enter your e-mail address (optional).

8. Click **Run DIALIGN**. The results will appear, showing the alignment of the four sequences. The expected result is shown in *Fig. 5(b)*. As you can see, the conserved motifs are only poorly aligned by the default implementation of DIALIGN.

9. Finally, we will use the *anchor* option to try to improve DIALIGN's performance, by pinning one single column in one of the conserved motifs so that the program is forced to align this column whilst the nonconserved flanking sequence is aligned around them. Repeat steps 5 and 6, but then scroll down the screen to **Optional anchor points**.

10. It is possible to upload a file containing the anchor point information, but in this example we will enter it manually. There is a table into which up to five different anchor points can be entered; each anchor point is entered in one row of the table.

11. Using the anchor option, we want to enforce the correct alignment of the first column of the third motif. To this end, we will enter three anchors, each anchoring one point on the first protein to one point on one of the other three proteins. Across the first row of the table, enter the numbers '1', '2', '27', '39', and '1' – this specifies that sequence 1 is anchored to sequence 2, that the anchored region starts at residue 27 in the first sequence and at residue 39 in the second, and that the anchored region is only one residue long. Similarly, across the second row, enter '1', '3', '27', '35', and '1' to anchor residue 27 of sequence 1 to residue 35 of sequence 3. Finally, enter '1', '4', '27', '34', and '1' to anchor residue 27 of sequence 1 to residue 34 of sequence 4.

12. Now click **Run DIALIGN**. The result this time will be the alignment shown in *Fig. 5(c)*. By forcing the sequences to align at one point (the start of the third conserved motif), we have improved the overall alignment of all motifs in the sequences.

13. Other combinations of anchor points can also be used. For example, if functional or structural evidence suggests the presence of several conserved motifs, then multiple points in one protein can be anchored to corresponding points in another.

(a) PROBCONS alignment

```
1r69     ------SIS---SRVKSKRIQLGLNQAELAQKVGT------TQQSIEQLENGK---TKRP
1au7A    GMRALEQFA---NEFKVRRIKLGYTQTNVGEALAAVHGSEFSQTTICRFENLQLSFKNAC
1neq     CSN-EKARDWHRADVIAGLKKRKLSLSALSRQFGY------APTTLANALE-----RHWP
1a04A    ERD-VNQLTPR-ERDILKLIAQGLPNKMIARRLDI------TESTVKVHVKHM---LKKM

1r69     RFLPELASALGVSVDWLLNGT
1au7A    KLKAILSKWLEEAEQKRRTTI
1neq     KGEQIIANALETKPEVIWPSR
1a04A    KLKSRVEAAVWVHQER---IF
```

(b) DIALIGN alignment (default)

```
1r69     ------SISS R---------- ----VKSKRI -------QLG LNQAELAQKV
1au7A    gmralEQFAN E---------- ----FKVRRI -------KLG YTQTNVGEAL
1neq     c-------SN Ekardwhrad viagLKKRKL SLSALSRQFG YAPTTLANAL
1a04A    -----ERDVN Q---------- ----LTPRER DILKLIAQ-G LPNKMIARRL

1r69     GTTQQSI--- ---------- ----------E Q-----LENG KTKRPRFLPE
1au7A    AAVHGSefsq tticrfenlq lsfknacKLK AILSKWLEEA EQKRrtti--
1neq     Erhwp----- ---------- -------KGE QIIANALETK PEv-------
1a04A    DITESTV--- ---------- ---------K V-----HVKH MLKKMKLKSR

1r69     LASALgvsvd WLlngt--
1au7A    ---------- ----------
1neq     ----I----- Wpsr----
1a04A    VEAAV----- WVhqerif
```

(c) DIALIGN alignment (with user-defined anchor points)

```
1r69     ------SISS RVK------- ------SKRI QLGLNQAELA QKVGT-----
1au7A    gmraleQFAN EFK------- ------VRRI KLGYTQTNVG EALAAvhgse
1neq     c-------SN EkaRDwh--- -RADVIAGLK KRKLSLSALS RQFGY-----
1a04A    e--------- ---RDvnqlt pRERDILKLI AQGLPNKMIA RRLDI-----

1r69     -TQQSIEQLE NGKTK-R--- ---PRFLPEL ASALgvsvdW Llngt----- --
1au7A    fSQTTICRFE NlqlsfK--- ---NACKLKA ILSK-----W LEEaeqkrrt ti
1neq     -APTTLANAL ERHWP-Kgeq iiaNALETKP EV-I-----W psr------- --
1a04A    -TESTVKVHV KHMLK-K--- ---MKLKSRV EAAV-----W VHQerif--- --
```

Figure 5. Comparison of PROBCONS with DIALIGN, with and without the use of anchors.
Each of the four sequences is known (from functional data) to contain conserved motifs (gray), but there is little primary sequence conservation. (*a*) PROBCONS successfully aligns the sequences, aligning the two larger motifs perfectly and the smaller one in three of the four proteins. (*b*) DIALIGN, in its default mode, fails to align the conserved motifs – the alignment is confused by other, random points of conservation throughout the sequences. (*c*) By providing anchor points (shown in bold) at which the sequences must align, DIALIGN can be coerced to produce a better overall alignment: the second motif (as well as the third) is now correctly aligned, and the small first motif is partially aligned.

3. ADDITIONAL WEB RESOURCES

- ALIGN-M: http://bioinformatics.vub.ac.be/research/research.html [11.15]. Versatile alignment tool combining sequence and structure information.

- DCA: http://bibiserv.techfak.uni-bielefeld.de/dca/ [11.16]. Efficient algorithm to calculate near-optimal global alignments according to the SP scoring scheme.
- MATCH-BOX: http://www.fundp.ac.be/sciences/biologie/bms/matchbox_submit.shtml [11.17]. Multiple protein alignment based on conserved blocks using statistical criteria.
- MULTALIN: http://bioinfo.genopole-toulouse.prd.fr/multalin/ [11.18]. Classical progressive approach to multiple alignment.
- MUSCLE: http://www.drive5.com/muscle/ [11.19]. Fast implementation of the progressive alignment strategy using k-mers for distance estimation and several other new heuristics.
- POA: http://www.bioinformatics.ucla.edu/poa2/ [11.20]. Multiple-alignment approach based on partial-order relations.
- PRODA: http://proda.stanford.edu/ [11.21]. Alignment method that can deal with repeats and shuffled motifs.
- PRIME: http://prime.cbrc.jp/ [11.22]. Progressive method with iterative refinement strategy and piecewise linear gap costs.
- SAGA: http://www.tcoffee.org/Projects_home_page/saga_home_page.html [11.23]. Multiple alignment based on genetic algorithms.

4. REFERENCES

1. **Notredame C** (2002) *Pharmacogenomics*, **3**, 131–144. – *Useful review article on multiple alignment.*
2. **Durbin R, Eddy SR, Krogh A & Mitchison G** (1998) *Biological Sequence Analysis.* Cambridge University Press, Cambridge.
3. **Dayhoff M, Schwartz RM & Orcutt B** (1978) In *Atlas of Protein Sequence and Structure*, vol. 5, pp. 345–352. National Biomedical Research Foundation, Washington, DC.
4. **Henikoff S & Henikoff JG** (1992) *Proc. Natl. Acad. Sci. U.S.A.* **89**, 10915–10919.
★ 5. **Needleman SB & Wunsch CD** (1970) *J. Mol. Biol.* **48**, 443–453. – *The classical paper describing the basic dynamic programming algorithm for pairwise alignment.*
6. **Smith TF & Waterman MS** (1981) *Adv. Appl. Math.* **2**, 482–489.
7. **Altschul SF, Gish W, Miller W, Myers EW & Lipman DJ** (1990) *J. Mol. Biol.* **215**, 403–410.
8. **Hogeweg P & Hesper B** (1984) *J. Mol. Evol.* **20**, 175–186.
★ 9. **Thompson JD, Higgins DG & Gibson TJ** (1994) *Nucleic Acids Res.* **22**, 4673–4680. – *A program description of* CLUSTALW.
10. **Lawrence CE, Altschul SF, Boguski MS, Liu JS, Neuwald AF & Wootton JC** (1993) *Science*, **262**, 208–214.
11. **Smith RF & Smith TF** (1990) *Proc. Natl. Acad. Sci. U.S.A.* **87**, 118–122.
12. **Schuler GD, Altschul SF & Lipman DJ** (1991) *Proteins*, **9**, 180–190.
13. **Bairoch A & Apweiler R** (2000) *Nucleic Acids Res.* **28**, 45–48.
14. **Morgenstern B, Dress A & Werner T** (1996) *Proc. Natl. Acad. Sci. U.S.A.* **93**, 12098–12103.
15. **Notredame C, Higgins DG & Heringa J** (2000) *J. Mol. Biol.* **302**, 205–217.
16. **Thompson JD, Plewniak F & Poch O** (1999) *Bioinformatics*, **15**, 87–88.
17. **Bahr A, Thompson JD, Thierry JC & Poch O** (2001) *Nucleic Acids Res.* **29**, 323–326.
18. **Thompson JD, Koehl P, Ripp R & Poch O** (2005) *Proteins*, **61**, 127–136.
19. **Mizuguchi K, Deane CM, Blundell TL & Overington JP** (1998) *Protein Sci*, **7**, 2469–2471.
20. **Van Walle I, Lasters I & Wyns L** (2005) *Bioinformatics*, **21**, 1267–1268.

★ **21.** **EdgarRC** (2004) *Nucleic Acids Res.* **32**, 1792–1797. – *The original description of* MUSCLE, *a highly accurate method for global multiple alignment.*

★ **22.** **Lassmann T & Sonnhammer EL** (2002) *FEBS Lett.* **529**, 126–130. – *Comprehensive evaluation of multiple-alignment methods. Whilst previous studies restricted themselves to globally related test data, this paper uses both global and local benchmark sequence sets.*

23. **Brown NP, Leroy C & Sander C** (1998) *Bioinformatics*, **14**, 380–381.

24. **Clamp M, Cuff J, Searle SM & Barton GJ** (2004) *Bioinformatics*, **20**, 426–427.

25. **Sammeth M, Griebel T, Tille F & Stoye J** (2006) *Bioinformatics*, **22**, 889–890.

26. **Sammeth M, Rothganger J, Esser W, Albert J, Stoye J & Harmsen D** (2003) *Bioinformatics*, **19**, 1592–1593.

27. **Morgenstern B, Goel S, Sczyrba A & Dress A** (2003) *Bioinformatics*, **19**, 425–426.

28. **Shih AC, Lee DT, Lin L,** *et al.* (2006) *BMC Bioinformatics*, **7**, 103.

29. **Delcher AL, Kasif S, Fleischmann RD, Peterson J, White O & Salzberg SL** (1999) *Nucleic Acids Res.* **27**, 2369–2376.

30. **Hohl M, Kurtz S & Ohlebusch E** (2002) *Bioinformatics*, **18** (Suppl. 1), S312–S320.

31. **Brudno M, Chapman M, Gottgens B, Batzoglou S & Morgenstern B** (2003) *BMC Bioinformatics*, **4**, 66.

32. **Bray N & Pachter L** (2003) *Nucleic Acids Res.* **31**, 3525–3526.

33. **Brudno M, Do CB, Cooper GM,** *et al.* (2003) *Genome Res.* **13**, 721–731.

★ **34.** **PollardDA, Bergman CM, Stoye J, Celniker SE & Eisen MB** (2004) *BMC Bioinformatics*, **5**, 6. – *Interesting benchmark study investigating the performance of alignment programs on noncoding functional genomic sequence.*

35. **Barton LM, Gottgens B, Gering M,** *et al.* (2001) *Proc. Natl. Acad. Sci. U.S.A.* **98**, 6747–6752.

36. **Chapman MA, Charchar FJ, Kinston S,** *et al.* (2003) *Genomics*, **81**, 249–259.

37. **Prakash A & Tompa M** (2005) *Nat. Biotechnol.* **23**, 1249–1256.

38. **Alexandersson M, Cawley S & Pachter L** (2003) *Genome Res.* **13**, 496–502.

39. **Batzoglou S, Pachter L, Mesirov JP, Berger B & Lander ES** (2000) *Genome Res.* **10**, 950–958.

40. **Gross SS & Brent MR** (2006) *J. Comput. Biol.* **13**, 379–393.

41. **Stanke M, Schoffmann O, Morgenstern B & Waack S** (2006) *BMC Bioinformatics*, **7**, 62.

42. **Chain P, Kurtz S, Ohlebusch E & Slezak T** (2003) *Brief. Bioinform.* **4**, 105–123.

43. **Fitch J, Gardner T, Kuczmarski T,** *et al.* (2002) *Proc. IEEE*, **90**, 1708–1721.

44. **Higgins DG & Sharp PM** (1988) *Gene*, **73**, 237–244.

45. **Gotoh O** (1996) *J. Mol. Biol.* **264**, 823–838.

46. **Morgenstern B, Prohaska SJ, Pohler D & Stadler PF** (2006) *Algorithms Mol. Biol.* **1**, 6.

★ **47.** **Subramanian AR, Weyer-Menkhoff J, Kaufmann M & Morgenstern B** (2005) *BMC Bioinformatics*, **6**, 66. – *A program description of* DIALIGN-T.

48. **Do CB, Brudno M & Batzoglou S** (2004) ProbCons: probabilistic consistency-based multiple alignment of amino acid sequences. In *Proceedings of the Nineteenth Conference on Artificial Intelligence*, p. 703. ISMB/ECCB.

★ **49.** **Do CB, Mahabhashyam MS, Brudno M & Batzoglou S** (2005) *Genome Res.* **15**, 330–340. – *The original description of* PROBCONS, *one of the best methods for global multiple alignment.*

50. **Katoh K, Kuma K, Toh H & Miyata T** (2005) *Nucleic Acids Res.* **33**, 511–518.

51. **Katoh K, Misawa K, Kuma K & Miyata T** (2002) *Nucleic Acids Res.* **30**, 3059–3066.

52. **Heringa J** (1999) *Comput. Chem.* **23**, 341–364.

53. **Simossis VA & Heringa J** (2005) *Nucleic Acids Res.* **33**, W289–W294.

54. **Altschul SF, Madden TL, Schaffer AA,** *et al.* (1997) *Nucleic Acids Res.* **25**, 3389–3402.

55. **Huang X & Miller W** (1991) *Adv. Appl. Math.*, **12**, 337–357.

56. **Siebert S & Backofen R** (2005) *Bioinformatics*, **21**, 3352–3359.

57. **Poirot O, Suhre K, Abergel C, O'Toole E & Notredame C** (2004) *Nucleic Acids Res.* **32**, W37–W40.

CHAPTER 12

Inferring phylogenetic relationships from sequence data

Peter G. Foster

1. INTRODUCTION

In molecular phylogenetics, we want to infer an underlying tree of relatedness from our gene sequence and its close relatives. We collect the same gene sequences, DNA, or protein from all of the taxa that we are interested in and align those sequences so that the columns of the alignment represent homologous sites of the gene, and then use that alignment to infer a phylogenetic tree. There are several different approaches to tree building. They have their strengths and weaknesses, and differ widely in the computational time required.

The most important tree-building methods are maximum parsimony (MP) and model-based methods, the latter including distance methods, maximum likelihood (ML), and Bayesian approaches. The distance methods are perhaps the simplest. In this approach, all possible pairwise evolutionary distances between sequences are calculated and these distances are then used to build trees that best explain those distances. The pairwise distances that this method use might simply be the percentage difference between sequences, but because superimposed mutations at the same site can mask the real extent of evolutionary divergence, we correct the observed distances with an explicit model of evolution. The other methods are character-based, and rather than first distilling the sequences to distances, the full information in the sequences is used in judging whether one candidate tree is better than another. The MP approach looks for the trees that can describe the inferred sequence changes over the tree in the smallest number of steps. ML and Bayesian approaches use models to find results that have highest probability. The methods have been tested on simulated data, where the true tree is known, and generally the ML and Bayesian methods perform best, but at a cost of increased computational complexity.

Most methods involve searching 'tree space' – which can be imagined as a landscape in which all possible trees are represented, with similar trees adjacent to each other, and in which the local height of the landscape represents the

Bioinformatics: *Methods Express* (Paul H. Dear, ed.)
© Scion Publishing Limited, 2007

goodness of the tree at that point. A comprehensive search of tree space is usually an impossible task because the number of trees is astronomical. For example, the number of unrooted trees with 100 taxa is about 1.7×10^{182}. It is not possible to evaluate every tree unless the number of taxa is less than a dozen or so, and so some sort of heuristic search must be used. It is here that compromises and shortcuts are made, and where the quality of the algorithms and the programming come into play. For example, if the tree space has multiple islands of good trees, then a 'greedy' search that only goes uphill and that starts near one of the suboptimal islands will tend to get stuck and never find the globally optimal tree. Better search strategies might use multiple starting points for searching, or have the ability to cross from one island to another, but would be more computationally expensive. Using any heuristic does not guarantee finding the best tree.

The result of our analysis is a phylogenetic tree. The tree is often drawn with parallel lines with the terminal taxa on the right (see *Fig. 1a, b*), which can perhaps lure us into thinking that the present-day taxa on the right evolved from the ancestor nodes on the left. However, in the methods that we generally use, there is no time-line information in the tree because the tree is unrooted, and we do not have the information to infer which taxa evolved from which ancestors. This idea is clearer in the tree shown in *Fig. 1(c)*, where there is no obvious ancestor to which the eye is erroneously drawn. The usual way to root a tree is to use an 'outgroup' (see *Fig. 2*). In this strategy, we call the taxa that we are interested in the 'ingroup' and we choose several additional taxa that we know are outside of that ingroup to use as an outgroup. Although the analysis as a whole is unrooted, we can infer that the place where the outgroup attaches to the ingroup roots the ingroup. The outgroup should be phylogenetically close to the ingroup, because if it is too distantly related it can distort the analysis.

It has become common to do phylogenetic analysis on more than one gene, perhaps a combination of mitochondrial and nuclear genes, using both ribosomal and protein-coding genes. This trend has arisen not only because of the increased

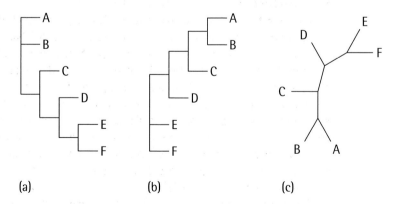

(a)   (b)   (c)

Figure 1. Unrooted trees.
The trees that we infer are generally unrooted and contain no information about ancestor–descendant relationships. The three trees in (*a*), (*b*), and (*c*) have identical topologies and are all unrooted. In particular, we should not infer from (*a*) or (*b*) that the ancestor of all of the taxa is the node on the far left.

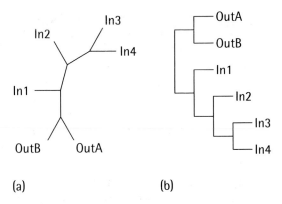

(a)   (b)

Figure 2. Outgroup rooting.
Here we use outgroup rooting, where taxa OutA and OutB are known to be outside the ingroup (taxa In1–In4) and so we can infer the root of the ingroup. As with most analyses, as a whole it is unrooted, as emphasized in (*a*). It is often drawn as in (*b*), where by convention a graphically introduced (but otherwise meaningless) bifurcating basal node is drawn to divide the ingroup from the outgroup.

availability of data, but also because it has been noted that the results of an analysis from one gene often differ from the results of another. Assuming that these differences are due to noise or different biases, it is the assumption and hope that in a combined analysis the effects of noise and bias will cancel out, but the true phylogenetic signal, perhaps weak in any one gene, will be the same in each gene and so will sum to give a well-supported tree. We might take any one of a few approaches in a multi-gene analysis. We might analyze each gene separately and then reconcile any differences between the respective trees afterwards; or we might simply concatenate the data and analyze it as if it were a single gene. However, the different genes may well have different evolutionary characteristics, and in a simultaneous analysis those differences can, in some software, be accommodated in a multi-partition model.

We want to find the best tree, but we also want to show how well supported our results are. Perhaps our single best tree is much better than any other tree, or perhaps, if our data are not decisive, then our best tree will not be significantly different from other similar trees. We may even find that we have many optimal trees, all equally good. Furthermore, some parts of the best tree might be more reliable than other parts. The usual way to determine these supports is with the bootstrap. In the bootstrap, we make pseudoreplicate datasets by resampling columns of the original alignment with replacement and then repeat the entire analysis on each. This process is repeated usually hundreds of times and the results summarized with a majority-rule consensus tree showing bootstrap support for the internal branches (see *Fig. 3*).

The methods should not be treated as black boxes that we throw our data into, from which a tree then emerges. Different methods have different assumptions that may or may not be met by your data. For example, in most common methods, there is an assumption that different sites in the alignment evolve independently. This may not be a good assumption, as for example in the

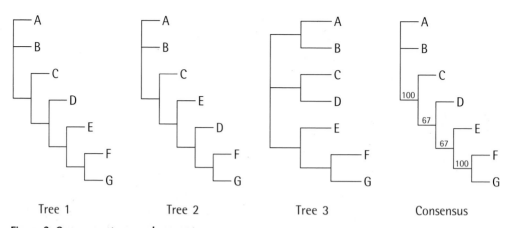

Tree 1   Tree 2   Tree 3   Consensus

Figure 3. Consensus trees and support.
The input trees might be bootstrap replicates or samples from a Bayesian Markov chain Monte Carlo. Here we make a majority-rule consensus tree from three input trees. The consensus shows the proportion of input trees that have the given split or tree bipartition (internal branches). Some splits, for example the grouping of taxa C and D together, were seen in only one of the three input trees and so do not appear in the consensus. Some supports are higher than others, and so we have 100% support for the groupings of A with B, and of F with G, as these splits were found in all of the input trees. However, we would have less confidence in other parts of the tree.

stem regions of structural RNA genes. Again, most methods assume compositional homogeneity, meaning that the sequences all have the same composition. If this assumption is not met, then the methods can fail and the recovered tree can be distorted. The MP method, in particular, is susceptible to long-branch attraction. Here, highly diverged but unrelated taxa tend to group together erroneously in the inferred trees. This artifact arises from the failure of MP to deal correctly with sequences that have many superimposed, and therefore hidden, mutations. You might think that using several different approaches would be safest, but this may not be so; if you get the same result from several different methods, it may be because you are getting the correct answer to an easy problem with clear data, or it may be because all the results are distorted in the same way by an unknown bias in your data.

There are, of course, many aspects of phylogenetics that are beyond the scope of this short chapter and many complex areas are glossed over. In particular, I will not be discussing molecular clocks and molecular dating, nor the identification of sites in a gene that are under selective pressure, both of which are important emerging applications of phylogenetics. These methods are changing quickly, and any attempt to make a recipe would immediately be out of date in these rapidly moving and often controversial fields (1). To get to grips with these, you need to dive into the primary literature in journals such as *Molecular Biology and Evolution* and *Systematic Biology*.

The field of phylogenetics has matured to a point that there is now a comprehensive textbook by Joe Felsenstein (2) (see also http://evolution.genetics. washington.edu/[12.1]). The classic chapter on phylogenetic inference by Dave Swofford and colleagues has not gone out of date in the decade since its writing

(3). The reviews written or co-written by Paul Lewis are especially readable (4, 5). The course material for the annual Workshop on Molecular Evolution is written by experts in the field and is a very useful resource, available online (http://workshop. molecularevolution.org/ [12.2]).

2. METHODS AND APPROACHES

2.1 Alignments

We start our phylogenetic analysis with an alignment of molecular sequences. We want to compare homologous positions in the genes, and the alignment is our statement of homology. Homologous sites in genes are kept in line in spite of insertions and deletions in the gene sequences by introducing alignment gaps, usually with the '–' character. As we do not know whether the gaps are due to deletions in one sequence or insertions in another, they are referred to as indels. You can identify a group of related sequences from public databases using BLAST. For some closely related genes, alignment is trivially easy and can be done by hand, often with few or no introduced gaps. However, for more diverged sequences, you will need to use multiple sequence alignment software. The standard program for making alignments is CLUSTALW (6), which has several web servers including one at the European Bioinformatics Institute (http://www.ebi.ac.uk/clustalw/ [12.3]). Another recommendable alignment program is MUSCLE, which also has a web server (http://www.drive5.com/muscle/ [12.4]). Making multiple alignments is a difficult computational problem and often a CLUSTALW alignment can be, and should be, improved by eye. This should be no surprise, as the human brain is very good at recognizing patterns. Use a graphic user interface multiple sequence editor with colored characters to help you to fix up the alignment. Often, it is not obvious what gap pattern to impose on the sequences to get the best alignment, and if that is the case those parts that you are not sure of should be masked out and not used in subsequent analyses. An attempt at automating the process of identifying unreliable parts of an alignment has been made (7). A comment should be made here concerning the alignment of DNA sequences for protein-coding genes: with the assumption that homology is determined by the protein sequence, you should translate the DNA into its protein sequence and align the protein first, and then back-align the DNA to the protein so that the gap pattern between codons is preserved.

2.2 File formats

There are many file formats for sequence data (see the section on formats at http://workshop.molecularevolution.org/ [12.5]), and most phylogenetic software is limited in the formats that it is able to read. In many areas of bioinformatics, the simple FASTA format is used, and it is commonly used as the input format for multiple-alignment programs. It is used occasionally in phylogenetics, even though it does not imply aligned sequences. A particularly common format for aligned sequence data is PHYLIP format, which might be either sequential (one sequence after another, like FASTA) or, more usually, interleaved, showing the alignment more clearly. Another important format is Nexus format, which is used by PAUP*, MRBAYES, and a few other programs (8). This extremely flexible format is able to accommodate information in addition to the sequence data, such as information about partitioning the data, about trees, and even commands for the programs that use this format. There is software that can convert formats, including READSEQ (http://iubio.bio.indiana.edu/soft/molbio/readseq/java/ [12.6]), with a web server (http://iubio.bio.indiana.edu/cgi-bin/readseq.cgi [12.7]). Trees can also be described in text files, almost always in Newick format (see http://iubio.bio.indiana.edu/cgi-bin/readseq.cgi [12.7]). Nexus format trees are Newick trees embedded in a Nexus structure. Newick format shows the tree topology using nested parentheses, and can incorporate branch lengths and labeling of interior nodes (often labeled with support values), as well as terminal taxa. Different programs more or less adhere to file format standards, but it is often the case that incompatibilities arise that force the user to make adjustments. If you are lucky, this can be automated or scripted, but often it cannot be changed easily and the data must be edited by hand. Format definitions are often modified and extended by software authors, which is an invitation to incompatibility when going from one program to another. The practicing phylogeneticist needs to be aware of this and to be able to debug problems.

In the case of a multi-gene analysis, if the data are simply concatenated and not kept separate, then there is no format problem. However, if we want to keep the different data partitions separate, perhaps to allow different overall rates in the different partitions, or if different model characteristics are to be applied to the different data partitions, then some way of indicating how the data are divided up into those partitions is needed. In Nexus files, there is a standard way of partitioning the data and that way is used by PAUP*. Other programs like MRBAYES and TREEFINDER that can do multi-partition data use their own ways of partitioning that are unfortunately not compatible with other programs.

2.3 Software

Joe Felsenstein maintains a valuable web page of all of the phylogenetic software that he knows about (http://evolution.genetics.washington.edu/phylip/software.html [12.8]), and many of the programs listed have web servers. The classic phylogenetic software includes the PHYLIP suite of programs by Joe Felsenstein, and PAUP*, written by Dave Swofford. Both are enormously capable and have been improved and debugged through wide use, and both are very well documented. PHYLIP includes a great many methods, but it can be slow and often other programs are able to do similar analyses faster. Many other programs use the PHYLIP data format and imitate the PHYLIP command-line menu-driven interface. PAUP* (http://paup.csit.fsu.edu/ [12.9]) is one of the few mainstream programs in phylogenetics that is not free – the source code is not available and there is a small license fee.

Nevertheless, it is worthwhile having, as it is excellent for parsimony, distance methods, and for ML of DNA.

PAUP* is in wide use for ML analysis of DNA. It is excellent for this, but is slow compared with newer ML programs such as PHYML (http://atgc.lirmm.fr/phyml [12.10]) and TREEFINDER (http://www.treefinder.de/ [12.11]). TREE-PUZZLE (previously called just PUZZLE; http://www.tree-puzzle.de/ [12.12]) is an ML program with good models for both DNA and protein that uses quartet puzzling to find ML trees. In this strategy, the problem is decomposed into finding the best trees from many different quartets of taxa sampled from the data and then combining (puzzling) the quartets to make the full tree. Although it is faster than PAUP* for ML, the resulting trees are generally not as good as more-thorough tree-based methods, and with the advent of newer ML programs, its niche is smaller.

Choices for analyzing multiple data sets simultaneously are limited. The major Bayesian program in wide use, MRBAYES, has a comprehensive set of models and is one of the few programs that is able to model multiple data partitions well. PAUP* is able to model separate DNA partitions with its site-specific model, by allowing the different data partitions to have their own overall rates; however, PAUP* otherwise assumes that the other model characteristics, rate matrix, and so on, are the same in all of the partitions. Modeling of different data partitions in MRBAYES is much more flexible, allowing combinations of datatypes and fitting of different complete models to the different partitions. TREEFINDER is also able to model different data partitions well under ML.

As larger analyses can be very computationally intensive, it would sometimes be appropriate to be able to run the analyses in parallel on clusters of computers. MRBAYES and TREE-PUZZLE have parallel versions. A bootstrap analysis is naturally amenable to running on a cluster, where different machines are each given a part of the task. Another up-and-coming strategy is cycle scavenging, where phylogenetic jobs flexibly migrate among idle desktop computers (http://www.cs.may.ie/distributed/multiphyl.php [12.13]).

The end result of a phylogenetic analysis is usually a tree with support values on the nodes, and often we will want to draw that tree. We will want to show the branch lengths accurately, perhaps with a scale bar, and we will want to show the support values on the nodes. We will want to use vector rather than bitmap graphics, and we will probably want the output of the tree-drawing program to be in a form that will allow further editing with your favorite graphics program. Unfortunately, software to draw trees can be frustrating and any given program may only satisfy some of these desiderata. Perhaps the best combination of ease of use and capabilities is TREEVIEW (see Rod Page's TREEVIEW page at http://taxonomy.zoology.gla.ac.uk/rod/treeview.html [12.14]). The PHYLIP programs DRAWTREE and DRAWGRAM can also be useful.

2.4 Tree-building methods

The main tree-building methods are MP and model-based methods. Parsimony is fast and able to handle many sequences and is excellent if divergences are small.

Indeed, if there is very little divergence, then attempting to fit a model will be difficult through lack of variation in the data. A search might find one MP tree, but often there are many such MP trees, each with the same minimum number of sequence changes over each tree; this might be considered an indication of the amount of ambiguity in the particular analysis. We usually will want to do a bootstrap analysis, in which case the result would be a consensus of the MP trees from each bootstrap replicate. A given bootstrap replicate might also have more than one MP tree, and these are weighted by the inverse of the number of these trees when the consensus is made.

Protocol 1

MP search using PAUP*

1. Save your data in Nexus format. (An example dataset is given in the Protocol_1 folder for this chapter on the book's web site, as <u>data.nex</u>[12.15].)

2. Make and save a text file containing the following:

```
#nexus

begin paup; [ Comments inside square brackets ]

[ Keep a log file. I assume the analysis will need to be repeated, ↻
  so I set replace=yes to overwrite the file ]

log file=log replace=yes;

set maxtrees=1000 increase=auto; [ appropriate for parsimony]

execute data.nex; [ read in your data ]

[ Do a parsimony bootstrap, with default settings, except that 1000 ↻
  bootstrap replicates are used. ]

bootstrap nreps=1000;

[ The following line saves the resulting consensus tree to a nexus ↻
  tree file. Using the from/to should not be needed, but it is a ↻
  workaround to a known bug. Bootstrap support values are saved ↻
  with the tree. If you repeat the analysis, the file is over-written. ]

savetrees from=1 to=1 file=bootTree.nex savebootp=nodelabels

maxdecimals=1 replace=yes;

log stop;

quit;

end;
```

(A copy of this text file is given in the Protocol_1 folder for this chapter on the book's web site, as <u>commands.nex</u>[12.16].)

3. Execute the file. On Mac Classic or Windows versions of PAUP*, you start the program and then use the execute command. On Unix-like machines, including Mac OS X and Linux, you can run it by saying, to your command line:

```
paup commands.nex
```

4. The example data file contains only 12 taxa and will run quickly. An alignment of 100 taxa might take a few hours to run.

5. At the end of the analysis, you will have a tree in a Nexus file, where internal nodes are decorated with bootstrap support values. The program will also make a log file as requested. Copies of the expected files (bootTree.nex[12.17] and log[12.18]) are in the Protocol_1 folder.

6. You can then import the result file into a tree-drawing program. If you want to import it into TREEVIEW X (http://taxonomy.zoology.gla.ac.uk/rod/treeview.html [12.14]), the file needs to be modified slightly – the zero just before the final semi-colon in the tree description needs to be removed. (A worked example is in the Protocol_1 folder: Treeview/bootTreeM.nex[12.19].)

7. In this analysis, branch lengths are meaningless and so a 'cladogram' format, where all of the leaf nodes line up on the right, is appropriate. TREEVIEW X can place the bootstrap support values on the internal nodes. You can print it to a file for further manipulation. (A copy of the output is in the Protocol_1 folder: Treeview/bootTreeM.pdf[12.20].)

Model-based methods tend to do better than MP methods when divergences are large. Models take into account the possibility of superimposed, and therefore hidden, mutations when calculating evolutionary distances, and branch lengths are more meaningfully expressed in terms of mutations per site. Model-based methods include distance methods, ML, and Bayesian methods. Of these, the distance-based methods do not do as well as the full tree-based ML and Bayesian methods, but distance methods are fast and so might be suitable for bootstrap analyses of many taxa. (However, newer ML implementations are getting faster (9, 10) and can now handle many taxa, so the reason for using distance methods is somewhat less compelling.) A distance analysis is a two-step process. First, a matrix of pairwise distances is made and then these distances are somehow made into a tree. The pairwise distances might be a very simple measure such as the number of changes between the sequences; these are uncorrected or p-distances and so are not really model based. Better distances can be obtained with models that take into account superimposed mutations. Maximum-likelihood distances can also be made, which would be recommended. A tree can then be made from the distance matrix using neighbor joining (11) or its relatives WEIGHBOR and BIONJ (12, 13). Neighbor joining is an algorithm and does not involve tree searching, so it is very fast. Another alternative to treeing distance matrices is to search tree space using something like the minimum tree length as the function for choosing the best tree.

As tested with simulated data, the best-quality tree-building methods are the full tree-based ML and Bayesian approaches, which both use the likelihood function. The likelihood is proportional to the probability of the data given the model and the tree. The likelihood of a tree is usually a very small number and so is expressed in its log form (and should not be confused with the probability of a tree). When we talk of the likelihood of a tree, we usually mean the maximized likelihood, where all of the nuisance parameters are optimized, maximizing the likelihood for that particular tree. The nuisance parameters are things like the branch lengths and model rate matrix parameters. The ML tree is the tree, found in a search, that has the highest maximized likelihood. As maximizing the likelihood uses numerical optimization and so is computationally intensive, searching for the ML tree can be slow. Because of this, phylogenetic programmers have, with some success, been pushing for increased speed via improved algorithms and workable compromises.

2.5 Choosing a model

Choosing a model of evolution is important, as it can greatly affect your analysis. You should not guess or assume a model, but rather choose the best available model, the one that best fits your data. Generally, we use ML to choose a model; the strategy is to evaluate the same tree with many different models and then to choose the best one. However, we do not simply choose the model that gives the highest likelihood. The process is analogous to fitting a polynomial curve to a scatter plot; we want our model to describe the important trends in the data, but without overfitting. Models differ in the number of parameters, and generally the more parameters there are in the model, then the better the fit of the model to the data and the higher the likelihood. On the other hand, with more and more parameters, the likelihood will rise, but we will see diminishing returns, and simple data described by an overly complex model might be modeling noise in the data – it would be overparameterized. We want to avoid that, and so when we evaluate our available models, we choose the one that gives the highest likelihood, but penalize by the number of parameters. This is formalized in the Akaike information criterion (AIC) (14). The AIC of a model is defined as $-2\log L + 2n$, where n is the number of free parameters in the model. We make a table of AIC values and the best choice of model is the one with the lowest AIC value.

Models for a single data partition can be described in terms of (see below):

- The rate matrix (e.g. GTR or HKY for DNA, JTT or WAG for protein)
- The composition (e.g. empirical or ML, +F in protein)
- The ASRV (none, +G, +I, +IG)

Models differ from each other in their free (i.e. adjustable) parameters. The parameters might be optimized by ML, or in a Bayesian analysis they can be free to be varied, or sometimes they can be fixed to reasonable numerical values. Models with variable numerical parameters can therefore be considered to be families of models. The rate matrix describes the rates of change between character states (for example, between any two of the four bases, A, C, G, and T). The most general rate matrix for DNA models in wide use is the GTR (general time reversible) matrix, and matrices for other models can be considered to be simplifications of it. Superimposed on the rate matrix is the composition of the model. Generally, we use the observed average composition of the data, i.e. the empirical composition, or we optimize the composition by ML. If the data are fairly homogeneous, the empirical and ML compositions will be about the same. However, sometimes whether you use empirical or ML compositions will make a difference to the analysis. If you have the computational time for it, it is recommended that ML-optimized compositions be used. On the other hand, optimization of parameters like this is computationally expensive and one reasonable way to save time is to use fixed empirical compositions. The third aspect of model descriptions is among-site rate variation (ASRV). This is an accommodation of the effects of selection on different sites and so would be especially important with biological sequences. It should definitely be taken into account (15). ASRV is usually modeled by using discrete Γ (gamma)-distributed among-site rates (+G), or a proportion of invariant sites (+I), or both together (+IG). The former strategy, using Γ-distributed rates, allows different sites to have different rates, where the relative rates can be described by the shape parameter of the Γ-distribution family. There is no biological basis for assuming that the ASRV is Γ distributed; it is simply a computational convenience to allow a wide range of different distributions all described by a single parameter. The other major strategy to accommodate ASRV is to allow a proportion of sites to be invariant (i.e. constant sites that are not allowed to be varied, in contrast to constant sites that have the potential to vary but have not yet done so). Such a proportion can be estimated by ML. There is some overlap (nonidentifiability) in the two strategies. For example, a Γ distribution with a low α value has an L-shaped distribution of site rates, implying many very slow sites. This would be describing approximately the same thing as a model with a sizable proportion of invariant sites, and so your estimate of one will affect your estimate of the other.

In protein models, we generally do not optimize the rate matrix parameters (as we would do with the DNA GTR model). Rather, we use one of the 'off-the-shelf' empirical models such as the JTT or WAG models. These have been made with large data sets and are meant to be generally applicable. It is assumed that your small alignment will behave in approximately the same way as the large data set. This is a good strategy because your small data set probably does not have enough information for an accurate estimation of all 189 rate matrix parameters. These large datasets have their own inherent composition, which may well differ from the composition of your data. Imposing your empirical composition on the chosen rate matrix is often called the +F model and is certainly recommended if available.

The strategy for choosing a model by ML is to optimize the same tree with different combinations of rate matrix, composition, and ASRV. The tree used for this purpose need not be the optimal tree; any reasonably good tree, such as a neighbor-joining tree, can be used. After evaluating the tree with different models, a table is made with the AIC (or its variants AICc or AIC2, or the Bayesian information criterion) and used to choose the best model. This can be done by hand, but it is tedious and so has been automated. The original automated model choice program is MODELTEST (16), which tests DNA models available in PAUP*, using PAUP* to do the likelihood calculations. The theory and practice of model selection can be found in the documentation accompanying MODELTEST. A version for protein models, PROTTEST (17), uses its own modified version of PHYML to do the likelihood calculations. A version of PROTTEST is available as a web server (http://darwin.uvigo.es/ [12.21]). A similar program, MODELGENERATOR (http://bioinf.may.ie/software/modelgenerator/ [12.22]), for both DNA and protein models, does not require external software for the likelihood calculations. Choosing a model tells you what model family to use (e.g.

HKY+G, or GTR+IG), but it also may tell you the optimized numerical parameters that were obtained. Whether you should use those numerical parameters depends on the context. Generally, for an ML search or a Bayesian analysis, the parameters should all be free (although we make an exception for empirical protein rate matrix parameters) and so we would not use those optimized parameter values. (Using PAUP* for ML searching is a special case, where we would use a successive approximation strategy for the best searches; 18.)

In a multi-gene analysis, the genes might be quite different in their evolutionary dynamics, in which case they should be modeled separately. If they are different data types, then they must be modeled separately. One strategy that can be used is to analyze the different genes completely separately. For this, you could use any ML program. You would need to evaluate (without searching) all of the possible candidate trees with all of the genes, and the ML tree is the tree that gives the highest sum. In this strategy, the analyses are completely separate and the branch lengths are unrelated. This has the disadvantage that there is a huge increase in the number of parameters because of all of the free branch lengths. Another reasonable strategy is to analyze the data together, but in separate data partitions, modeled separately. Available software for this is more limited. The partitions would be able to have overall rates, and so we would have fast genes and slow genes. These rates could be considered branch length multipliers, forcing the corresponding branch lengths in the different partitions to be proportional to each other. This has far fewer parameters.

Protocol 2

Automated model choice

The program MODELGENERATOR will be used here because it is applicable to both DNA and protein, and it does not require additional software to do the likelihood calculations (19). MODELGENERATOR will recognize whether the data are DNA or protein and test a full complement of rate matrices and ASRV. It constructs a neighbor-joining tree to make model comparisons. For protein models, both the inherent model composition and your empirical data composition (the +F model) are tested. For DNA models, the composition is estimated by ML, where applicable. The models can be compared and chosen by the AIC (or its variant AICc, used for short data; 20).

1. Get the software from its website (http://bioinf.may.ie/software/modelgenerator/ [12.22]) and unpack it.

2. Install the program on Mac OS X or Linux by making the following text file:

```
java -jar /path/to/your/modelgenerator.jar $1 $2
```

Name the file modelgenerator[12.23], make it executable, and put it in your path.

3. Put the data in FASTA or PHYLIP format (data for a worked example is given in the Protocol_2 folder for this chapter on the book's web site, named data.phy[12.24]).

4. Run the following command:

```
modelgenerator data.phy 4
```

Here the '4' is the number of discrete Γ-distributed among-site rate categories; four is generally sufficient. The results with the suggested model are saved in a log file.

5. The output for the worked example is given in the Protocol_2 folder as modelgenerator0.out[12.25]; a README[12.26] file is also provided.

Protocol 3

ML with PHYML

PHYML (9) is a fast implementation of ML with a good set of models for both DNA and protein. It is fast enough that bootstrapping can be done in a reasonable amount of time. A local installation can be run using command line options (for a complete list, use the command 'phyml -h'), but the PHYLIP-like interface allows more options. PHYML also has a web interface, which we will use in this protocol.

1. Choose a model for your data (see *Protocol 2*).

2. Put your data into PHYLIP format. A worked example is given in the Protocol_3 folder for this chapter on the book's web site as data.phy[12.24]; note that this file is the same as the one used in *Protocol 2*).

3. Call up the web page (http://atgc.lirmm.fr/phyml [12.10]) and, in the upper-right part of the screen, change the setting for the input file from **example file** to **file** and use the **choose file** button[a] to upload the data.phy[12.24] data file. Also check the **Perform bootstrap** box, and set the number of bootstrap replicates ('Number of data sets') required (200 for the worked example; generally 100 at a minimum and usually more if time allows).

4. The model suggested by the AIC in *Protocol 2* for these data was the GTR+G model, so choose the **GTR** substitution model. Make the proportion of invariant sites **fixed** and set to zero, but turn on Γ-distributed ASRV by setting the number of substitution rate categories to 4, leaving the Γ-distribution parameter in its default setting of **estimated**. Leave the other settings at their default values, fill in your e-mail address, and submit the job.

5. When the job finishes, the results are e-mailed to you. A copy of the expected results is given in the Protocol_3 folder as phyml_online_results[12.27].

6. The final line of the results gives the tree structure with its bootstrap values. To view the tree in TREEVIEW X, put this line into a new text file (the worked example file in the Protocol_3 folder is Treeview/tree.phy[12.28]).

7. Use this as an input into TREEVIEW X to create a graphical file (the expected result is given in the Protocol_3 folder as Treeview/tree.pdf[12.29]). Note that bootstrap values are given as numbers out of 200 (the number of bootstraps used in this example – see step 3 above), rather than as percentages.

Note

[a]Depending on which web browser you are using, this button may instead be called **Browse**.

2.6 A Bayesian approach to phylogenetics

The past decade has seen the emergence of Bayesian methods into the repertoire of the phylogeneticist (4, 21). Bayesian methods are also based on the likelihood of trees, but the likelihood is not optimized as in ML. Rather the result is expressed in terms of the posterior probability distribution of the trees or splits, and takes into account the variance of all of the nuisance parameters. The posterior probability distribution cannot be calculated directly, but it can be approximated by means of the Markov chain Monte Carlo (MCMC) method. This is a computational process that samples tree topologies, branch lengths, and model parameters in proportion to their posterior probability; those trees with a higher posterior probability are sampled more often. It is a chain that goes from state to state, typically for many thousands of state changes or generations. At the end, when we have collected enough samples, we need to summarize those samples for our result, and generally that summary will be the consensus tree topology from the sampled trees. Typically, we start our MCMC from a random tree topology, with arbitrary model parameters and, after a while, the chain converges to the posterior probability distribution. Because of this delay in reaching convergence, samples taken from the beginning of the chain are not representative of the posterior and are discarded as 'burn-in'. Assessing whether the chain has converged is difficult, and the investigator needs to pay careful attention to the available diagnostics. The posterior distribution is proportional to the likelihood of trees with their models, but it is also proportional to prior probabilities, which is what you think the results should be before you see the data. Generally, the prior is chosen so that it does not influence the result much and so, for example, the prior probability of any one tree topology is the same as any other. Generally, when there is a sizable amount of data, the influence of the prior will be small and the posterior will mostly be driven by the likelihood, and so ML and Bayesian analyses will be quite similar.

A valid MCMC involves many practical considerations. We want the MCMC to show good mixing, i.e. we want our samples from the MCMC to cover the whole of the posterior distribution, and there are strategies to promote that. One simple strategy is to sample the chain rarely, for example every 1000 generations, rather than every generation. Another strategy uses Metropolis-coupled MCMC, or MCMCMC; the interested reader is invited to consult the MRBAYES manual for an explanation (Huelsenbeck and Ronquist's MRBAYES web page at http://mrbayes. csit.fsu.edu/[12.30]). We also want some assurance that our MCMC has reached convergence. We cannot have absolute assurance, but there are diagnostics that we can use. A reasonable assumption is that if we run the analysis more than once, with each run starting from a different random starting tree, and if we find that each run converges to the same consensus topology, then we can have more confidence that they have converged. Doing more than one run is highly recommended and two runs is now the default in MRBAYES. Perhaps the simplest diagnostic is to plot the likelihood of a run and check that it has reached the more or less noisy plateau that is consistent with convergence. This is a widely used quick-and-dirty method, but it is not reliable. Certainly, if the likelihood plot has not reached a plateau, then we can say that the chain has not reached convergence, but we cannot argue the reverse. Other parameters besides the likelihood can be examined, and the 'sump' command in MRBAYES aids this. Generally, we are especially interested in the topology, and we want to make sure that the topology has converged. To aid this, MRBAYES now examines how similar the sampled trees are from different simultaneous runs in the form of the average standard deviation of split frequencies. When this reaches a low level, such as 0.01, then the two runs are similar to each other and we can have more confidence that the chains have converged. Whilst many investigators use a fairly short burn-in and so get a few more samples, it is recommended that a long burn-in be used to give a better chance of sampling truly converged chains.

Protocol 4

Simple Bayesian analysis with MRBAYES

MRBAYES has a comprehensive wiki manual and tutorial introduction on its web site (http://mrbayes.csit.fsu.edu/ [12.30]) and this should be consulted. There is also extensive online help within the program itself and there are example files that come with the software. It is recommended that more than one run be made, and doing two simultaneous runs is the default in MRBAYES. Instructions for doing multi-partition analyses are given in the wiki manual.

1. Install MRBAYES from its web site (http://mrbayes.csit.fsu.edu/ [12.30]). MRBAYES is generally installed under the name 'mb[12.31]', and you can start the program for interactive use by typing 'mb' at your prompt.

2. Choose a model for your data[a].

3. Put your data into Nexus format, using a Nexus data block (8). (Data for this worked example is given in the Protocol_4 folder for this chapter, as data.nex[12.15].)

4. Run the analysis for a short time, perhaps 10 000 generations, interactively (as in the tutorial). That way you can get some idea of how long a longer analysis will take. When running interactively, you can continue running more generations after the first lot has finished.

5. When you have an idea of how long to run the MCMC, you can collect your commands in a Nexus file. Such a commands file might contain:

```
#nexus
begin mrbayes;
log filename=mbout.log append start;
execute data.nex;
lset nst=6 rates=invgamma;
mcmc ngen=1000000 samplefreq=500 printfreq=10000 filename=mbout;
sump filename=mbout burnin=1001;
sumt filename=mbout burnin=1001;
log stop;
quit;
end;
```

(A copy of a similar worked example is given in the Protocol_4 folder as commands.nex[12.16].)

The 'lset' line, which specifies the model, should specify your chosen model. Here, I have specified a million generations, which might take overnight for a medium-sized data set. I collect samples every 500 generations ('samplefreq'), which means that I collect 2001 samples for each of the two runs. The burn-in is given by the number of samples, not the number of generations (in this case, 1001 samples, or half of the run)[b]. The 'sump' command summarizes the likelihood and parameter values, and the 'sumt' command summarizes the sampled trees and makes a consensus tree; these commands also specify the names for the respective output files.

6. Run the command file from the command line as 'mb commands.nex'.

7. At the end of the run, examine the output to determine whether topological convergence has been reached, and whether you have convergence of the model parameters, good chain swapping, and good proposal acceptance rates. (The output files for this worked example are given in the Protocol_4 folder as mbout.*[12.32], along with a README[12.26] file describing their interpretation.) If everything went well (as in this worked example), then the analysis is finished, but, if not, then you need to make adjustments and run it again, perhaps for a longer run.

8. Bayesian analysis results are very straightforward to interpret. The split supports, given in the consensus tree, are the posterior probability that the split is true, given the data and model.

Notes

[a]MRBAYES does not implement the many subsets of the GTR model such as the TN and K3P models, and so if those models are chosen, you can use the GTR model. In this case, it is probably better to err on the side of slightly overparameterizing rather than underparameterizing.

[b]A long burn-in is recommended. Some diagnostics, such as a likelihood plot, might reach apparent convergence quickly, but other aspects, such as topology, might be slower to converge. You can use AWTY ('Are we there yet?') for more convergence diagnostics (http://king2.csit.fsu.edu/CEBProjects/awty/awty_start.php [12.33]).

3. TROUBLESHOOTING

The process of evolution has been complex, and the present-day gene sequences that we use might easily contain biases or other aspects that can confound our phylogenetic methods; a few will be mentioned here. For example, if there is compositional heterogeneity in the sequences, and there are unrelated sequences that have a similar composition, those sequences might erroneously be attracted to each other in the recovered tree. When we model ASRV, one assumption is that the rate of individual sites is constant over the tree. However, there is evidence that the rate of sites can change in diverged taxa, and even the spectrum of invariant sites can change over time. These covarion and heterotachy effects are difficult to model. In early evolution, horizontal (or lateral) gene transfer appears to have been more common and of course that will cause gene trees to be different from organism trees. Perhaps such effects might better be described by networks rather than trees. Long-branch attraction in parsimony has long been appreciated, and model-based methods are less susceptible to this. However, we can also see long-branch effects in model-based methods when the model does not fit well. Such problems are not easily identified. At the least, you might want to identify very long branches or compositionally divergent taxa, and as an experiment remove them for a repeat of your analysis to see whether it affects your conclusions. A similar problem with difficult data occurs when the outgroup is highly diverged from the ingroup to the extent that its presence distorts the ingroup. As a workaround, you can remove the outgroup to get a better supported but unrooted ingroup topology, and then as a separate analysis attempt to see where the outgroup attaches.

Phylogenetic software can be frustrating. It is often written by academics whose main expertise is not programming, and it is often user-hostile and buggy. Something as simple as differences in line endings in files (Unix versus old Mac versus DOS line endings) might cause a program to fail. These otherwise invisible characters can be seen with the 'od -c' command in Unix. Perhaps you have long names for taxa and you intend to do an analysis with a program that only takes short names. It might be worthwhile changing the names to arbitrary short names to do the analysis and then changing the names back to the long versions at the end. A script using a dictionary or hash in Perl or Python can facilitate that. You can save time debugging methods if you use small datasets. Debugging is often helped by example files provided with the program, and trial and error debugging is made easier when you use very small files.

Bayesian analysis requires more expertise than other methods in order to adjust the MCMC and to assess convergence. It also has its own problems (see, for example, (39)). You will often find that, when the same data are analyzed by both ML and Bayesian approaches, posteriors are higher than bootstraps; these inflated posteriors should be kept in mind when interpreting results. You should be especially suspicious of high support on short internal branches.

4. REFERENCES

1. **Graur D & Martin W** (2004) *Trends Genet.* **20**, 80–86.
★ 2. **Felsenstein J** (2004) *Inferring Phylogenies.* Sinauer Associates, Sunderland, MA. – *Standard comprehensive textbook.*
3. **Swofford DL, Olson GJ, Waddell PJ & Hillis DM** (1996) In *Molecular systematics.* Edited by DM Hillis, G Moritz & BK Mable. Sinauer Associates, Sunderland, MA, pp. 407–514.
★ 4. **Holder M & Lewis PO** (2003) *Nat. Rev. Genet.* **4**, 275–284. – *A very readable explanation of Bayesian methods and the MCMC.*
★ 5. **Lewis PO** (2001) *Trends Ecol. Evol.* **16**, 30–37. – *A paper describing some of the new directions in phylogenetics.*
6. **Chenna R, Sugawara H, Koike T, et al.** (2003) *Nucleic Acids Res.* **31**, 3497–3500.
7. **Castresana J** (2000) *Mol. Biol. Evol.* **17**, 540–552.
8. **Maddison DR, Swofford DL & Maddison WP** (1997) *Syst. Biol.* **46**, 590–621.
9. **Guindon S & Gascuel O** (2003) *Syst. Biol.* **52**, 696–704.
10. **Jobb G, von Haeseler A & Strimmer K** (2004) *BMC Evol. Biol.* **4**, 18.
11. **Saitou N & Nei M** (1987) *Mol. Biol. Evol.* **4**, 406–425.
12. **Bruno WJ, Socci ND & Halpern AL** (2000) *Mol. Biol. Evol.* **17**, 189–197.
13. **Gascuel O** (1997) *Mol. Biol. Evol.* **14**, 685–695.
14. **Akaike H** (1974) *IEEE Trans. Autom. Contr.* **19**, 716–723.
15. **Sullivan J & Swofford DL** (1997) *J. Mamm. Evol.* **4**, 77–86.
16. **Posada D & Crandall KA** (1998) *Bioinformatics,* **14**, 817–818.
17. **Abascal F, Zardoya R & Posada D** (2005) *Bioinformatics,* **21**, 2104–2105.
18. **Sullivan J, Abdo Z, Joyce P & Swofford DL** (2005) *Mol. Biol. Evol.* **22**, 1386–1392.

19. **Keane TM, Creevey CJ, Pentony MM, Naughton TJ & McInerney JO** (2006) *BMC Evol. Biol.* **6**, 29.

20. **Posada D & Buckley TR** (2004) *Syst. Biol.* **53**, 793–808.

21. **Huelsenbeck JP, Ronquist F, Nielsen R & Bollback JP** (2001) *Science,* **294**, 2310–2314.

APPENDIX
Additional useful bioinformatic resources

It would be futile and impossible to try to provide an up-to-date list of all bioinformatic databases and web-based tools. Instead, I urge you to refer to the annual database and web-server issues of *Nucleic Acids Research*. Both of these issues, freely available online, provide compilations of direct links to the sites they describe:

Nucleic Acids Research Database Issue: http://nar.oxfordjournals.org/. Updated annually, this is the most comprehensive listing of databases relevant to molecular biology. The 2007 Database Issue (*Nucleic Acids Res.* 35, Database issue; follow link from the NAR front page) describes almost a thousand web-accessible databases covering every imaginable (and the occasional improbable) topic.

Nucleic Acids Research Web Server Issue: http://nar.oxfordjournals.org/. Complementing the annual Database issue, this is a well-nigh exhaustive list of web-accessible tools for molecular biologists. The 2007 Web Server Issue (*Nucleic Acids Res.* 35, Web Server issue; follow the link from the NAR front page) lists over a thousand tools in fields from *ab initio* protein folding to Z-DNA prediction.

What follows is a short, selective, and arbitrary miscellany of particularly useful bioinformatic sites. I have tried to avoid overlap with the topics covered elsewhere in this book, and have rated sites (one to three stars) on the basis of their general interest and ease of use. Many thanks to my colleagues at LMB for recommending their own favorite sites, many of which are included here.

1. Collections of bioinformatic tools

There are many sites that act as gateways, providing links to collections of bioinformatics resources of general utility or focusing on a specific field. Several of these gateways are discussed in the main text, and a few more examples are given here.

★★BioWeb Online Analysis Tools: http://molbiol-tools.ca/. A compendium of links to numerous useful bioinformatic tools ranging from DNA sequence alignment through transcription analysis and RNA secondary structure predictions to polysaccharide structure prediction.

★★Bioinformatics and Research Computing at Whitehead Institute: http://jura.wi.mit.edu/bio/. The **Bioinfo Tools** link from this page will take you to a variety of handy tools for analyzing and manipulating DNA and protein sequences, as well as to a collection of databases.

★★Sequence Manipulation Suite: http://www.bioinformatics.org/sms2. A collection of extremely useful, easy-to-use tools for basic manipulation and analysis of DNA, RNA, and protein sequences. The 50 or so tools cover basic tasks such as DNA sequence format conversion, reverse complementation, translation, codon usage, motif prediction, restriction site identification, PCR primer analysis, prediction of protein properties, mutagenesis, and many more.

★★CBS Prediction Servers: http://www.cbs.dtu.dk/services/. A suite of web-interfaced prediction programs by the Centre for Biological Sequence Analysis. Most of these tools are devoted to the prediction of genes from DNA sequences and of protein features, including the identification of signal peptides and other functional motifs in protein sequences.

2. Specialist tools

These are largely single-purpose sites, generally covering topics not addressed elsewhere in this book.

Restriction maps and *in silico* digests

★★★NEBcutter: http://tools.neb.com/NEBcutter2/index.php. Identifies open reading frames and restriction sites in sequences. A wide variety of options is available for predicting the results of single or multiple restriction digests, including the option to display the expected fragment sizes or a simulated gel image of the expected result.

★Web Map: http://pga.mgh.harvard.edu/web_apps/web_map/start. Produces restriction maps, six-frame translations, and predicted open reading frames from pasted or uploaded DNA sequences, with a wide range of user-selectable options.

PCR

★★PRIMER3: http://primer3.sourceforge.net/. Web-based PCR primer design, also available at a number of other sites.

★★Electronic PCR: http://www.ncbi.nlm.nih.gov/sutils/e-pcr/. Follow the **Forward e-PCR** link to search a candidate DNA sequence for the presence of previously characterized sequence-tagged sites. More useful is the **Reverse e-PCR** link, which allows you to predict what your PCR primers will amplify from a given genome (seven species, including human, are catered for at the time of writing); useful for checking that your chosen primers will amplify a unique product from the target genome.

DNA packaging

★Online Nucleosomes Position Prediction: http://genie.weizmann.ac.il/pubs/ nucleosomes06/index.html. Predicts the location of nucleosomes on eukaryotic DNA based on nucleotide sequence. Follow the link to **Predict your sequence,** paste your sequence into the window, and the software returns a prediction showing the likeliest phasing of nucleosomes.

DNA/RNA secondary structure prediction

★ZHUNT web server: http://gac-web.cgrb.oregonstate.edu/zDNA/. Predicts likely regions of Z-DNA formation in a DNA sequence.

★★MFOLD and related secondary-structure software: http://www.bioinfo.rpi.edu/applications/mfold/. A range of software by Michael Zuker, for predicting secondary structures of RNA and DNA sequences. Other software on the site predict melting temperatures and hybridization between nucleic acids.

★Quadruplex-forming G-rich sequence mapper: http://bioinformatics.ramapo.edu/QGRS/index.php. Predicts regions of potential quadruplex structure in DNA sequence.

CpG islands

★CpG island prediction: http://www.ebi.ac.uk/emboss/cpgplot/. Part of the EMBOSS suite; identifies probable CpG-islands in DNA sequence.

DNA repeats

★★TANDEM REPEATS FINDER web server: http://tandem.bu.edu/cgi-bin/trf/trf.300.pl. Finds and reports simple tandem repeats in nucleotide sequence data.

★★REPEATMASKER web server: http://www.repeatmasker.org/. Identifies both simple and complex repetitive regions in nucleotide sequence, and can mask the sequence in preparation for database searches, to avoid numerous spurious matches to the repeats.

3. Resolving synonyms

Few things are as frustrating as the variety of names by which a single gene or protein may be known. The sites below will help you to find synonyms:

★★★The Gene and Protein Synonym Database (GPSDB): http://biomint.pharmadm.com:80/protop/bin/bmstaticpage.pl?userType=guest&p=gpsdb. Created by the BioMinT consortium, GPSDB allows you to search for synonyms of gene or protein names across many databases. For example, a search for 'polynucleotide kinase' reveals 23 synonyms ranging from 'DNA Kinase' to 'EC 2.7.1.78', and provides links to each entry in its respective database. UniProt and Entrez Gene accession numbers can also be searched, and PubMed queries against some or all of the synonyms can be launched.

★★HUGO Gene Nomenclature Committee: http://www.gene.ucl.ac.uk/nomenclature/. The HUGO Gene Nomenclature Committee approves the official nomenclature (name and abbreviation) for each human gene. This site gives the official gene names, allows searches for synonyms and provides links to sequence databases. It also allows researchers to request nomenclature for new genes.

★★ ID Converter: http://idconverter.bioinfo.cnio.es/. Accepts a variety of types of identifiers for genes, proteins or clones, and returns the equivalent identifiers from other data resources, as well as direct links to those entries.

4. Sequence formats

Nucleotide and protein sequences can be represented in a bewilderingly wide range of different formats. The sites below explain these formats, and offer interconversion between them.

★★EBI Help File – Sequence Formats: http://www.ebi.ac.uk/help/formats_frame. html. Explains most of the common sequence formats and provides annotated examples.

★★★READSEQ format conversion: http://www.ebi.ac.uk/cgi-bin/readseq.cgi. There are many web servers providing interconversion between DNA sequences in different formats, but Readseq is probably the most versatile and straightforward to use.

5. Graphic representation of data

These sites help with visualization and presentation of data in graphic formats.

★BOXSHADE server: http://www.ch.embnet.org/software/BOX_form.html. Produces graphic representations, in a variety of formats, from multiple sequence alignment files (which can also be in a variety of formats). Output formats include Postscript, EPS, and PICT, making them ideal for further editing for publications and presentations.

★★Chaos plot: http://bioweb.pasteur.fr/seqanal/interfaces/chaos-simple.html. Chaos plots give a simple graphical 'fingerprint' of the composition of a DNA sequence. This EMBOSS version has only limited flexibility, but the **Advanced chaos form** provides a few additional options.

★★★DOTLET dot plotter: http://myhits.isb-sib.ch/cgi-bin/dotlet. Dot plots are a simple way to represent regions of similarity between two sequences. DOTLET is a very pretty implementation of this, and lets you analyze both nucleotide and protein sequences.

★★Protein Colourer: http://www.ebi.ac.uk/cgi-bin/proteincol/ProteinColourer.pl. Renders protein (or nucleotide) sequences in color, according to a user-changeable color key (for example, hydrophobic residues in red). Useful for visualizing regions with a particular amino acid or base composition, the output can be copied and pasted directly into Microsoft Word and other applications.

6. Other utilities

★★★Biobar: http://biobar.mozdev.org/. If you're using Firefox, Netscape, or certain other web browsers, installing BioBar gives an additional toolbar for direct and rapid access to all of the major biological databases. From the home page, click the **Install/Uninstall** link and follow the instructions. When you restart your browser, the new database access toolbar will appear.

Index

کامران خان

Biography